Laser Science
and Technology

ETTORE MAJORANA
INTERNATIONAL SCIENCE SERIES
Series Editor:
Antonino Zichichi
European Physical Society
Geneva, Switzerland

(PHYSICAL SCIENCES)

Recent volumes in the series:

A Continuation Order Plan is available for this series. A continuation order will bring delivery of each
new volume immediately upon publication. Volumes are billed only upon actual shipment. For further information please contact the publisher.

Laser Science and Technology

Edited by

A. N. Chester
Hughes Research Laboratories
Malibu, California

V. S. Letokhov
Institute of Spectroscopy
Troitsk, Moscow Region, USSR

and

S. Martellucci
The Second University of Rome
Rome, Italy

Plenum Press • New York and London

Library of Congress Cataloging in Publication Data

International School of Quantum Electronics on Laser Science and Technology (1987: Erice, Sicily)
  Laser science and technology / edited by A. N. Chester, V. S. Letokhov, and S. Martellucci.
     p.   cm.—(Ettore Majorana international science series. Physical sciences; v. 35)
  Proceedings of the International School of Quantum Electronics on Laser Science and Technology, held May 11–19, 1987, in Erice, Italy"—T.p. verso.
  Includes bibliographical references and index.
  ISBN 0-306-43033-9
  1. Lasers—Congresses. 2. Quantum electronics—Congresses. I. Chester, A. N. II. Letokhov, V. S. III. Martellucci, S. IV. Title. V. Series.
TA1623.I575 1987   88-25582
621.36′6—dc19   CIP

Proceedings of the International School of Quantum Electronics on Laser Science and Technology, held May 11–19, 1987, in Erice, Italy

Printed in the United States of America

PREFACE

The conference "Laser Science and Technology" was held May 11-19, 1987 in Erice, Sicily. This was the 12th conference organized by the International School of Quantum Electronics, under the auspices of the "Ettore Majorana" Center for Scientific Culture. This volume contains both the invited and contributed papers presented at the conference, covering current research work in two areas: new laser sources, and laser applications.

The operation of the first laser by Dr. Theodore Maiman in 1960 initiated a decade of scientific exploration of new laser sources. This was followed by the decade of the 1970s, which was characterized by "technology push" in which the discoveries of the 1960s were seeking practical application. In the 1980s we are instead seeking "applications pull," in which the success and rapid maturing of laser applications provides both inspiration and financial resources to stimulate additional work both on laser sources and applications.

The papers presented in these Proceedings attest to the great vitality of research in both these areas:

New Laser Sources. The papers describe current developments in ultraviolet excimer lasers, X-ray lasers, and free electron lasers. These new lasers share several characteristics: each is a potentially important coherent source; each is at a relatively short wavelength (below 1 micrometer); and each is receiving significant development attention today.

Laser Applications. The papers span the most rapidly growing areas for laser applications. Some areas are already of commercial importance, involving biomedicine, materials working, laser-induced chemistry, industrial process monitoring, and inspection: other applications areas remain at this time more research-oriented: nuclear fusion, nuclear physics, optical computing, and frequency standards.

It is hoped that these proceedings will be a useful guide to research in both these areas, which are forming a strong basis for advances in electro-optics technology.

The editors wish to express their sincere appreciation to Prof. A.M. Scheggi, who served as Scientific Secretary and organizer for this conference. Thanks are also due to the organizations who sponsored the conference, especially the E. Majorana Centre for Scientific Culture, whose support made this conference possible.

Before concluding, the editors acknowledge the invaluable help of Miss R. Colussi, who volunteered to retyping and revising the entire manuscript, as well as the continuous assistance of the editorial staff of Plenum Publishing Co. in London and New York in the preparation of this volume.

A. N. Chester
Hughes Research Center
Malibu, California (USA)

V. S. Letokhov
Institute of Spectroscopy
Troitsk, Moscow Region (USSR)

S. Martellucci
The Second University of Rome
Rome (Italy)

June 23, 1988

CONTENTS

NEW APPLICATIONS OF LASERS

PROGRESS IN LASER TECHNIQUES: TOWARD INDUSTRIAL U.V. LASER TOOLS

FOR THE NINETIES

M. L. Gaillard

Laboratoires De Marcoussis, Centre de Recherche de la
Compagnie Générale d'Electricité
Route de Nozay, 91460 Marcoussis, France

INTRODUCTION

After twenty five years of active laser research, several thousands of
laser systems have been identified and tested in laboratories throughout
the world. Though many such systems found a niche in scientific instrumentation, the list of the laser devices which are apt to survive in the harsh
environment of industry remains nevertheless amazingly short. No one over-
looks the inroads made by the gas (mainly CO_2) and solid (YAG) lasers in
material processing workshop. During the last decade however, only two serious new contenders have emerged: the excimer and the copper vapor lasers.
From an industrial point of view both are still very much in the testing
stage despite an impressive body of scientific and technical evidence which
clearly speaks in favor of U.V. and visible lasers for industrial use.

All in all, current industrial laser research focuses its attention
on four basic laser systems and four only. Such a narrow field of interest
may be left as an insult to scientist's flare for innovation and creativity.
The truth, however, is that the industrial development of even the oldest of
the four industry favorites is not yet complete. Full engineering at the
level required in order to win a wide acceptance on sound demonstrated
economical basis is still a major endeavor of industrial laser laboratories. New scientific and technical issues are being raised which will be
answered only through long and costly developments.

The purpose of this set of lectures is to give an appreciation of
current development trends in the fields of the U.V. industrial laser
systems as well as a hint of some of the technical problems still requiring basic research efforts. It may be worth pointing out from the out-
set that industrial laser sources, including U.V. lasers, have already
lost a lot of their early technical glamor. In industrial laser systems
or equipments, the laser is nowadays just a mere component, which development is driven by the "iron law" of the market: cost reduction. Keeping
down investment cost is achieved by miniaturisation and improvement of performance at a given market price ("more bang for the buck"). Running costs

1

are reduced by gains in efficiency, reliability and ease of maintenance. Quite often, this is achieved by careful addition of painful, minute, detailed improvements in widely different fields of technology: aerodynamics, plasma physics, electronics, thermodynamics...

But breakthroughs do occur in industrial research as well, although they may be a lot more difficult to spot than pure scientific discovery. In order to become a true contribution to progress, technical innovation has to occur at the proper time during product development and to win a wide market acceptance. According to such criteria, Galois' Group theory would hardly ever qualify! With all due caution, we will nevertheless try to point out several major contributions in the industrial U.V. laser field which are likely to speed up the development of U.V. laser machining centers and have been or will soon be implemented as market standards for laser optics, laser circulators, power supplies, preionization sources.

1. FUNDAMENTALS

The industrial excimer lasers belong to a class of pulsed lasers which operate on electronic transitions of essentially diatomic molecules, whose ground state potential energy curve is either purely repulsive or only very weakly bound. Population inversion and hence gain is easily achieved on these transitions by the rapid dissociation of the ground state molecules. Subsequently, the free constituents can enter the pumping cycle again. Most commercially available excimer lasers operate with the rare gas halides (Table I). Excited state densities sufficient for laser action have been produced by using intense pulses of excitation by electron beams or electrical discharges. The basic similarity in design to CO_2 pulsed TEA lasers has played a fundamental role in this development. For all practical purposes, the avalanche discharge excitation method has emerged as the only one able to cope with industrial requirements such as: repetition rate, compactness, reliability, low maintenance... We will thus focus our attention on rare gas halide excited devices using either KrF or XeCl as model systems. Since their discovery in 1975, excimer lasers have been the subject of several comprehensive reviews[1,5]. The reader is referred in particular to reference 5 for a complete introduction to excimer laser physics.

Table 1. Rare gas halide excimer lasers (wavelength in nm)

	F	Cl	Br
A	193		
Kr	249	223	
Xe	352	308	282

A. Molecular structure of rare gas halides

Continuum emission from diatomic rare gas halides was observed for the first time in 1974[4],[6] during the course of flowing afterglow experiments devoted to atomic rare gas metastable state quenching by halogen rich compounds. The possibility of laser action is based on the specific properties of the potential energy curves of the ground and excited states of the molecular system which is formed during the close approach of rare gas and a halogen atom. If both partners are initially in their ground states with the halogen in the $^2P_{3/2}$ state, two different quasimolecules can be formed which differ according to the projection of the electron angular momentum on the axis joining the nuclei. This corresponds to two terms $(A)^2\pi_{3/2}$ and $(X)^2\Sigma_{1/2}$ which are characterized by weak energy dependence at large internuclear distances (Van der Waals interaction) and a strong short range (exchange) repulsion over distances smaller than or of the order of the atomic size (Fig. 1). In the case when a halogen atom is in a fine structure state $^2P_{1/2}$, its approach to an inert gas atom produces a $(A)^2\pi_{1/2}$ term which essentially as the $(A)^2\pi_{3/2}$ state. The ground electronic state which results is either purely repulsive (KrF case[8]) or very weakly bound (XeCl case) and thus unstable at room temperature[9] but with a finite dissociation rate.

Quite to the contrary, the first excited electronic states of the rare gas halides are strongly bound by electrostatic interaction between ion pairs. Indeed, rare gas atoms in their first excited electronic state $np^5(n+1)s^1$ are similar to the alkalis and thus prone to ionic bonding with the halogen. Again, when the internuclear distance is reduced, three closely spaced molecular states appear (B, C, D). But whereas the interaction

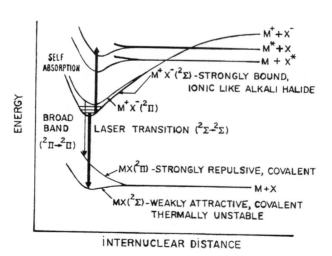

Fig. 1. Schematic potential energy diagram illustrating the electronic the structure of the rare gas monohalides[5].

between neutral atoms is significant only for distances smaller than ten angstroms (and then essentially as a repulsion in the ground stare rare gas halide case), Coulomb forces have long range of interaction and lead to a large decrease of the potential energy at short internuclear distance. Ionic potential energy curves thus fall below the covalent curves correlated to either X* (excited rare gas) + H (halogen) or X + H*.

From the first stable excited state $(B)^2\Sigma_{1/2}$, a radiative transition can take place either toward the weakly bound or dissociative $(X)^2\Sigma_{1/2}$ or toward the unstable $^2\pi_{3/2}$ and $^2\pi_{1/2}$ states giving rise to two dissociative continua (Fig. 2). Only the $\Sigma - \Sigma$ transition has given rise to useful laser emission due to its higher cross section, related in turn to the narrower natural emission bandwith (with the exception of the (C-A) $^2\pi-^2\pi$ transition also observed in XeF).

B. Pumping mechanism in discharge excitation

Production of excited state rare gas halide molecules by pulsed discharge excitation relies on complex plasma processes occurring on the short time scale characteristic of the upper laser state natural lifetime:

$$\sim 9 \text{ ns for } (Krf)*$$

$$\leq 40 \text{ ns for } (XeCl)*.$$

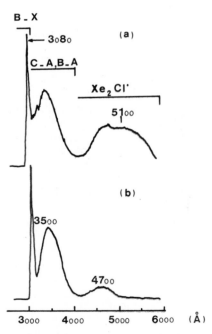

Fig. 2. Emission spectra of XeCl and Xe_2Cl excited by electron bombardament[19] at two typical temperature values : (a) 120 K , (b) 200 K.

The discharge excitation mechanism differs significantly from electron beam excitation, which has historically received the largest amount of attention due to high peak power laser research for fusion.

In the excimer laser, the active medium consists mainly of a buffer gas; neon, which has been shown to improve the performances of both XeCl[10] and KrF lasers, is now the favorite choice of laboratory users, but reduced operations costs still justify industrial use of Fe or Ar in some cases. The buffer gas plays an important role. By its high pressure (up to several atmospheres), it permits high electrical energy deposition in the medium while the pressure of the excimer rare gas itself is limited by the loss channels which would otherwise arise from the formation of trimers such as Kr_2F and Xe_2Cl. The buffer also acts as a diluant which speeds up the deexcitation rate of non-lasing levels, including the vibrationally excited levels of the B state. It further prevents bottlenecking in the lowel laser level of XeCl by collision increase of the dissociation rate of the weakly bound X state. Of course, care must be taken in order to avoid accidental coincidence of laser emission with buffer gas absorption. This is partly responsible for the superior behavior of neon[10] compared with other rare gases, Ar in particular.

Despite wide attention during more than ten years, the plasma chemistry of self sustained discharge pumped rare gas halide excimer lasers still remains a rich field of investigation and controversy. Such lasers tend to be difficult to instrument because minor alterations of their geometry, circuitry and inner material composition are known to drastically alter their performance and thus modify their operation in uncontrolled ways. In-situ time resolved diagnostics are scarce, and the electron density in a discharge pumped XeCl has been measured for the first time only recently[11]. The dynamics of transient species population and of their electronic or vibrational state of excitation still needs to be experimentally measured for many important discharge-formed species. Most of the abundant excimer literature provides only oscillograms of discharge current and voltage, laser gain and laser pulse. Sophisticated complete device theoretical models, including electrical circuits, optical cavity coupling and plasma kinetics have been built[12]. They are quite able nowadays to reproduce the experimental data. But the poor information content of the available experimental data make such comparison relatively insensitive to the finer details of the plasma kinetics.

It is by now firmly established that in the (Xe,HCl,Ne) discharge, a very large fraction of the upper laser level excitation mechanism is provided by three body ion-ion recombination (Fig. 3). The main contribution comes from the rapid reaction[13]:

$$Cl^- + Xe^+ + Ne \longrightarrow (XeCl)^* + Ne$$

The rate constant is maximum with neon as the third body, another distinct advantage associated with the use of neon as buffer gas rather than He. The "precursor" halogen negative ions are formed by dissociative attachment of the discharge free electrons on vibrationally excited states of the halogen donor, essentially the v = 1 level of the electronic ground

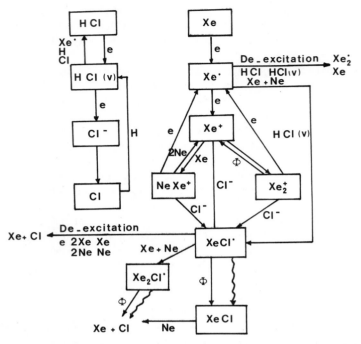

Fig. 3. Most important kinetic reactions used in IMFM XeCl model according
to Gevaudan et al.[12]

state of HCl[14]:

$$e^- + HCl \ (v = 1) \longrightarrow Cl^- + H.$$

Among the identified loss mechanisms, the main saturable losses arise
from:

- electron quenching
- two body quenching by HCl, Ne and Xe
- trimer formation according to[15]

$$(XeCl)* + Xe + Ne \longrightarrow (Xe_2Cl)* + Ne.$$

According to Gevaudan et al[12], trimer formation not only lowers the
excimer upper laser level population but also contributes predominantly
to absorption by the discharge at 308 nm. Non-radiative deexcitation limits the instantaneous excited state density to about 10^{16} cm^{-3}. Thus,
the stored energy on the upper laser level cannot easily exceed about
5 J/l. Since the excimer formation feeds on HCl and laser action leads to
atomic chlorine, the recombination process to reconstitute the HCl stock
is an important process. Unfornately it is rather slow, so that HCl is
consumed as the discharge proceeds, leading to halogen depletion instabilities.

In the (Kr, F_2, Ne) mixture, the "ionic" channel of formation of the
upper laser level is superseted by the "metastable" channel which plays
only a secondary role in XeCl. According to this new mechanism, also called the "neutral" channel, (KrF)* formation results from metastable krypton
atom collisions with the halogen donor molecule fluorine F_2 (Fig.4). The
reaction is known to proceed with a branching ratio close to one toward
the upper laser B state. It is representative of a class of reactions called
"harpooning" reactions[16]. Due to the high electronegativity of the
halogen molecule and the low ionization potential of the rare gas metastable level, the excited electron acts as a harpoon and leads to the creation of an intermediate triatomic complex:

$$Kr* + F_2 \longrightarrow Kr^+ F_2^-$$

which then dissociates upon return of the harpooning electron toward the
rare gas atom:

$$Kr^+ F_2^- \longrightarrow (KrF)* + F.$$

The very short lifetime of the B state, further reduced by quenching,
results in poor energy storage in the upper laser level and precludes
Q-switching operation of both KrF and XeCl. Non-saturable losses are most
troubling for laser design. They occur either by premature quenching of
the "precursor" species in the discharge (via metastable quenching or atomic ion interception before excimer formation) or by photon absorption by
transient species. Photoabsorption by atomic and molecular ions is an important source of non saturable losses. These losses prevent the storage

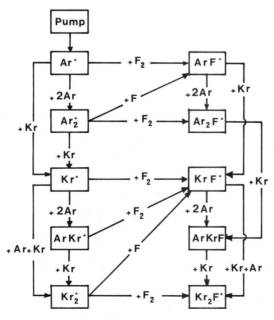

Fig. 4. Neutral channel (KrF)* formation according to Huestis and
Lorents, cited in reference 67

of a high power electromagnetic wave in the cavity and thus role out cavity dumping operation. Since one cannot keep the electromagnetic wave in the laser for a large number of transits through the gain medium, the overall length of interaction available in order to reach laser threshold is short. Thus excimer lasers have to be operated in a high gain mode and the necessary pumping energy is high: >100 kW/cm^3. For the same reason, there is a limit to the power scaling which is achievable by increasing the length of the laser medium. Amplification along the laser axis is given by

$$\frac{dI}{dz} = \left[\frac{g_o - \alpha_s}{1 + \dfrac{I}{I_s}} - \alpha_{ns} \right] I$$

where g_o is the small signal gain, α_s the saturable loss coefficient, α_{ns} the non-saturable loss coefficient, I_s the saturation intensity which is given by:

$$I_s = \frac{h\upsilon}{\sigma\tau}$$

where h is Planck's constant, υ the laser line central frequency, τ the collisional lifetime of the laser upper state, σ the stimulated emission cross section.

Assuming for simplification that the laser line is homogeneously broadened, the stimulated emission cross section can be approximated near the peak of the band by

$$\sigma = \frac{1}{4\pi} \sqrt{\frac{\text{Log } 2}{\pi}} \; \frac{A\lambda^4}{c.\Lambda\lambda}$$

where A is Einstein's spontaneous emission coefficient, λ the emission wavelength, c the speed of light, $\Lambda\lambda$ the FWHM of the laser emission band.

For KrF, with $\lambda = 248$ nm, $\Lambda\lambda \cong 2$ nm and $A \cong 10^8$ s^{-1}, one obtains

$$\sigma \cong 3 \; 10^{-16} \text{ cm}^2$$

which leads to

$$I_s \cong 350 \text{ kW/cm}^2 \text{ assuming } \tau \cong 2 \text{ ns.}$$

For large intensities, $I = 3 \times I_s \cong 1$ MW/cm^2, the net saturated gain goes to zero. With g_o close to 10% cm^{-1} which is achievable at high pumping energy, the maximum useful laser length is limited to a few tens of centimeters, typically less than a meter.

The principal parameter which governs the efficiency of an electric discharge rare gas halide laser is the ratio of the electric field E to the buffer gas density N. At atmospheric pressure, the optimal E/N is of the order of 10^{-15} V. cm^2 and corresponds to electric fields in the range of 10^4 - 10^5 V/cm. At higher applied voltages, the fraction of energy deposited in the gas that results in rare gas atom excitation decreases because of two-step excitation and ionization. Since for industrial purposes it quickly becomes awkward to handle voltages much higher that 25 to 35 kV, this means that only relatively small electrod gaps amounting to a few centimeters can be used. As a result, one has to create an electric field applied transversely across the laser tube. The formation on a nanosecond time scale of a transverse homogeneous discharge of high electron density in a highly electronegative gas mixture at high pressure presents a technical challenge which has been the subject of a vast majority of the publications concerning electric discharge excited excimer lasers.

C. Discharge analysis

In a high pressure rare gas halide discharge, true steady-state operation is never achieved: the electron number density increases (by neutral species ionization) at the same time as the electron loss rates decrease (by depletion of the halogen donor which operates as electron sink via attachment). There is however a very large difference in the time constant for depletion (microsecond) and for ionization (tens of nanoseconds). Transient equilibrium can thus be achieved on a short time scale even though long term equilibrium cannot be reached in the attachment dominated regime. Like all transient phenomena, the quasi steady-state equilibrium of the excimer discharge depends strongly on the initial conditions. Control of the discharge initiation has been the key to many recent excimer laser research progress.

It has been appreciated since Townsend that the single most important physical process for plasma formation is electron avalanche. Uncontrolled electron leads to arcs or filamentarly discharges rather than to the homogeneous discharges which are necessary for proper laser action. Since each avalanche is formed from a primary seeding electron, the spatial distribution of the electron avalanche and thus the discharge homogeneity is directly connected to the seed or "preionization" electron distribution. According to the "streamer" model applied in 1975 to the TEA CO$_2$ laser by Palmer[17], the initial electron density must be sufficiently high to insure that the electron avalanche overlaps from the very beginning of the discharge (Fig. 5). This appears to be the only way to avoid space charge field gradients which develop into arcs or filaments. Levatter[18] improved the model by taking care of the applied electric field variation, the voltage rise time and the gas property changes during the early stage of the discharge. Further refinements were recently applied by Gevaudan[19]. Initial homogeneous electron densities between 10^6 to 10^8/cm^3 have been found necessary in order to achieve homogeneous discharge formation in the rare gas halide lasers[20]. Such high "seed" electron densities can be supplied only by rather efficient auxiliary sources of ionizing radiation able to create electron-ion pairs at a rate of 10^{15} - 10^{17} cm^{-3}. s^{-1}.

Most commercial lasers currently rely on U.V. preionization (Fig. 6) achieved with ultraviolet light from sparks or corona discharges that are

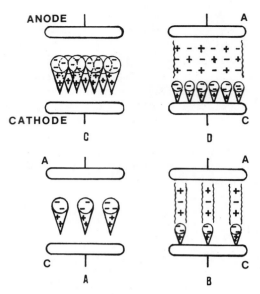

Fig. 5. Evolution of a discharge according to the initial preionization electron density[19]. (A, C) avalanche distribution arising from the initial electrons. (B, D) established discharge (A-B unstable cases, C-D stable situations).

generated inside the laser gas before the main discharge is fired. Because U.V. Light is strongly absorbed by the laser medium, only a limited depth of gas can be preionized in that manner. This sets one of the most stringent limit on the size and thus power increase of the excimer laser head. A large body of empirical evidence has been gathered about the relevance of:

- threshold preionization electron density
- duration of preionization
- preionization electron loss process
- time delay between preionization and main voltage application.

The mechanism of preionization in fluorine rich mixtures remains vertheless poorly understood since free electrons undergo extremely rapid dissociative attachment with the halogen donors to form negative F^- ions. The resulting homogeneous F^- distribution may act as a volume electron source when the discharge voltage is turned on, but this remains to be further demonstrated convincingly[21].

In the laboratory, U.V. preionization has been superseded by X-ray preionization which offers much more flexibility and is much easier to control for diagnostic purposes[22]. With laboratory X-ray guns, it become feasible to preionize large volumes with a good homogeneity, thus lifting some of the most severe constraints on power scaling[23]. Although several attempts are underway in order to introduce X-ray preionized excimer lasers on the commercial market, it must be pointed out that muche remains to be done in order to improve the reliability of the pulsed X-ray guns using cold, hot or plasma cathodes. There has been a marked shift in current

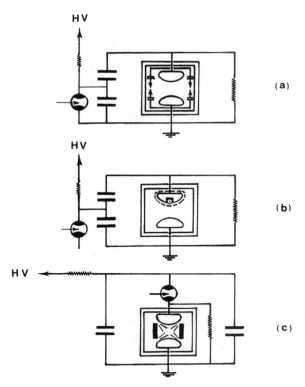

Fig. 6. Examples of U. V. preionized laser head and schematic
electrical circuits[3]. (a) parallel spark (LC inverting circuit);
(b) corona preionization; (c) semiconductor preionization.

industrial research which now has to improve the technology of two largely
independent susystems: the laser head on one side and the X-ray source on
the other side, both being separated by a thin metallic window (a few mm
of aluminium for example; Fig. 7). Fortunately, it has been demonstrated
that as little as 10^{-4} rad of 20 to 40 KeV X-rays is sufficient for effectively stabilizing the discharge of an excimer laser[24]. Such a low requirement leaves a good possibility that the technical problem raised by the
design of an industrial X-ray gun will be solved reliably.

Efficient gas mixture preionization is a necessary condition for a
homogeneous discharge but under the normal mode of operation of discharge
lasers, it is not a sufficient one. Since the main discharge electric field
is turned on after preionization, care must be taken so that during the
transient regime resulting from the interaction of the growing E field and
the growing space charge, not enough time is left for the development of
transverse inhomogeneities in the interelectrode space: in particular,
electron depletion in the cathode space must be avoided. An instantaneous
stratification of the electron density between the electrodes may locally
favor arc or filament formation. Practical as well as theoretical considerations lead to the conclusion that the voltage rise time should be made
as short as possible with upper theoretical limits of: 250 ns for (Ne,Xe,
HCl) at 3 bars as against 25 ns for (He,Xe,HCl), another advantage in favor

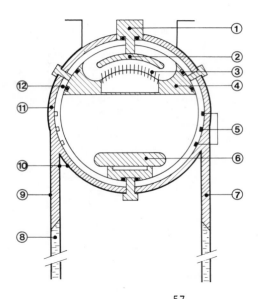

Fig. 7. X-ray preionized laser schematic[57]. (1) X-ray gun connector; (2) X-ray gun anode; (3) X-ray cathode and spatial filter; (4) laser cathode; (5) field electrodes; (6) laser anode; (7) solid dielectric line; (8) water line; (9) ground return; (10) anode connection; (11) glass-epoxy body; (12) teflon chemical shield.

of neon buffer gas[19]. For KrF, this requirement is even more stringent. This is enough to explain why KrF discharge lasers have regularily been found so far to display poorer performance than XeCl laser. Time constant considerations have a large impact on laser design. In order to assure such sharp voltage increase, the inductance of the energy deposition circuit has to be made as small as possible, with important consequence on laser head and electrical circuit layout.

D. Device design

Laboratory devices have been built so far with the purpose of improving our understanding of the discharge behavior and to investigate the peak power and efficiency limits of the rare gas halide excimer laser. Low repetition rates are sufficient for that objective, thus most reported laboratory lasers did not address the issues raised by the high repetition rates and high average powers which are necessary for industrial use. Single shot devices using either U.V. or more recently X-ray preionization must then be considered purely as test beds and researchs tools necessary for proof of concept demonstrations.

a. Low repetition rate lasers

U.V. preionization is limited to low and medium peak pulse power laser heads. The corona discharge has been found to provide enough useful U.V. light and is a lesser contaminant to the laser gas mixture than sparks or arcs. It is best applied through a screen electrode using the old trigger electrode discharge concept[25] (Fig. 6).

A wide variety of X-ray gun concepts have recently been tested, and development work on new design is underway. The main drawback from a practical point of view is the introduction of expensive vacuum equipment. As mentioned before, the reliability of the cathode is also a critical issue (Fig. 7).

Low repetition rate laser heads (<10 Hz) generally operate with little gas circulation, slow axial flow being sufficient to insure gas purity when combined with a cryogenic trap according to reference 26.

Under such circumstance, it becomes geometrically easier to couple the laser head and the main discharge circuitry with low inductance connections (a few nH). Best results have been obtained through the use of an electrical circuit based on a pulse forming line (normally, a deionized water line) connected to the laser head by a rail gap[10,27].

In combination with X-ray preionization (Fig. 7), one achieves then full flexibility for the independent control of two crucial parameters:

- the initial electron density
- the main voltage rise time.

Incorporation of these design principles into high repetition rate devices is by no means trivial. Water lines for example, which have been intensively used as cheap storage and pulse forming laboratory devices, become very difficult to operate at high repetition rate because of water conductivity problems. Spark gaps which offer interesting switching capability for laboratory use evidently have to be given up in the industrial world.

b. High repetition rate device

The efficiency of the conversion of electrical energy deposited in the discharge to optical energy extracted from the laser medium remains low ($\leq$ 4%). Excess energy appears as thermal and acoustic waste and strongly perturbs the laser medium. Even at medium repetition rates (> 10 Hz) diffusion alone does not provide a sufficient drain and the return to equilibrium is too slow in the laser cavity (several hundreds of milliseconds[28]). Circulation of the gas through the electrodes at high speed is a solution which has been carried over from years of previous research on TEA CO_2 lasers. Diffusion of the heated gas as well as purging of the recirculation zones imply that the clearing ratio (the ratio of volume of gas passing between the electrodes between two pulses to the laser head volume) is larger than three in order to insure full gas renewal and return to equilibrium.

All current commercial devices use the internal flo-loop configuration which integrates the components into a compact cylindrical pressure vessel. A transverse fan is used to match the high aspect ratio flow geometry (Fig. 8). Such circulators still have a large potential for development, and their operating characteristics at high speed and high pressure need to be further investigated[29] in order to better appraise the limits of the internal flow laser designs.

The circulator bearings are critical components under all circumstances. Their exposure to corrosive halogen rich mixtures is difficult to avoid in conventional set-ups without endangering the rotating shaft seal. Magnetic bearings and magnetic couplings are possible solutions, however expensive ones at the moment.

The electrical switch of the fast rise time ($> 10^{12}$ A/s) electrical pulse power supply is the other component which suffers most from operation at high repetition rates. Air or oil cooled thyratrons have long been limited to 10^{6} shots lifetime with wide statistical fluctuations under the severe requirements of excimer discharges. Magnetic assist and magnetic pulse compression[30] have considerably relaxed the switching requirements and the thyratron performances have improved accordingly, up to a point were thryratron lifetime does not seem to be a problem any more at current conservative commercial levels ($> 10^{8}$ shots)[31].

2. STATE OF THE ART

A. Commercial product survey

Until very recently, commercially available rare gas halide lasers have been designed with the scientific instrumentation market in mind. Specifications of current products neverheless give a good picture of the level of performance which can be achieved today on a routine basis, even if the operation of most existing devices still requires daily assistance of expert scientific laser users.

For future industrial users, the maximum average power and the pulse repetition frequency are useful parameters in order to evaluate the pro-

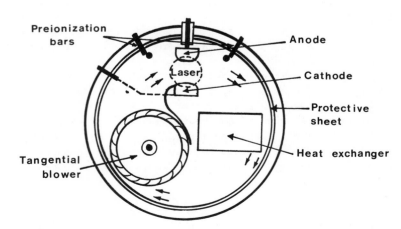

Fig. 8. Cross-section of a typical high repetition rate U.V. preionized avalanche discharge excimer laser[4].

ductivity of a potential U.V. laser process. A record breaking 150 W average power laser has been offered to prospective buyers at the beginning of 1986. Middle of the road products tend to achieve significantly lower values so far, between 10 an 100 W. Concerning the pulse repetition rate, 500 Hz remains the highest reliable commercial offering. High P.R.F.'s are associated with lower pulse energy and efficiencies, with the best available compromises ranging between 1 J/100 Hz and 300 mJ/500 Hz. At very low repetition rates, a maximum pulse energy of 5 J is available for KrF as well as XeCl. Under all circumstances, the wall plug efficiency is never much higher than 1%, very often much lower than that.

With short pulse operation (typically 20 ns), discharge driven excimer lasers have limited time to develop good spatial and temporal coherence. Beam quality is thus likely to become a major issue, with emphasis on low beam divergence, good beam homogeneity and good pulse to pulse reproducibility. For high performance devices, best results are obtained with Master oscillator / Power amplifier configuration (Fig. 9) at the cost of a significant increase in complexity, which will raise difficult technical issues for an industrial product[32].

Maintenance and reliability have already received considerable attention. Twenty four hour testing is underway at all major producer plant and some national laboratories[33]. Laser gas lifetimes are limited by slow consumption of the halogen component and by discharge generated impurities. Thus, the excimer laser power has a natural tendency to decline during

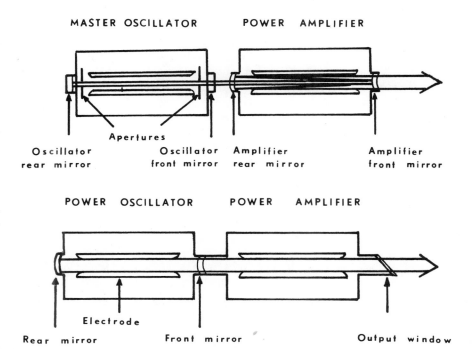

MASTER OSCILLATOR   POWER AMPLIFIER

Apertures

Oscillator   Oscillator   Amplifier   Amplifier
rear mirror   front mirror rear mirror   front mirror

POWER OSCILLATOR   POWER   AMPLIFIER

Electrode

Rear mirror   Front mirror   Output window

Fig. 9. (A) Typical MOPA (Master Oscillator – Power Amplified) and (B) POPA (Power Oscillator – Power Amplifier) layout for excimer lasers according to reference 32.

operation. XeCl lifetimes exceeding 10^8 shots have been achieved by a combination of better gas handling practice[26] and computer controlled long term stabilization of the laser output[31,34] which is now a market standard.

At the repetition rates which are required for high industrial process throughput, component lifetimes become of importance. Maintenance including major component replacement: switch, electrode, optics... is still required at least every 3×10^8 shots with XeCl. Electrode wear and thus optics degradation is somewhat larger for KrF.

B. Recent laboratory achievements

Laboratory results, even more than current commercial product specifications, are quite remote from the performance which should be expected from an industrial laser. They are useful however to set the limits of the achievable results under the best possible conditions and should be used in order to plan realistic targets for industrial development. As usual, ultimate performances concerning pulse energy, efficiency... have never been achieved simultaneously, but rather with devices purposefully biased in order to achieve record breaking results in one direction.

Highest peak energy is recorded at 60 Joules[35] for XeCl (at 0,9% efficiency) and average powers beyond 1 kW at 1 kHz have been reached with KrF in Los Alamos as cited in reference 36. Long pulse operation (which is a key to efficiency emprovement) has been demonstrated up to 1.5 microsecond[37] while a record 4.2% efficiency has been reached at an energy extraction of 3.5 J/1 in XeCl[38] with FWHM of 120 ns using the double discharge concept. According to its inventors, at the cost of added complexity of the electrical circuitry, this is the easiest way to achieve an optimized energy transfer to the discharge during both the early high impedance stage and the late low impedance stage of the discharge (Fig. 10). The best specific energy extraction reported so far, but at much lower efficiency, is 5.8 J/1 at 1.8% efficiency[39].

Very promising results concerning the reliability of a new X-ray gun concept have recently been obtained at ONERA[40] and IMFM[41]. In this secondary emission electron gun, ions are generated in a low pressure He ionization chamber ($\sim$50 m Torr). They are extracted by a grid and accelerated toward a permanently biased high voltage cathode ($\sim$120 kV). The secondary emission electrons produced on the cathode are accelerated toward the grid, go through the ionization chamber and are X-ray converted while going across the laser thin metallic side-wall (Fig. 11).

This type of electron gun is switched on by application of a relatively low voltage ($\sim$15 kV) pulse on a thin thermoionic wire at the center of the ionization chamber. Using a 50×4 cm^2 gun at 70 mA/cm^2 peak current density, a 50 mrad X-ray dose per pulse was obtained in the component wear. This patented design[42] clearly represents a potential solution of great interest for the industrial X-ray preionized rare gas halide discharge laser.

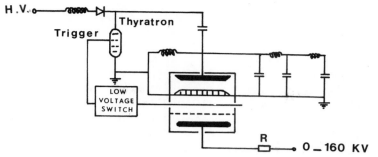

Fig. 10. Electrical configuration for double discharge operation[38].

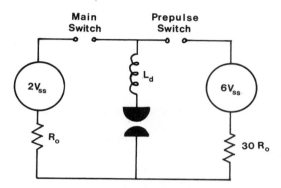

Fig. 11. Diagram of the IMFM laser head fitted with the patented X-ray gun by D. Pigache[41].

3. CURRENT TRENDS

It should be clear from the introductory survey just given, that rare gas halide discharge driven laser technology still has a large potential for improvement. Most observer agree that a mature excimer technique has a good chance to meet the requirements of the industrial production line, at least for high added value material processing.

In a system which reaches the optimization stage, integration, and thus cross correlations between various subparts, becomes more and more important. Every component change has to be carefully evaluated for its direct benefit against a background of detrimental side effects on other parameters of the complete equipment. There remain nevertheless critical areas where progress would almost always justify doing away with carefully crafted system trade offs. For the rare gas excimer laser three such points are currently under severe scrutiny:

- high pulsed power electrical circuitry
- high speed laser gas recycling
- preionization and long pulse kinetics.

Success in any one of these three areas has led or will lead to major changes in the current architecture of the excimer laser.

We will only briefly discuss the first issue, since the techniques which have been used in order to improve the reliability and the efficiency of the excimer laser main discharge circuit, mainly magnetic assist switching[43,44] and magnetic pulse compression[30], are already implemented in various commercial products and are being pushed to new extremes with the development of the vapour laser. Furthermore, the stringent requirements imposed on the electrical circuit by the classical operating procedure of the discharge laser may be soon considerably relaxed by some of the new operating modes which result from our better understanding of the preionization role.

Problems raised by gas recycling have a more fundamental character and will be carefully addressed in the following.

A) Closed cycle excimer fluid dynamics

Under the high repetition rates and high peak power per pulse conditions which are characteristics of future industrial lasers, the fluid dynamics of the gas loop becomes a central issue. "Natural damping" of the aerodynamic perturbations due to the heavy excess energy deposit in the discharge volume which occurs during each pulse is not fast enough to insure the density uniformity in the laser cavity required by optical beam quality[45] and discharge stability[46].

Attenuation of pressure disturbances from levels of the order of 1 atm to levels of 0.0001 atm is typically required, in less than 10 ns. A complete survey of aerodynamical problems connected with excimer laser operation was recently given by Cassady[29]. Only the principal conclusions will be summarized in the following, with examples taken from the paper by Forestier et al.[47]

Two types of pressure waves can be differentiated by their propagation direction: longitudinal and transverse waves.

Transverse pressure disturbances that propagate along the optical axis have little effect on the optical quality of the medium but must be kept below 1% to avoid discharge instabilities. Unsteady pressure waves also propagate between the electrodes. They are generated by spatially non uniform energy loading and structural vibration. Attenuation can be achieved by dampers behind the anode, by tilting the reflecting sidewalls to "walk" the waves out of the cavity or by diffraction from the electrode contours themselves.

More troublesome are the longitudinal waves which travel in the upstream or downstream direction (Figs. 12 - 14). One dimensional, unsteady, non-linear analysis has been used with some success to model complete closed-cycle laser flow loops. Two dimensional modeling has also been applied locally[48,49].

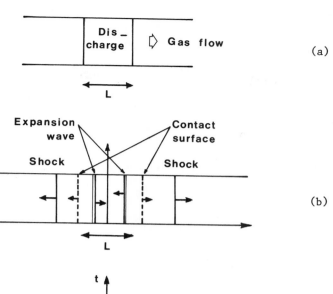

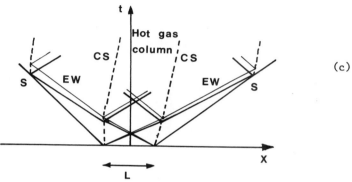

Fig. 12. Simplified aerodynamic perturbations in the laser cavity[3].
(a) discharge geometry cross section; (b) wave front evolution
after the discharge; (c) one dimensional time evolution of the
perturbation.

Broad band, passive damping that minimally affects the average flow
is best achieved with sidewall mufflers (see Fig. 15). Such devices have
been found to stretch the wavelength of the longitudinal waves so that
the net change of index of refraction across the optical aperture width
becomes acceptably small. In order to reduce their length, which scales
with the diameter of the flow channel, dampers should be located as close
to the cavity or upstream wall generated disturbances that are convected
into the cavity.

The interaction of pressure waves with regions of nonisentropic flow
(i.e., through screens, honeycomb, heat exchangers...) gives rise to late
time density disturbances known as entropy waves:

$$\left(\frac{\Lambda\rho}{\rho}\right)_{entropy} = K\,C_D\,M\left(\frac{\Lambda P}{P}\right)_{acoustic} \quad ;$$

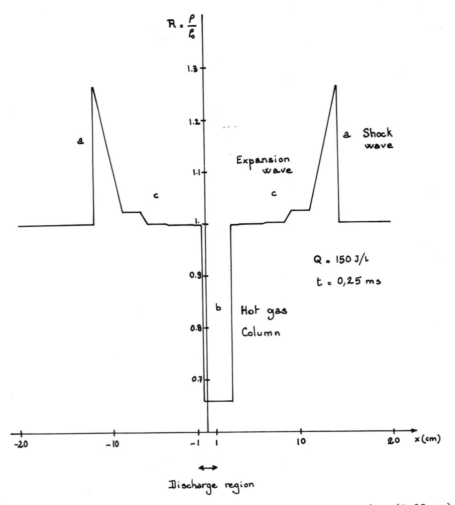

Fig. 13. Instantaneous density gradient in the laser cavity (0,25 ms) after a single excitation pulse as calculated according to the one dimensional model reported in reference 3.

with $K \sim (\gamma - 1)/ \gamma$ for a simple screen, C_D is the drag coefficient of the flow element, M the local Mach number.

Fortunately, the entropy disturbance arising within upstream mufflers is confined to the vicinity of the sidewalls. On the other side, the boundaries of the hot gas volumes produced by the preceeding pulses move downstream at the flow speed and create discontinuities that reflect pressure waves back upstream into the cavity region. The current return grid which is usually required to minimize discharge circuit impedance can be designed to promote enhanced axial mixing of the hot/cold gas interfaces by turbulent transverse wawes. The flow expansion which usually follows the cavity can also efficiently reflect the entropy waves downstream[49].

The fluid dynamic problem of high repetition rate excimer lasers still needs further consideration in order to provide designers with the neces-

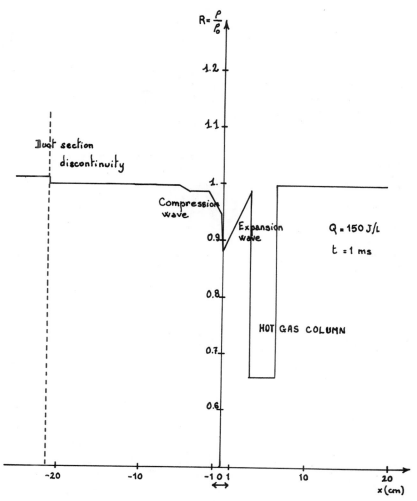

Fig. 14. Same situation 1 ms after the pulse and taking into account
the converging section before the laser cavity[3]

sary tools to achieve optimal trade off between conflicting system requirements:

- fast gas renewal
- low turbulence flow
- fast acoustic wave damping
- efficient circulator operation.

Although internal flow loop configuration may in the long run be used for fully developed high power industrial lasers, research is currently carried out with wind tunnels using centrifugal blowers[47]. They readily provide the flow speed necessary for good clearing ratio at high repetition rate (> 100 m/s) and they are easy to instrument. A closed laser flow

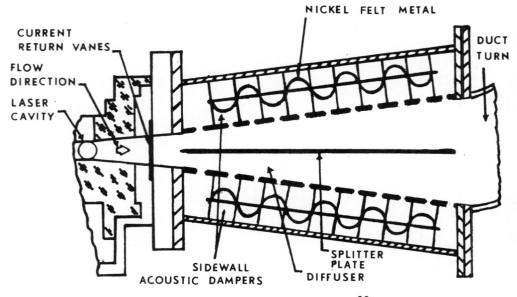

NICKEL FELT METAL

CURRENT
RETURN VANES

FLOW
DIRECTION

LASER
CAVITY

DUCT
TURN

SIDEWALL
ACOUSTIC DAMPERS

SPLITTER
PLATE
DIFFUSER

Fig. 15. Sidewall acoustic damper[29].

loop that can achieve an RMS medium homogeneity of about 10^{-4} within 18 acoustic transits after each lasing pulse, for an initial cavity overpressure of 0.4, has recently been operated successfully.[48]

B. Preionization and long pulse kinetics

Despite the fact that preionization has long been recognized to play an essential role in discharge excited rare gas halide lasers, it has been very difficult so far to obtain experimental information on the many parameters which control homogeneous discharge initiation and short term discharge equilibrium. Work is underway in various laboratories in order to gather more physically significant measurements on the time resolved behavior of the discarge and on the kinetics of discarge created atomic and molecular species[50,53]. This long term endeavor still suffers from the lack of precision of number density and reaction rate measurements. Empirical discharge research thus remains of value at least as a guide for more quantitative plasma kinetic work.

a. Preionization with U.V. laser light

Introduced by Taylor, Alcock and Leopold[54], the laser-induced preionization of XeCl laser discharges is a valuable tool for the precise investigation of the discharge properties. In a recent paper[55], Taylor presented an important summary of the experiments from the Ottawa group using the Krf laser as a flexible preionization source for the XeCl laser discharge (Fig. 16). Although three different laser structures were operated during the course of the Ottawa experiments, a common feature was their electrical circuit operation mode with a fast (10 - 20 ns) high voltage spike to break down the gas mixture. Even within the limits of this conventional mode of operation, laser induced preionization can be used to control the time and space resolution of the preionization as well as its intensity. The main conclusions are:

(a) large preionization gradients in the electric field direction result in only modest changes of laser beam uniformity
(b) beam uniformity depends approximately linearly on preionization gradients transverse to the field direction
(c) discharge uniformity down the length of the laser exhibits the same sensitivity as case b to preionization gradients.

The most unusual behavior is obtained when the preionization beam is blocked over part of the interelectrode space (Fig. 17). This result, reproduced with other preionization sources[56] and other operation modes[57] indicates that: "regions in the discharge with insufficient preionization electrons undergo a local change in discharge uniformity over the entire duration of the discharge despite efforts of the electric field to eliminate such gradients".

This can be explained only through a better understanding of the time behavior of the discarge. Systematic recordings of the discharge fluorescence with fast image converters proved to be very valuable during the Ottawa experiments. A rather complete phenomenological picture of the instability processes which develop during a long pulse ($\sim$400 ns FWHM) of energy deposition in the discharge is given by Taylor[55]. On a qualitative basis, the observations seems to confirm that two basic mechanisms are at work:

1. The halogen depletion instability[58,59]. The halogen donor molecule which serves to buffer to growth of electron number density by attachment dissociates during the discharge. Direct electron impact, dissociation reactions with rare gas excited states, attachment of secondary electrons and collisional quenching of rare gas halide excited states all act to decrease the HCl or F_2 density, hence leading to instabilities when HCl or F_2 density falls locally be below that required to balance two step ionization;

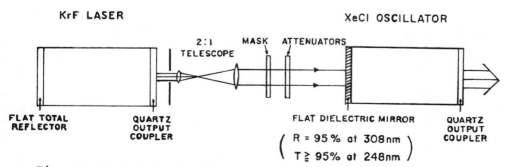

Fig. 16. Laser induced preionization experimental layout in Otawa[55].

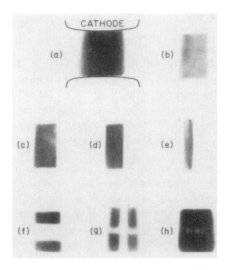

Fig. 17. Laser burn pattern on U.V. sensitive paper for masked preionization beam[55]. (a) with full aperture preionization; (b - h) with
various blocking patterns.

2. The thermally driven cathode sheath instability[60,61]. A local increase in E/N (which usually occurs in the cathode sheath region)
leads to an increased electron production by ionization which
produces an increase in local gas temperature, which further increases E/N locally and thus initiates the instability. According
to Taylor's experimental evidence, hot spots develop at the cathode.
The alogen gas in the mix promotes initial growth of a well defined filament from each hot spot. Filament development leads to
glow discharge collapse and discharge pinching in the cathode
region. Although electrical power can still be deposited into the
discharge in fully developed filament mode (without arcing), strong
induced discharge inhomogeneities lead to premature lasing disrup-
tion.

Due to the much longer time constants involved in the XeCl laser
kinetics, it seems nevertheless possible to achieve long pulse operation[37]
under the following conditions:

(1) low and uniform electrical energy deposition (≤ 200 kW cm^{-3})
(2) low Xe partial pressure (≤ 15 Torr)
(3) low HCl partial pressure (≤ 1 Torr)

This will be considerably more difficult with KrF, as confirmed by
recent results in Marcoussis which we will now discuss.

b. Laser self-switching

As mentioned before, discharge excited lasers have always been operated through a combination of a gas preionization followed by a fast high
overvoltage triggering the main discharge. Such a scheme implies that a

fast low inductance switch is introduced between the energy storage element and the discharge. A new operation mode was put forward in 1982[62] which solved the problem raised by the fast electrical switch (Fig. 18). The laser head is directly connected to the storage capacitor and loaded to its design voltage with a relatively slow rise time pulse. The triggering of a stable discharge is then obtained by rapid creation in the interelectrode spacing of the required electron density using photoionization either by a corona U.V., X-ray or KrF laser photon flux.

In this case, electron-ion pairs are injected in the presence of the applied electric field. The avalanche process begins as soon as the electrons are created, with an e-folding time as short as 4 ns. As a result, the minimal preionization density has to be reached in less than 4 ns, since pairs produced at a lower rate have a negligible effect compared to the electrons generated by the multiplication process in the discharge.

Recording the time lag between triggering and voltage drop across the discharge, Lacour and Vannier[57] obtained a rather sensitive measurement of the avalanche time with a fixed applied electric field across the discharge as a function of gas mixture composition, reduced field and gas pressure (Fig. 19). Under most circumstances, the avalanche process is faster in KrF than in XeCl, thus making the fluorine rich discharge more difficult to control than the corresponding chlorine rich one, and of course that the pure rare gas discharge.

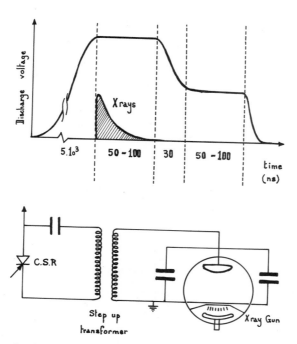

Fig. 18. Photoionization triggering principle and experimental layout.

Photoionization self switching has been actively investigated in Marcoussis due to its obvious potential in terms of circuit simplicity and laser compactness. High repetition rates (up to 350 Hz) and high efficiencies (2.2%) have been achieved as well as high peak pulse power: 1.7 Joules with an energy extraction of 1.7 J/l in XeCl. This mode of operation seems to require during the first few nanoseconds of the avalanche a much higher level of preionization electron density than traditional excimer laser operation. Later on during the energy deposition pulse, the set up is favored by the impedance matching between the energy storage circuit and the laser discharge, which is easily achieved without the need for an expensive duplication of the electrical circuitry as required by the original double discharge concept[38].

In contradiction with some statements presented by recent users of double discharge triggering, the efficiency of the self triggered laser with pulse forming line is not intrinsically limited by impedance mismatch during the early breakdown phase. As pointed out by Long at al.[38], with a line charged at twice the discharge steady-state voltage, nearly complete energy transfer to the gas is theoretically possible in a long discharge mode of operation ($\geq$ 100 ns).

Self triggering is best suited for XeCl discharge lasers and stil requires progress to be made in the design of fast rise time high flux preionization sources. Double discharge operation with magnetic isolation[44] remains an interesting option for short laser pulse $\leq$ 50 ns and fluorinated compounds if full solid state power switching can be implemented on high repetition rate designs without excessive losses in the saturable magnetic elements.

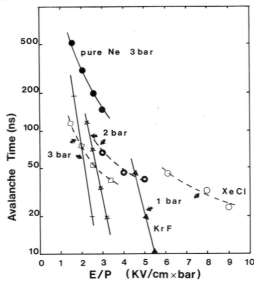

Fig. 19. Avalanche time as a function of reduced field for typical XeCl mixture (doted lines), KrF mixtures (full lines) and pure neon[57].

CONCLUSIONS

A detailed understanding of the rare gas halide discharge laser slowly emerges from ten years of applied and fundamental research. The transient behavior of high pressure transverse discharges has been a challenging test ground for the integration of concepts arising from many domains of physics.

Despite the fact the main laser features are by now understood, there still remain major puzzles in the kinetics of high pressure discharges. Collins et al.[63], working within the framework of a US-France cooperative project in high pressure reaction kinetics, have established the significance of three-body effects that enhance the reaction rates for energy transfer and for ion-molecule reactions that occur in atmospheric pressure of inert gas diluent event in the absence of any formation of product clusters. Unfortunately, in the rare gas halide case few of the necessary three-body reaction rates have been reliably measured. Charge transfer in the reaction

$$He_2^+ + Xe + He \longrightarrow products$$

has recently been found[64] equal to the bimolecular rate at 1 atm of He partial pressure and higher beyond. For several years, the three body charge transfer[65]

$$Ne_2^+ + Xe + Ne \longrightarrow Xe^+ + 3Ne$$

has been expected to play a key role in XeCl kinetics at several atmospheres of neon buffer gas, but it is curiously missing from several prevailing models.

Photoabsorption, photodetachment and/or photoionization of many of ionized species present in the discharge are also poorly known. It recently came as a major surprise that the photoionization cross section of Kr_2F was in fact five times smaller than anticipated[66]. This brings down the absorption losses due to trimer formation and leaves unexplained the observed strong U.V. absorption by the laser medium, late in the pulse when F^- absorption also vanishes. For the physicists, work is not over yet!

However, for the engineers, it is already possible to build simple, efficient and powerful industrial tools based on existing gas halide discharge principles. Industrial (and medical) applications call for an array of distinct specialized device designs. Construction of multikilowatt U.V. lasers is one of the most challenging tasks currently offered in gas laser development. It is actively pursued within government sponsored programs in the U.S. and in Japan. With the Eureka Eurolaser project, Europe is also in a position to develop efficiently its own industrial excimer laser technology.

AKNOWLEDGEMENTS

The author is in debt to his coworkers B. Lacour, P. Pinson and L. Torchin for numerous discussions and useful criticisms while preparing this survey. Thanks are also due to all participants in the Eurolaser project definition phase.

REFERENCES

1. A. V. Eletskii, Excimer Lasers, Sov. Phys. Usp., 21:502 (1978)
2. J. J. Ewing, Excimer Lasers, in: "Laser Handbook", M.L.Slitch, ed. North Holland, Princeton (1979)
3. B. L. Fontaine, Lasers à eximères, in: "Summer School: rectivity in plasmas, application to lasers and surface treatments", Editions de Physique, Les Ulls (1984)
4. P. Burlamacchi, Excimer laser: Pratical excimer laser sources, in: "Physics of new laser sources", N.B. Abraham, F.T. Arecchi, A. Mooradian and A. Sona ed. Nato Asi Series, Plenum Press, New-York (1985)
5. C. A. Brau, Rare gas halogen excimers, in: "Excimer lasers", C.H.K. Rhodes, ed. 2nd Edition, Springer-Verlag, Berlin (1984)
6. M. F. Golde and B. A. Thrush, Vacuum U.V. emission from reactions of metastable inert gas atoms: chemiluminescence of $A_r O$ and ArCl, Chem. Phys. Lett., 29:486 (1974)
7. J. E. Velazco and B. A. Setser, Bound-free emission spectra of diatomic Xenon halides, J. Chem. Phys., 62:1990 (1975)
8. T. H. Dunning and P. J. Hay, The covalent and ionic states of the rare gas monofluorides, J. Chem. Phys., 69:134 (1978)
   The electronic states of KrF, J. Chem. Phys., 66:1306 (1977)
9. P. J. Hay and T. H. Dunning Jr, The covalent and ionic states of the Xenon halides, J. Chem. Phys., 69:2209 (1978)
10. R. S. Taylor, P. B. Corkum, S. Watanabe, K. E. Leopold and A. J. Alcock, Time dependent gain and absorption in a 5 J U.V. preionized XeCl laser, IEEE J. Quant. Electron. QE-19:416 (1983)
11. M. Hiramatsu, M. Furuhashi and T. Goto, Determination of electron density in discharge pumped excimer laser using stark broadening of H line, J. Appl. Phys., 60:1946 (1986)
    R. C. Hollins, D. L. Jordan and J. Coutts, Time-resolved electron density measurements in rare gas halide laser discharges, J. Phys. D., 19:37 (1986)
12. M. Ohwa and M. Obara, Theoretical analysis of efficiency scaling laws for a self sustained discharge pumped XeCl laser, J. Appl. Phys., 59:32 (1986)
    A. Gevaudan, B. L. Fontaine, B. M. Forestier and M. L. Sentis, Modelling of the X-Ray preionized XeCl self-sustained discharge laser in: "Gas flow and Chemical lasers", S. Rosenwaks, ed. Springer-Verlag, Berlin (1987)
13. M. R. Flannery and T. P. Tang, Ionic recombination of rare gas ions X^+ with F^- in a dense gas X, Appl. Phys. Lett.,32:327 (1978)

14. M. Allan and S. F. Wong, Dissociative attachment from vibrationally and rotationally excited HCl and HF, J. Chem. Phys., 74:1687 (1981)

    W. L. Nighan and R. T. Brown, Efficient XeCl (B) formation in an electron beam assisted Xe/HCl laser discharge, Appl. Phys. Lett., 36:498 (1980)

15. D. C. Lorents, Excited state kinetics for XeCl* in: "Proceeding of the International Conference on Laser 84", STS Press (Mc Lean, Va) (1985)

    J. Lecalve, M. C. Castex, B. Jordan, G. Zimmerer, T. Moller and D. Haaks, Time resolved studies of the RgCl (B-X) emission of a synchrotone radiation state selective excitation of Cl_2 by Rg (Xe, Kr, Ar) mixture, in: "Proc. 38th International Meeting of the French Chemical Society", F. Lahmany, Ed. p. 639, Elzenez Scientific Publishers (1985)

16. M. G. Prisant, C. T. Rettiner and R. N. Zare, A direct Interaction model for chemiluminescent reactions, J. Chem. Phys., 81:2689 (1984)

17. A. J. Palmer, A physical model on the initiation of atmospheric pressure glow discharge, Appl. Phys. Lett., 25:138 (1974)

18. J. I. Levatter, Necessary conditions for the homogeneous formation of a volume avalanche discharge with specific application to rare gas halide excimers, Ph. D. Thesis, University of California, San Diego (1979)

19. A. Gevaudian, Modélisation d'un laser à excipiexe XeCl à décharge par avalanche préionisée par faisceau de rayons X, Thèse, Université d'Aix-Marseille III (1986)

20. R. C. Sze and T. R. Loree, Experimental study of a KrF and ArF discharge laser, IEEE J. Quant. Elect., QE-14:944 (1978)

21. S. C. Lin, C. E. Zheng, D. L. J. Matsumoto and S. B. Zhu, Attachment kinetics and life time of preionization electrons in rare-gas halogen laser mixtures, Appl. Phys., B 40:15 (1986)

22. J. I. Levatter and S. C. Lin, Necessary conditions for the homogeneous formation of pulsed avalanche discharges at high gas pressures, J. Appl. Phys., 51:210 (1980)

23. A. V. Kozyrev, Y. D. Korolev, G. A. Mesyats, Y. N. Novselov, A. M. Prokhorov, V. S. Skakun, V. F. Tarasenko and S. A. Genkin, Use of X-ray radiation to preionize the active medium in high pressure gas lasers, Sov. J. Quant. Elec., 14:356 (1984)

24. C. R. Tallman and I. J. Bigio, Determination of the minimum X-ray flux for effective preionization of an XeCl laser, Appl. Phys. Lett., 42:149 (1983)

25. R. Dumanchin and J. Rocca-Serra, Augmentation de l'énergie et de la puissance fournie par unité de volume dans un laser à CO_2 en régime pulsé, C.R. Acad. Sci. Paris, B269:916 (1969)

26. R. Tennant, Control of contaminants in XeCl lasers, Laser Focus, 17, 10:65 (1981)

27. K. Midorikawa, M. Obara and T. Fujioka, X-ray preionization of rare-gas halide lasers, IEEE J. Quant. Elect., QE-20:198 (1984)

28. R. Buffa, P. Burlamacchi, M. Matera, H. F. Ranea Sandoval and R. Salimbeni, High repetition rate effects in XeCl TEA lasers, Optic. Comm., 40:288 (1982)

29. P. E. Cassady, Fluid dynamics in closed-cycle pulsed laser, AIAA J, 23:1922 (1985)

30. I. Smilanski, S. R. Byron and T. R. Burkes, Electrical excitation of an XeCl laser using magnetic pulse compression, Appl. Phys. Lett., 40:547 (1982)

31. L. Holmes, Excimer lasers take on a practical look, Laser Focus, 22, 7:72 (1986)

32. G. Klauminzer, Oscillator-amplifier approach in excimer lasers, Lasers and applications, 5, 9:75 (1986)

33. G. Balog and R. C. Sze, Lifetime studies of a commercial XeCl excimer laser, p.713, in:"Proceeding of the International conference lasers 85", C. P. Wang ed. STS Press, Mc Lean, Va (1986)

34. H. Pummer, The excimer laser: 10 years of fast growth, Photonics Spectra, 19, 5:73 (1985)

35. L. F. Champagne, A. J. Dudas and B. L. Wexler, Large volume X-ray preionized XeCl laser, Paper ThR81, Cleo 1984, Anahelm, California, June 19 - 22 (1984)

36. B. Fontaine and B. Forestier, Rapport sur l'état des recherches sur les lasers à ecxiplexes de grande puissance moyenne et perpectives à moyen terme, Rapport SGDN n° 16 du 9 Octobre 1984

37. R. S. Taylor and K. E. Leopold, Microsecond duration optical pulses from a U.V.-preionized XeCl Laser, Appl. Phys. Lett., 47:81 (1985)

38. W. H. Long Jr, M. J. Plummer and A. E. Stappaerts, Efficient discharge pumping of an XeCl laser using a high voltage prepulse, Appl. Phys. Lett., 43:737 (1983)

39. K. Miyazaki, Y. Toda, T. Hasama and T. Sato, Efficient and compact discharge XeCl laser with automatic U.V. preionization, Rev. Scl. Instrum., 56:201 (1985)

40. D. Pigache, J. Bonnet and D. David, A short pulse secondary emission electron gun for high pressure gas lasers and plasma chemical reactors, XVI Internat. Conf. on "Phenomena in ionized Gases", Dusseldorf (RFA), 29 August (1983)

41. M. L. Sentis, B. Forestier, B. Fontaine, P. Issarties and D. Pigache, High-pulse repetition rate limitations in a high average power XeCl laser, in: Gas Flow and Chemical Lasers", S. Rosenwaks Ed. Springer-Verlag, Berlin (1987)

42. D. Pigache, French Patent, 72-38368 (1972)

43. R. R. Butcher and T. S. Falhen, Magnetically switched 150 W XeCl laser, CLEO 84 Technical Digest THP1, p. 202 (1984)

44. C. H. Fisher, M. J. Kushner, T. E. Dehart, J. P. Mc Daniel, R. A. Petr and J. J. Ewing, High efficiency XeCl laser with spiker and magnetic isolation, Appl. Phys. Lett., 48:1574 (1986)

45. M. D. Hogge and S. C. Crow, Flow and acoustic in pulsed excimer systems, AIAA Conf. on fluid dynamics of high power laser, Paper n° II-4 Cambridge (MA), Oct. (1978)

46. M. Baranov, D. D. Malyuta, V. S. Mezhevov and A. P. Napartovich, Influence of gas density perturbations on the ultimate characteristics of pulse periodic lasers with U.V. preionization, Sov. J. Quant. Elect., 10:1512 (1980)

47. B. M. Forestier, M. L. Sentis, S. M. Fournier and B. L. Fontaine, Flow and acoustics in a closed-loop high pulse rate frequency XeCl laser, _in_: "5th GCL Symposium, Oxford", Ed. Adam Hilger, London (1985)

48. E. Baum , C. G. Koop, V. A. Kulberny, K. R. Magiawala and J. Shwartz, Density homogeneity control in repetitively pulsed gas laser, _in_: "Gas flow and chemical lasers", S. Rosenwaks, Ed. Springer Verlag, Berlin (1987)

49. C. J. Knight, Sidewal muffler design for pulsed exciplex lasers, AIAA 23rd Aerospace Sciences Meeting, Jan. 14-17 (Reno, Nevada) (1985)

50. H. Shields and A. J. Alcock, XeCl$_2$ fluorescence and absorption in self sustained discharge XeCl lasers, _Appl. Phys._, B 35:167 (1984)

51. J. E. Andrew and P. E. Dyer, Gain measurements in ArF and FrF excimer discharges using axial and sidelight fluorescence detection, _Opt. Commun._, 54:117 (1985)

52. P. Kh. Mirdla, V. E. Peet, R. A. Sorkina, E. E. Tamme, A. B. Treshchalov and A. V. Sherman, Theoretical and experimental investigations of an electric discharge plasma of an XeCl laser, _Sov. J. Quantum Electr._, 16:1438 (1986)

53. A. V. Dem'Yanov, V. S. Egorov, I. V. Kochetov, A. P. Napartovich, A. A. P tor, N. P. Penkin, P. Yu. Serdobintsev and N. N. Shubin, Investigation of the dynamics of the populations of electronic states of atoms and ions in a self-sustained discharge in an HCl -Xe-He mixture, _Sov. J. Quant. Electr._, 16:817 (1986)

54. R. S. Taylor, A. J. Alcock and K. E. Leopold, Laser induced preionization of a rare-gas halide discharge, _Opt. Lett._, 5:216 (1980)

55. R. S. Taylor, Preionization and discharge stability study of long optical pulse duration U.V. preionized XeCl lasers, _Appl. Phys._, B 41:1 (1986)

56. S. Sumida, K. Kunitomo, M. Kaburagi, M. Obara, T. Fujioka and K. Sato, Effect of preionization uniformity on a KrF laser, _J. Appl. Phys._, 52:2682 (1981)

57. B. Lacour and C. Vannier, Photo triggering of a 1 J excimer laser using either U.V. or X-rays, _Appl. Phys. Lett._ to be published

58. W. L. Nighan, Plasma processes in electron beam controlled rare gas halide lasers, _IEEE J. Quant. Electr. QE_, 14:714 (1978)

59. J. Coutts and C. E. Webb, Stability of transverse self-sustained discharge excited long pulse XeCl lasers, _J. Appl. Phys._, 59:704 (1986)

60. E. P. Velikhov, _in_: "Molecular gas lasers", MIR Edition Moscow (1981)

61. R. Turner, The glow to arc transition in a pulsed high pressure gas discharge, _J. Appl. Phys._, 52:681 (1981)

62. O. de Witte, B. Lacour and C. Vannier, Photoionization switching of the gas lasers, _in_: "Conference on laser and Electro Optics", CLEO 82, Technical Digest, paper WD6, Phoenix, Ariz. (Apr. 1982)

63. C. B. Collins, Z. Chen, V. T. Gylys, H. R. Jahani, J. M. Pouvesle and J. Stevefelt, "The importance of three body processes to reaction kinetics at atmospheric pressures", _IEEE J. Quant. Electron. QE_ 22:38 (1986)

64. J. M. Pouvesle, Thesis-Universitè d'Orléans (1986)

65. C. B. Collins and F. W. Lee, "Measurement of the rate coefficients for the bimolecular and termolecular ion-molecule reactions of Ne_2^+ with selected atomic and molecular species", <u>J. Chem. Phys.</u>, 72:5381 (1980)

66. K. Hakutaq, H. Komori, N. Mukai and H. Takuma, Absolute photoabsorption cross-section measurement of the Kr_2F excimer at 248 nm, <u>J. Appl. Phys.</u>, 61:2113 (1987)
    A. W. Mc Cown, Absorption at 248 nm by Kr_2F^*, <u>Appl. Phys. Lett.</u>, 50:804 (1987)

67. T. H. Johnson and A. M. Hunter, Physics of the krypton fluoride laser, <u>J. Appl. Phys.</u>, 51:2406 (1980)

X-RAY LASER WORLD WIDE PROGRESS

D. L. Matthews

Lawrence Livermore National Laboratory
University of California
Livermore, CA 94550 USA

The quest for the demonstration of an x-ray wavelength laser took $\sim$20 years with successful, unambiguous results coming only recently from research efforts in the USA[1-3], Great Britain[4], and France[5]. Numerous reviews[6-8] have been written on the subject, but the most current are actual accounts of various groups' research in "conference proceedings"[9-11]. In this manuscript, I will describe the various types of successful lasing schemes employed to date, comment on their status and speculate on their future development. I will not be describing all attempts to make the x-ray laser, nor listing an account of various theoretical designs, or inversion schemes that are not currently used. Successful x-ray lasers have to date, only been demonstrated by forming a plasma with a high power optical or infrared wavelength laser.

What is an x-ray laser and why is it (or has it been) so difficult to make? An x-ray laser is the same in priciple, as its UV, optical, infrared, and microwave (MASER) predecessors; in that it is a source of extremely bright monochromatic radiation of small divergence (i.e., represented by a beam formed by source geometry, not by optics) which has been produced by the stimulated radiative decay of atoms (or ions, in practice) excited enough to produce x-ray wavelength transitions. In present day schemes, these transitions are due to L-N or L-M transitions in H, He-, or Li-like low z ions or M-M or N-N transitions in Ne- or Ni-like, moderate z ions. In practice, it has been impossible to produce the necessary inversion density among excited levels in atoms without first generating a plasma. X-ray lasers are difficult to make because enormous power is required to produce an inverted amplifier medium, x-ray wavelength oscillators are not yet practical, and the fact that plasma amplifiers are difficult to form with desired characteristics over large dimensions.

To understand the large gap between optical and x-ray wavelength inversion power requirements, consider the simple three-level laser shown in Fig. 1. We know from solving the radiation transport equation for a plasma that the output intensity along a given direction is given by

$$\vec{I} = S \ (e^{GZ} - 1),$$

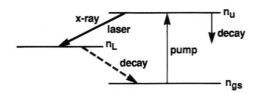

Fig. 1. Generic x-ray laser grotrian diagram which shows paths for pumping of upper state (n_u) and decay of lower state (n_L).

where I is the intensity at a given frequency ν and for an amplifying medium of source function

$$S \equiv \frac{n_n A_{uL}}{(n_u - \frac{g_u}{g_L} n_L) B_{uL}} \quad , \text{ gain } G \equiv \frac{\lambda^2}{8\pi} A_{uL} \, \Delta n \, d(\nu),$$

where $\Delta n = (n_u - \frac{g_n}{g_L} n_L)$, and length Z. Here

and length Z. Here A and B are the Einstein coefficients for radiative rate or absorption and $\phi(\nu)$ is the larger transition's line shape, usually dominated by doppler broadening.

In order to produce gain we require that $N_u/g_u > N_L/g_L$ from the expression above for gain G and in order to maintain it, we need to always pump at a rate faster than the total decay rate of the upper laser level, $A_T = A_{uL} + A_{ug}$. This means that $N_u A_T \leq p$, where p is the volumetric pump rate $(cm^3 - sec^{-1})$. Therefore, the gain, a line center λ_0, can be expressed as

$$G(\lambda_0) \leq \frac{\lambda_0^3 A_{uL}}{8\pi \Delta \nu_D} \Delta n \cong \frac{\lambda_0^3 P}{8\pi v_t}$$

where $\Delta \nu_D = 2 \nu_0 (2kT/mc^2 \, \ln 2)^{1/2} \sim v_t/\lambda$, v_t is the thermal ion velocity. This relation gives our first significant problem with x-ray lasers, namely, the gain scales at the 3rd power of wavelength. Furthermore, if we define the volumetric pump power at $P = p \, (hc/\lambda_0)$, then the gain

$$G(\lambda_0) \leq \frac{\lambda_0^4}{8\pi v_t} \frac{P}{hc} \text{ for } P \, (W/cm^3).$$

Therefore, the pump power (W/cm^3) needed to maintain a constant gain scale as the inverse 4th power of laser wavelength or the 0th power of the atomic number of the atom. Thus, an x-ray laser operating at 100 Å requires a factor 10^8 more power than, e.g., the Nd-glass laser operating at 1.053

μm. As a consequence, due the large power density, it can deliver in brief time intervals, the "conventional" high power optical laser such as is used for inertial confinement research[12] has been predominantely chosen by current researchers[1,3-5] as the pump source for x-ray lasers. Again, due to the exorbitant power requirements necessary to produce a single amplifier, x-ray lasers to date have been made from amplified spontaneous emission (ASE)[13] in the amplifier and not from injecting a single mode beam from a high quality oscillator (which also would require significant power to pump) into a high gain large aperture amplifier. This method of amplification limits beam coherence properties if reasonable power is to be obtained.

The next roadblock to x-ray laser development has been the lack of mirrors operable at normal incidence as well as the production of long-lived amplifiers to take advantage of multiple pass geometries. Current research[14] on multilayer mirrors can possibly overcome part of this problem, at least to provide multipass (perhaps 3 or 4 passes) capability. However, for now, it is out of the question to consider using the conventional procedure of producing a low gain x-ray amplifier ($G \sim 0.001 - 0.010$ cm^{-1}) and then depending on 10^3-10^4 passes to build up significant amplification.

Another problem concerns the use of high temperature, dense plasma as the amplifier. As discussed by Rosen and London[15] and Sobel'man and Vinagradov[8], these plasmas can contain steep electron density gradients and spatial inhomogeneities which can seriously effect the transport of the x-ray laser beam through the amplifier. For example, the distance traveled by a photon emitted in a direction parallel to the longitudinal axis of a cylinder (which represents the amplifier) is given by $L_R \sim L_\perp (2n_c/n_0)^{1/2}$ where $L_\perp$ is the electron density scalelength (distance for density to decrease by $1/e$) transverse to the longitudinal axis of the cylinder, and n_c/n_0 is the ratio of the critical density ($n_c = 10^{13} \lambda^{-2}$ cm^{-3}) for total reflection of x rays wavelength to the peak electron density achieved in the amplifier (normally along the longitudinal axis). Typical $L_\perp$ values for these types of plasmas are from 10-100 μm which, if the density is high enough (10^{21} cm^{-3} e.g.), can lead to propagation distances of only 0.14 to 1.4 cm; far less than desirable.

The last problem unique to x-ray amplification limits the usable diameter of the amplifier and thus reduces the total output power that can be achieved. Typical successful inversion schemes (which are to be discussed in the next section) rely on large radiative rate (A) transitions to empty the lower state of the x-ray laser transition. Necessarily, transitions with such large A values are highly susceptible to radiation trapping, a phenomenon which, if large enough, reduces the A value and ultimately destroys the population inversion. In general, maintaining several optical depths on these lower state dump transitions will prevent the problem. In practice, this limits plasma amplifier diamters (apertures) to dimensions of order 100-300 μm.

How have researchers managed to develop the x-ray laser given the above conceptual difficulties? The answer lies in cleverly utilizing advantageous, population mechanisms and excited-state decay characteristics. Figure 2 illustrates the tricks required to produce population inversions

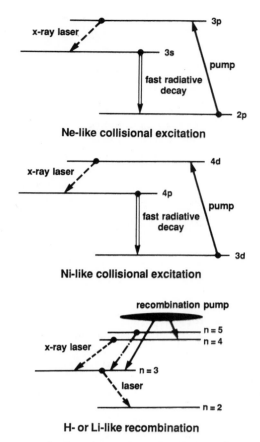

Ne-like collisional excitation

Ni-like collisional excitation

H- or Li-like recombination

Fig. 2. Diagram of three most popular laser inversion schemes.

in highly excited x ray emitting ions. The collisional excitation scheme
for Ne-like, and now Ni-like ions was first described by Elton[16]. It uses
electron collisions to populate n=3s, n=3d orbitals with electrons from
the 2p shell of a Ne-like atom or in the case of Ni-like, 4p, 4d, and 4f
orbitals from the 3d shell. Rapid x-ray decay of some but non all of the
levels then automatically set up the inversion. Non-dipole coulomb interactions of electrons and ions excite the 3p or 4d levels from the corresponding 2p or 3d levels. Once populated, these levels cannot decay by
dipole x ray emission back to the ground-state from which they were formed.
Instead, these orbitals are either collisionally mixed with nearby 3s, 3p
(or 4p and 4f levels) or are stimulated to decay in the form of 3p-3s or
4d-4p transitions. Actually, the most important ingredient in producing the
population inversion is not the population since it is non-selective, but
the fast decay of either of the 3s or 4p levels back to the ground state.
This rapid radiative depletion of the lower level of the potential x-ray
laser transitions makes it relatively simple to maintain the inversions so
long as the upper levels are continuously populated and the density is low
enough so that the 3s-2p or 4d-4p levels are not equilibrated. In the case
of Ne-like, gains of order 5-10 cm^{-1} have been predicted using reasonable
plasma conditions.

The other important method to get a population inversion is to rely on three-body and dielectronic recombination processes. If the ion is quickly ionized to one ionization state beyond that which is desired and then rapidly cooled (to prevent collisional population of the lower stat of a potential lasing transition), then these recombination processes strongly favor population of high n states over lower ones (by a factor of n^4) thus automatically producing a population inversion. These techniques for producing lasers were first introduced by Gudzenko and Shelepin[17] for hydrogen lasers. These ideas have been widely extrapolated today to H-, He- and Li-like ions[2,4,5].

Presuming we can achieve amplification at x-ray wavelengths with the methods introduced above, we conclude our introduction to x-ray lasers with a compilation of their characteristics at saturation. Table 1 gives the linewidths, power in MW, coherence lengths, beam divergence, and spectral brightness as a function of the wavelength and/or doppler width of the x-ray laser transitions. The brightness that can be achieved for a 4.4 nm laser is a factor of 10^{10} times that which is achievable on present day synchrotons. Furthermore, the extremely narrow linewidths (~0.3 times the doppler with) leads to large temporal coherence lengths of order 300-500 μm. Finally, we hope to obtain output power levels of a GW or more, i.e. provided we can achieve saturation at sub 4.4 nm which should be possible based on scaling the power levels already obtained at 20 nm. Achieving this much power would lead to a production efficiency of order 10^{-5} or 10^{-4} (x-ray laser output energy/pump laser energy). Typical x-ray laser pulsewidths to date have been in the 10 nsec to 0.1 nsec range and methods are planned to obtain pulse lengths perhaps a factor of 10x shorter.

GOALS, PRESENT INVERSION SCHEMES AND RESULTS

Most programs of research are now satisfied that they can design and demonstrate amplification at relatively long x-ray wavelengths. The current problem is to develop a powerful, coherent and short-wavelength x-ray laser. In essence, our formidable task is now to develop a good laser that will be useful and efficient. At Livermore, we have set the lofty goal of developing a 4.4 nm wavelength, fully-coherent, 1 GW laser. Most of our colleagues at other facilities share our desire to develop the short wavelength laser as soon as possible. The following summary of research efforts world wide hopefully demonstrates how much emphasis is being placed on winning this coveted "wavelength derby".

RECOMBINATION SCHEMES

The freely and rapidly expanding plasma formed by a high power laser irradiating a solid target has become one of the most popular methods[2,4,5] of producing a population inversion, and if line focus geometries are used, of forming amplifier. The Grotrian level diagram depicting the ions and levels normally used in these schemes was shown in Fig. 2. G. Pert[18] of Hull University, has pioneered a novel and efficient approach to this method of producing amplification. By using both small mass and small initial radius

Table I. Satured X-ray Laser Characteristics

$\Delta\nu$ (frequency width)　　　　　　　　$\sim 0.3 \, \Delta\nu_D$

Power (MW)

$\sim (\dfrac{D^2}{10^{-4} \, cm^2}) \, (\dfrac{200 \, \overset{\circ}{A}}{\lambda})^4$　at constant inversion density

Transverse coherence length　　$\sim L\lambda/D$

Longitudinal coherence lenght　$\sim \lambda^2/D\lambda$

Beam divergence either D/L or λ/D, whichever is greater

Spectral brightness

$$\sim \frac{\text{power (MW)} \times (6 \times 10^{24}) \times \dfrac{1}{h\nu(eV)} \, \text{photons}}{|D(mm)|^2 \, mrad^2 \, (0.01\% \, \text{bandwidth})}$$

---

fibers of pure carbon, (typically 7 to 8 μm in diameter) or most recently, carbon coated with Al or LiF, this method optimizes the coupling of the available laser energy to the target. The ensuing plasma then rapidly strips the atoms to bare (C and F) or He-like (Al) ionization stages and then, by "superadiabatic expansion" quickly cools the plasma to foster strong re-combination excitation into upper levels of CVI, FIX or AlXI without promot-ing collisional excitation into lower levels.

Researchers[4] at Rutherford-Appleton Labs (RAL) in England, have demon-strated substantial gain ($\sim$3-4 cm^{-1}) using this exploding fiber concept for producing amplification of the Balmer alpha transition in CVI at 18.22 nm and FIX at 8.1 nm. They use their Vulcan laser facility to produce up to 1 TW, 70 ps pulses of 0.53 μm of pump laser light. In collaboration with P. Jaegle and colleagues[5], this same group has also measured gain on the 5f-3d transition in Li-like Al. Their well-diagnosed experiments have pro-vided ample evidence of gain, illustrated the plasma kinetics and established gain length products of up to 4 for CVI, i.e., they have obtained implying 30 times amplified x-ray signals. In the future, the RAL group intends to use isoelectronic scaling of both hidrogenic and Li-like schemes to reach even shorter wavelengths. This is a formidable problem especially in the case of H-like where achieving gain on the Balmer alpha line is predicted[19] to require very short heating pulses (10 to 20 psec), some transient electron heating phenomena, and even the enhanced cooling rates that can be provided by use of efficiently radiating dopants. Demonstration of longer (perhaps saturated) amplifiers using double or multipass mirror geometries will also be pursued in the case of CVI.

Particularly for the case of Li-like Al ions, P. Jaegle[5] and colleagues have shown that amplification can take place for 4f-3d (15.4 nm) and 5f-3d (10.57 nm) transitions in the freely expanding coronal plasma of a simple slab target that has been irradiated by a high power laser. This scheme relies on dielectronic instead of three-body recombination. Gains of up to 2/cm at 10.57 nm have been reported[5] in this geometry. In November, 1987, the Orsay group plans to perform new experiments to obtain 12 to 15 gain length on the AXI scheme by using a newly completed facility at Palaiseau.

Since the scheme does not have to ionize the atom as extensively as in the case of H-like systems it is predicted to be more efficient. A.Sureau[20] at Orsay has predicted that reasonable gains can be obtained for transitions shorter than 4.4 nm by a simple isoelectronic scaling of this scheme provided the inversion density is increased 3 to 4 fold and temperature is also raised enough to provide the proper ionization stage.

A clever variation on these freely expandig schemes, which provides good efficiency and long time durations for the gain has been developed by Suckever et al.[2], at Princeton. The concept here is to use a CO_2 laser to produce a carbon or aluminum plasma (sometimes doped with higher z radiators to enhance radiative cooling) and to allow the free streaming ions to diffuse into a solenoidal magnetic field ($\sim$9 Tesla) which suppresses radial expansion (suppresses rate of density drop) which then prolongs the inversion period. Anomalous transport of CVI or AlXI ions from the axis of the magnetically confined plasma to the radiatively cooled outer regions results in both strong recombination and population inversion. For the CVI balmer alpha line at 18.22 nm the Princeton group has reported gain length products of $\sim$8 (implies 500 fold amplification), absolute divergence of 5 mrad, power of 100 kW and pulse lengths of 10-30 nsec. These impressive results yield an efficiency of 10^{-5} for the system.

Future experiments at Princeton include a new two-laser approach[9] in order to achieve lasing action significantly below 100 Å. In this new system, a high power CO_2 laser will produce a magnetically-confined, highly ionized plasma column (as before) in which a second powerful 10^{16} (W/cm^2) short pulse ($\sim$1 psec) laser will produce a population inversion and gain. The CO_2 will produce a plasma of Kr- or Ar-like ions whose outer shell transition will then be pumped through multiphoton processes by using the short pulse high power laser. Using the Kr isoelectronic sequence as an example, Suckewer et al.[2], predict the achievement of gain at wavelengths below 8.9 nm using Cd[12] ions and $4s^24p^45s^2 - 4s^24p^55s$ transitions. They also predict that the multiphoton process can enhance the gain they have already achieved in their CVI plasma experiments.

COLLISIONAL EXCITATION SCHEMES

The Ne-like n=3p to 3s, and recently the Ni-like n=4d to 4p exploding foil amplifier schemes have been extensively studied by the group at Lawrence Livermore National Laboratory[1,9-11] and at the Naval Research Labs[3]. This scheme uses the gentle plasma density and temperature gradients formed in an axploding foil amplifier (see Rosen et al.[1]) plus large collision rates in a highly collisional plasma to invert the n=3p and 3s levels for Ne-like ions or the n=4d and 4 p in the case of Ni-like systems. Progress with this scheme has been rapid with gain lengths reported to as high as $\sim$ 17 for the Ne-like Se j=2 to 1 transitions at 20.69 nm, gain coefficients up to 5.5 cm, power to at least 2 MW, divergence of 10 mrad and wavelengths as short as 6.6 nm (Ni-like Eu) with a gainlength product of $\sim$ 3. Preliminary evidence for isoelectronic scaling of the gain in Ni-like ions has just been observed by MacGowan and co-workers[21] at 5.0 nm. Future work with the thin schemes includes scaling to sub 4.4 nm with Ni-like W, demonstration of single mode operation[22], demonstration of multipass laser amplifier (double pass has already been demonstrated by Ceglio et al.[10]), and x-ray holographic imaging of in vitro microbiological organisms[23].

For those interested, a critical comparison of these various types of amplification techniques has been given by Matthews et al.[11]. In particular, they discuss their suitability for achieving the goal of saturated lasing at sub 5.4 n with fully coherent beams.

The status of x-ray laser development world wide is summarized in Fig. 3. Here I illustrate the gain length product or degree of amplification ($e^{gain\ length}$) vs the laser wavelengths achieved. We are delighted to have achieved such short wavelengths and high power in such a short time interval (first reports of amplification were in November, 1984). Much work remains, though, to achieve saturated amplification at shorter wavelengths and fully-coherent beams. More efficient laser schemes which embody gain enhancement techniques must be considered. These enhancement techniques include selective radiative pumping of performed plasmas (see Matthews et al.[11]), the multiphoton excitation scheme mentioned with the Princeton work and formation of long-lived amplifiers suitable for use with multilayer mirrors cavities.

CONCLUSIONS

It is now possible to produce, in theory, amplification at sub 4.4 nm wavelengths utilizing either a recombination pumped H-, Li-like or collisionally pumped Ni-like system. Fully-coherent, high power x-ray lasers at these same wavelengths must await testing of new ideas for more efficient schemes or await the availability of larger pump sources. Possibilities for significant applications of the x-ray laser are now being planned and will be demonstrated in the very near future.

ACKNOWLEDGEMENTS

This work was performed under the auspices of the U.S. Department of Energy by the Lawrence Livermore National Laboratory under contract number W-7405-ENG-48.

The author presented the lecture and wrote this manuscript, but much credit for the development of the x-ray laser as well as contribution to this work belongs to C. Keane, R. London, B. MacGowan, S. Maxon, M. Rosen, J. Trebes, and E. M. Campbell (Livermore), as well as the work of other

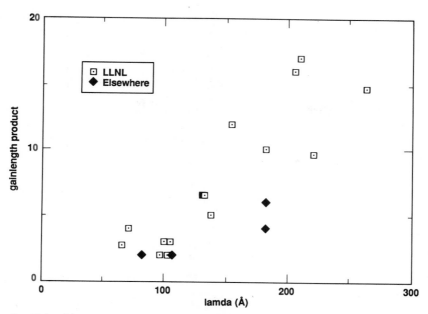

Fig. 3. This figure represents our progress to date in the degree of amplification (gain or length of amplifier) achieved vs wavelength of x-ray laser transitions.

groups that have contributed to the present body of knowledge of x-ray lasers; M. Key, O. Willi and colleagues (Rutherford Appleton Labs); S. Suckewer, C. Skinner and co-workers (Princeton Plasma Physics Labs); P. Jaegle, G. Jamelot and co-workers (LSAI Orsay, France); T. Lee, E. Mclean and R. Elton (Naval Research Labs).

REFERENCES

1. D. L. Matthews, P. L. Hagelstein, M. D. Rosen, M. J. Eckart, N. M. Ceglio, A. U. Hazi, H. Medecki, B. J. MacGowan, J. E. Trebes, B. L. Whitten, E. M. Campbell, C. W. Hatcher, A. M. Hawryluk, R. L. Kauffman, L. O. Pleasance, G. Rambach, J. Scofield, G. Stone, and T. A. Weaver, Phys. Rev. Lett. 54, 110 (1985)
   M. D. Rosen, P. L. Hagelstein, D. L. Matthews, E. M. Campbell, A. U. Hazi, B. L. Whitten, B. MacGowan, R. E. Turner, and R. W. Lee, Phys. Rev. Lett. 54, 106 (1985)
2. S. Suckewer, C. H. Skinner, H. Milchberg, C. Keane, and D. Voorhees, Phys. Rev. Lett., 55, 1753 (1985)
   S. Suckewer, C. H. Skinner, D. Kim, E. Valeo, D. Voorhees, and A. Wouters, Phys. Rev. Lett., 57, 1004 (1986)
   S. Suckewer, C. H. Skinner, D. Kim, E. Valeo, D. Voorhees, and A. Wouters, J. de Physique (Paris) 47, C6, 23 (1986)
3. R. Elton, private communication and R. Elton, T. N. Lee and W. A. Molander, JOSA B4, 539 (1987)
4. M. H. Key, private communication and M. Key, J. E. Boon, C. Brown, C.

Chenai-Popovics, R. Corbett, A. R. Damerell, P. Gottfeld, C. J. Hooker, G. P. Kiehn, C. L. S. Lewis, D. A. Pepler, G. J. Pert, C. Reagan; S. J. Rose, I. N. Ross, P. T. Rumsby, S. Sadaat, R. Smith, T. Tomi, and O. Willi, J. Phys., Paris,, 47, C6, 71 (1986)

5. P. Jaegle, A. Carillon, A. Klisnick, G. Jamelot, H. Guennou, and A. Sureau, Europhys. Lett., 1, 555 (1986)
    P. Jaegle, G. Jamelot, A. Carillon, and A. Klisnick, J. Phys., Paris, 47, C6, 31 (1986)
    P. Jaegle, G. Jamelot, A. Carillon, A. Klisnick, A. Sureau, and H. Guennou, JOSA B4, 563 (1987)

6. M. Key, Nature, 316, 314 (1985)

7. R. Waynant and R. Elton, Proc. IEEE 66, 1059 (1976)

8. I. I. Sobel'man and A. V. Vinagradov, Adv. in Atomic and Molecular Physics, 20, 327 (1985)

9. Proceeding of the International Colloquium on X-ray Lasers, Aussois France, April 14-17, 1986. Published in J. de Physique, 47, C6 (1986)

10. "Multi-layer Structures and Laboratory X-ray Laser Research", Ed. by N. Ceglio and P. Dhez, Proc. Soc. Photo-Opt. Instr. Eng., 688 (1987)

11. "The generation of Coherent XUV and Soft X-ray Radiation", special issue, JOSA B4 (1987)

12. R. Craxton, R. L. McCrory and J. Soures, 255 (1986)

13. G. Peters and L. Allen, J. Phys. A4, 238 (1971)

14. See N. Ceglio et. al., in Ref. 10

15. M. Rosen and R. London, private communication, Lawrence Livermore National Laboratory (1986)

16. R. C. Elton, App. Opt., 14, 97 (1975)

17. L. I. Gudzenko and L. A. Shelepin, Sov. Phys., JETP 18, 998 (1964)

18. G. J. Pert in Ref. 11, p. 602, (1987)

19. M. Key, private communication

20. A. Sureau, private communication

21. B. MacGowan et al., Proc. of IQEC/CLEO, Baltimore, MD (1987)

22. M. Rosen, J. Trebes and D. Matthews, Comm. Plasma Phys. Controlled Fusion 10, 245 (1987)

23. J. Trebes, private communication and Ref. 9.

ELECTRON WIGGLING INFLUENCE ON OPTICAL GUIDING IN A FEL

S. Solimeno* and Yu-Juan Chen**

* Dept. of Nuclear Physics, Structure of the Matter and
  Applied Physics, University of Naples, Italy
**Lawrence Livermore National Laboratory, University of
  California, Livermore, CA 94550, U.S.A.

INTRODUCTION

When a light beam travels along a free electron laser amplifier, the
beam cross section is modified by the counteracting effects of diffraction
and interaction with the electron beam. In particular, diffraction can be
overcome by a combination of refraction and gain. In a sufficiently long
undulator unstable beam shape, independent of the initial distribution
(guided mode), may be eventually reached.

Tang and Sprangle[1] first recognized that light in a FEL may be re-
fracted toward the axis. The existence of guided modes for undulators
operating in the linear exponential regime was discovered by Scharlenann,
Sessler and Wurtele[2] (see also La Sala et al.[3]) by simulating the propagation through the Livermore FEL with the numerical code FRED, and was
anticipated analytically by Moore[4]. Recently, some indirect evidence of
optical guiding in the Mark III FEL has been reported by Madey et al.[5].

The guided modes are in general investigated by extending the one-
dimensional analysis of normal waves in a FEL developed by Kroll[6] to the
three dimensional case. By combining the cubic dispersion equation valid
for the exponential growth regime and the wave equation, a second order
differential equation for the transverse distribution of the guided modes
is obtained.

Apart from the solution obtained by Moore for a uniform current distribution by following an approach similar to that used for a step-index
fiber, for a more general profile only approximate solutions can be found.
In particular, the transverse wave equation can be approximately integrated
by applying the WKB method as shown by Luchini and Solimeno[7].

Pantell, Fontana and Feinstein[8] have argued that the bending of the
electron beam due to the undulator field can produce a strong diffraction
of the guided beam, which eventually leads to the loss of the guided modes.
Their argument is based on the assumption that a beam in a FEL amplifier

43

can be analyzed by assimilating the FEL to an optical fiber with suitable complex refractive index bent periodically along the z-axis. This periodicity simulates the periodic bending of the e-beam.

Some authors have opposed the Pantell arguments by claiming that the e-beam bending is implicitly accounted for in the expression of the complex refractive index of the equivalent medium used for representing the influence of the wiggling electrons on the propagating laser beam.

DERIVATION OF DRIVING CURRENT

The transverse canonical momentum $\vec{P}_\perp$ is related to the transverse velocity $\vec{v}_\perp$ by the relation

$$\vec{P}_\perp = m\gamma\vec{v}_\perp - e\vec{A} \tag{1}$$

written for MKS quantities, γ being the Lorentz factor and $e(>0)$ the electron charge. The transverse current density $\vec{j}_\perp$ can be obtained from the electron distribution function F by using the integral expression

$$\vec{j}_\perp (\vec{x}_\perp, t; z) = - e \int \frac{\vec{p}_\perp + \vec{a}}{\gamma} \; f \; d^2\vec{p}_\perp \; d\gamma \tag{2}$$

having indicated with $\vec{p}_\perp = \vec{P}_\perp/mc$ and $\vec{a} = e\vec{A}/mc$ adimensional quantities and having replaced F with $f = m^2c^2F$.

The distribution function f at the abscissa z can be derived from the same function and its partial derivatives at the entrance of the undulator by using the Boltzmann equation. Having chosen z as the independent variable, it can be shown that the Hamiltonian coincides with minus the z-component $-p_z = - P_z/mc$ of the adimensional momentum $\vec{p}$. Consequently, the evolution of f is described by the Boltzmann equation

$$\frac{\partial f}{\partial z} + \frac{\partial p_z}{\partial \gamma}\frac{\partial f}{\partial t} - \frac{\partial p_z}{\partial \vec{p}_\perp} \cdot \frac{\partial f}{\partial \vec{x}_\perp} - \frac{\partial p_z}{\partial t}\frac{\partial f}{\partial \gamma} + \frac{\partial p_z}{\partial \vec{x}_\perp} \cdot \frac{\partial f}{\partial \vec{p}_\perp} = 0 \tag{3}$$

For ultrarelativistic electrons and in absence of space charges the Hamiltonian $-p_z$ can be approximated by

$$-p_z = - \gamma + \frac{1 + (\vec{p}_\perp + \vec{a})^2}{2 \gamma} \tag{4}$$

$\vec{a} = \vec{a}_w + \vec{a}_1$ being the superposition of the dimensionless potential vectors of the undulator $(\vec{a}_w)$ and laser $(\vec{a}_1)$ fields, supposed to be perpendicular to the z-axis.

PERTURBATIVE SOLUTION OF THE BOLTZMANN EQUATION

The Boltzmann equation can be solved perturbatively by expanding f into power series in a_1,

$$f = f_o + f_1 + f_2 + \ldots \tag{5}$$

f_i being a function of the i-th power of a_1.

Introducing the retarded time τ at the average electron speed

$$\tau = t - \int_0^z \left(1 + \frac{1 + a_w^2}{2\gamma_0^2} \right) dz' \tag{6}$$

we obtain

$$\tilde{L}_0 f_0 = 0$$
$$\tilde{L}_0 f_1 = \tilde{L}_1 f_0$$
$$\tilde{L}_0 f_2 = \tilde{L}_1 f_1 + \tilde{L}_2 f_0$$
$$\tilde{L}_0 f_i = \tilde{L}_1 f_{i-1} + \tilde{L}_2 f_{i-2} \tag{7}$$

where $\tilde{L}_0$, $\tilde{L}_1$ and $\tilde{L}_2$ are the differential operators

$$\tilde{L}_0 = \frac{\partial}{\partial z} + \left(\frac{1 + (\vec{p}_\perp + \vec{a}_w)^2}{2\gamma_0^2} - \frac{1 + a_w^2}{2\gamma_0^2} \right) \frac{\partial}{\partial \tau} + \frac{\vec{p}_\perp + \vec{a}_w}{\gamma} \frac{\partial}{\partial \vec{x}_\perp}$$

$$\tilde{L}_1 = - \left(\gamma^{-2} \vec{a}_1 (\vec{p}_\perp + \vec{a}_w) \frac{\partial}{\partial \tau} + \gamma^{-1} \vec{a}_1 \frac{\partial}{\partial \vec{x}_\perp} \right)$$
$$- \left\{ \frac{\vec{p}_\perp + \vec{a}_w}{\gamma} \cdot \frac{\partial \vec{a}_1}{\partial \tau} \frac{\partial}{\partial \gamma} - \gamma^{-1} \frac{\partial}{\partial \vec{x}_\perp} (\vec{a}_1 \cdot (\vec{p}_\perp + \vec{a}_w)) \frac{\partial}{\partial \vec{p}_\perp} \right\}$$
$$\equiv \tilde{L}_{1,1} + \tilde{L}_{1,2}$$

$$\tilde{L}_2 = - \frac{a_1^2}{2\gamma^2} \frac{\partial}{\partial \tau} - \frac{\partial a_1^2}{\partial \tau} \frac{\partial}{\partial \gamma^2} + \frac{1}{2\gamma} \frac{\partial a_1^2}{\partial \vec{x}_\perp} \cdot \frac{\partial}{\partial \vec{p}_\perp} \tag{8}$$

DISPLACEMENT OPERATOR

By introducing the displacement operator

$$\tilde{D}(z,z') = \exp \left(- \frac{1}{\gamma} \int_{z'}^z (\vec{p}_\perp + \vec{a}_w) \, dz'' \cdot \frac{\partial}{\partial x_\perp} \right.$$
$$\left. - \frac{1}{2} \int_{z'}^z \left(\frac{1 + \vec{p}_\perp + \vec{a}_w)^2}{\gamma^2} - \frac{1 + a_w^2}{\gamma_0^2} \right) dz'' \frac{\partial}{\partial \tau} \right) \tag{9}$$

45

and the functions f'_i

$$f'_i(z) = \tilde{D}(z,0)f'_i(z) \tag{10}$$

a system of evolution equations is immediately obtained,

$$\frac{\partial}{\partial z} f'_0 = 0$$

$$\frac{\partial}{\partial z} f'_1 = \tilde{L}'_1 f'_0$$

$$\frac{\partial}{\partial z} f' = \tilde{L}'_1 f'_1 + \tilde{L}'_2 f'_0$$

$$\frac{\partial}{\partial z} f'_i = \tilde{L}'_1 f'_{i-1} + \tilde{L}'_2 f'_{i-2} \tag{11}$$

with

$$\tilde{L}'_i = \tilde{D}(0,z)\tilde{L}_i\tilde{D}(z,0). \tag{12}$$

Integrating the above system we obtain

$$f'_0 = f_{in}$$

$$f'_1 = \int_0^z \tilde{L}'_1 dz' \, f_{in}$$

$$f'_2 = (\int_0^z \tilde{L}'_1 dz' \int_0^{z'} \tilde{L}'_1 dz'' + \int_0^z \tilde{L}'_2 dz') \, f_{in} \tag{13}$$

f_{in} being the distribution function at the undulator entrance.

DRIVING CURRENT

The transverse component $\vec{J}_\perp$ of the current density produced by the interaction of the laser beam with the wiggling electrons is given by

$$\vec{J}_\perp = -e \, \frac{\vec{p}_- + \vec{a}_w + \vec{a}_1}{\gamma} \, (f_0 + f_1 + f_2 + \ldots)d^2 p_\perp d\gamma$$

$$J_0 + J_1 + J_2 + \ldots \tag{14}$$

with

$$\vec{J}_0 = -e \, \frac{p_- + a_w}{\gamma} \, f_0 d^2 p_\perp d\gamma$$

$$\vec{J}_1 = -e\vec{a}_1 \int \frac{f_0}{\gamma} d^2 p_\perp d\gamma - e \int \frac{\vec{p}_\perp + \vec{a}_w}{\gamma} f_1 d^2 p_\perp d\gamma \equiv \vec{J}'_1 + \vec{J}''_1$$

$$\vec{J}_2 = -e\vec{a}_1 \int \frac{f_1}{\gamma} d^2 p_\perp d\gamma - e \int \frac{\vec{p}_\perp + \vec{a}_w}{\gamma} f_2 d^2 p_\perp d\gamma \tag{15}$$

46

In the following we will concentrate the attention on the field radiated by $\vec{J}_1$, that is from the superposition of a current $\vec{J}_1'$, proportional to a_1, and a component $\vec{J}_1''$, depending on a_1 through f_1. While $\vec{J}_1'$ accounts for the well known negative contribution of the electrons of a nonmagnetized plasma to the dielectric constant, $\vec{J}_1''$ represents the response of the electrons of a FEL to the laser field. Combining the above equation yields

$$\vec{J}_1 = - e \int \left(\frac{\vec{p}_\perp + \vec{a}_w}{\gamma} \int_0^z \tilde{D}(z,z') \tilde{L}_1(z') \tilde{D}(z',0) dz' \right.$$

$$\left. + \vec{a}_1(z) \tilde{D}(z,0) \right) f_{in} d^2 \vec{p}_\perp d\gamma \qquad (16)$$

In particular, the operators $\partial/\partial\gamma$ and $\partial/\partial\vec{p}_\perp$ appearing in $\tilde{L}_{1,2}$ can be replaced in the first integral on the RHS of (16) by their adjoints $-\partial/\partial\gamma$ and $-\partial/\partial\vec{p}_\perp$ operating on the functions on their left, so that after some algebra we obtain

$$\vec{J}_1 = e \int \left\{ \frac{\vec{p}_\perp + \vec{a}_w}{\gamma} \int_0^z \tilde{D}(z,z') \left[\frac{1}{2} \vec{a}_1 \cdot (\vec{p}_\perp + \vec{a}_w) \frac{1}{\gamma} \frac{\partial}{\partial\tau} + \frac{1}{\gamma} \vec{a}_1 \cdot \frac{\partial}{\partial\vec{x}_\perp} \right. \right.$$

$$+ 2 \frac{\vec{p}_\perp + \vec{a}_w}{\gamma^2} \cdot \frac{\partial\vec{a}_1}{\partial\tau} - \frac{1}{\gamma} \frac{\partial \ln\tilde{D}}{\partial\gamma} (\vec{p}_\perp + \vec{a}_w) \cdot \frac{\partial\vec{a}_1}{\partial\tau}$$

$$+ \frac{1}{\gamma} \frac{\ln\tilde{D}}{\partial\vec{p}_\perp} \cdot \left(\frac{\partial}{\partial\vec{x}_\perp} [\vec{a}_1 \cdot (\vec{p}_\perp + \vec{a}_w)] \right) + \frac{1}{\gamma} \left(\frac{\partial}{\partial\vec{x}_\perp} \cdot \vec{a}_1 \right) \right] \tilde{D}(z',0) dz'$$

$$\left. + \frac{1}{\gamma^2} \int_0^z \tilde{D}(z,z') \left(\frac{\partial}{\partial\vec{x}_\perp} [\vec{a}_1 \cdot (\vec{p}_\perp + \vec{a}_w)] \right) \tilde{D}(z',0) dz' - \vec{a}_1 \tilde{D}(z,0) \right\} f_{in} d^2 \vec{p}_\perp d\gamma$$

$$(17)$$

MONOENERGETIC AND COLLIMATED E-BEAM, AND TIME-HARMONIC LASER FIELD

For a monoenergetic, collimated and continuous e-beam, characterized by a current density J, the distribution function f_{in} takes the form

$$f_{in} = \frac{J(\vec{x}_\perp)}{e} \delta(p_x) \delta(p_y) \delta(\gamma - \gamma_0) \qquad (18)$$

In addition, for a time-harmonic laser field

$$\vec{a}_1(\vec{x}_\perp, z, t) = e^{i\omega t} \vec{a}(\vec{x}_\perp, z)$$

$$= e^{i\omega\tau + i\delta(z,0)} \vec{a}(\vec{x}_\perp, z) \qquad (19)$$

where

$$\delta(z,0) = \omega \int_0^z \left(1 + \frac{1+a_w^2}{2\gamma_0^2} \right) dz' \qquad (20)$$

For the above expression of f_{in},

47

$$\tilde{D}(z,z') \quad -\frac{1}{\gamma_o}\tilde{X}(z,z') + \frac{i\omega}{\gamma_o^3}\int_{z'}^{z}(1+a_w^2)dz'' \tag{21}$$

$$\frac{\partial \ln \tilde{D}}{\partial \vec{p}_\perp} = -\frac{z-z'}{\gamma_o}\frac{\partial}{\partial \vec{x}_\perp} - i\frac{\omega}{\gamma_o^2}\int_{z'}^{z}\vec{a}_w dz'' \tag{22}$$

being

$$\tilde{X}(z,z') = -\frac{1}{\gamma_o}\int_{z'}^{z}\vec{a}_w dz'' \frac{\partial}{\partial \vec{x}_\perp} \tag{23}$$

Now, integrating with respect to $\vec{p}_\perp$ and γ we obtain

$$\vec{J}_1 = e^{i\omega t}\{\frac{\vec{a}_w}{\gamma_o}\int_0^z e^{\tilde{X}-i\Delta}[i\frac{2\omega}{\gamma_o^2} + i\frac{2\omega\tilde{X}}{\gamma_o^2}$$

$$+\frac{\omega^2}{\gamma_o^4}\int_{z'}^{z}(1+a_w^2)dz'' - \frac{z-z'}{\gamma_o^2}\nabla_\perp^2)\vec{a}\cdot\vec{a}_w$$

$$-\frac{z-z'}{\gamma_o^2}(\frac{\partial}{\partial \vec{x}_\perp}\vec{a}\cdot\vec{a}_w)\cdot\frac{\partial}{\partial \vec{x}_\perp} + \frac{1}{\gamma_o}\frac{\partial}{\partial \vec{x}_\perp}\vec{a} + i\frac{2\omega}{\gamma_o^2}\vec{a}\cdot\vec{a}_w\tilde{X}]dz'$$

$$+\frac{1}{\gamma_o^2}\int_0^z e^{\tilde{X}-i\Delta}(\frac{\partial}{\partial \vec{x}_\perp}\vec{a}\cdot\vec{a}_w)dz' - \vec{a}\} e^{\tilde{X}(z,0)}J(\vec{x}_\perp)$$

$$e^{i\omega t}\frac{1}{\gamma_o}\tilde{P}e^{\tilde{X}(z,0)}J(\vec{x}_\perp) \tag{24}$$

being $\Delta = \delta(z,z')$. The operator $\tilde{P}$ contains $\partial/\partial\vec{x}_\perp$ in combination with vector functions of $\vec{a}$ and $\vec{a}_w$, and can be conveniently split as

$$\tilde{P} \equiv \vec{P}_0 + \frac{\vec{a}_w}{\gamma_o k_w}\vec{P}_1\cdot\frac{\partial}{\partial x_\perp} \tag{25}$$

where

$$P_0 = a_w\int_0^z e^{\tilde{X}-i\Delta}(i\frac{2\omega}{\gamma_o^2} + i\frac{2\omega}{\gamma_o^2}X$$

$$+\frac{\omega^2}{\gamma_o^4}\int_{z'}^{z}(1+a_w^2)dz'' - \frac{z-z'}{\gamma_o^2}\nabla_\perp^2)\vec{a}\cdot\vec{a}_w - \frac{1}{\gamma_o}(\frac{\partial}{\partial x_\perp}\cdot\vec{a})\ dz'$$

$$+\frac{1}{\gamma_o}\int_0^z e^{\tilde{X}-i\Delta}(\frac{\partial}{\partial \vec{x}_\perp}\vec{a}\cdot\vec{a}_w)dz' -\gamma_o\vec{a} \tag{26}$$

$$\vec{P}_1 = k_w\int_0^z e^{\tilde{X}-i\Delta}[\vec{a} - \frac{z-z'}{\gamma_o}\frac{\partial}{\partial x_\perp}(\vec{a}\cdot\vec{a}_w)$$

$$- i\frac{2\omega}{\gamma_o^2}\vec{a}\cdot\vec{a}_w\int_{z'}^{z}\vec{a}_w dz''] dz' \tag{27}$$

UNIFORM HELICAL UNDULATOR

For an uniform helical undulator with the field rotating clockwise for increasing $z, \vec{a}_w$ is given by

$$\vec{a}_w(z) = -\frac{a_w}{2} \sum_q (\hat{x} + iq\hat{y})e^{-iq_wz}$$

$$\equiv -\frac{a_w}{\sqrt{2}} \sum_q q>e^{-iqk_wz} \tag{28}$$

with $q \pm 1$ and $a_w = |\vec{a}_w|$ constant and $|q >$ the polarization vector.

Now, if we approximate ω with the resonance frequency ω_r,

$$\omega \cong \omega_r = \frac{2\gamma_o^2 k_w}{1 + a_w^2} \tag{29}$$

and introduce the normalized vectors $\vec{a}' = \vec{a}/a_w$, $\vec{x}' = \vec{x}_\perp/\delta r$, with $\delta r = a_w/\gamma_o k_w$ the radius of the wiggling electron orbit, and replace z with the adimensional coordinate $\zeta = k_w z$, $\vec{P}_0$ and $\vec{P}_1$ will be given by

$$\frac{\vec{P}_0}{a_w} = \hat{a}_w \int_0^\zeta e^{X-i\Delta} \left[4\alpha(i + \zeta - \zeta' + i\tilde{X}) - (\zeta - \zeta') \nabla'^2\right] \hat{a}_w \cdot \vec{a}' \, d\zeta'$$

$$+ \int_0^\zeta e^{X-i\Delta} (\frac{\partial}{\partial \vec{x}'} \hat{a}_w \cdot \vec{a}') \, d\zeta' - \gamma_o \vec{a}' \tag{30}$$

with $\alpha = a_w^2/(1 + a_w^2)$, $\nabla'^2 = \partial^2/\partial x'^2 + \partial^2/\partial y'^2$ the normalized transverse Laplacian, while

$$\frac{\vec{P}_1}{a_w} = \int_0^\zeta e^{X-i\Delta} \left[\vec{a}' - 2i\alpha(\int_{\zeta'}^\zeta \hat{a}_w d\zeta'')\hat{a}_w \cdot \vec{a}' - (\zeta - \zeta')(\frac{\partial}{\partial \vec{x}'} \hat{a}_w \vec{a}')\right] d\zeta' \tag{31}$$

EXPANSION OF THE DRIVING CURRENT UP TO THE SECOND ORDER IN THE ELECTRON WIGGLING RADIUS

Owing to the slow variation of $J(\vec{x}')$ across the transverse section of the electron beam, the operator $\hat{P}$ can be conveniently expanded into power series of $\partial/\partial \vec{x}'$, namely

$$\frac{\vec{P}_0}{a_w} = \vec{P}_0^{(0)} + \epsilon \vec{P}_0^{(1)} + \epsilon^2 \vec{P}_0^{(2)} + \ldots$$

$$\frac{\vec{P}_1}{a_w} = \vec{P}_1^{(0)} + \epsilon \vec{P}_1^{(1)} + \ldots \tag{32}$$

where ϵ is the smallness parameter which will be chosen equal to $\delta r/\sigma$, σ being the width of the gaussian modes used as basis of expansion of the field. In particular, observing that the operators $X, \partial/\partial \vec{x}'$ are first order in ε, while ∇'^2 is second order in ε, after some algebra we obtain the string of equations

$$\vec{P}_0^{(0)} = 4\alpha\hat{a}_w (iI(1) + I(\zeta - \zeta')) - \gamma_\circ a'$$

$$\vec{P}_0^{(1)} = 4\alpha\hat{a}_w (2iI(\frac{\tilde{X}}{\varepsilon}) + I((\zeta - \zeta')\frac{\tilde{X}}{\varepsilon})) + I(\frac{1}{\varepsilon}\frac{\partial}{\partial\vec{x}'})$$

$$\vec{P}_0^{(2)} = 2\alpha\hat{a}_w (3iI(\frac{\tilde{X}^2}{\varepsilon^2}) + I((\zeta - \zeta')\frac{\tilde{X}^2}{\varepsilon^2}))$$

$$- \hat{a}_w I((\zeta - \zeta')\frac{\nabla'^2}{\varepsilon^2} + I(\frac{\tilde{X}}{\varepsilon^2}\frac{\partial}{\partial\vec{x}'})$$

$$\vec{P}_1^{(0)} = I'(1) - 2i\alpha I(\int_{\zeta'}^{\zeta} \hat{a}_w d\zeta'')$$

$$\vec{P}_1^{(1)} = I'(\frac{\tilde{X}}{\varepsilon}) - 2i\alpha I(\frac{X}{\varepsilon}\int_{\zeta'}^{\zeta} \hat{a}_w d\zeta'') - I(\frac{\zeta - \zeta'}{\varepsilon}\frac{\partial}{\partial\vec{x}'}) \tag{33}$$

having used the shorthands

$$I(\tilde{O}) \equiv \int_0^\zeta e^{-i\Delta}\tilde{O}\hat{a}_w \cdot \vec{a}' \, d\zeta'$$

$$I'(\tilde{O}) \equiv \int_0^\zeta e^{-i\Delta}\tilde{O} \vec{a}' \, d\zeta' \tag{34}$$

WAVE EQUATION

To obtain the 3-D form of the laser field we employ the wave equation

$$(\nabla'^2 + \frac{a_w^2}{\gamma_\circ^2}\frac{\partial^2}{\partial\zeta^2} + \omega'^2) \vec{a}' = -\frac{4\pi(\delta r)^2}{I_A a_w}\vec{J}_1 e^{-i\omega t} \tag{35}$$

where $\omega' = \omega\delta r$ and $I_A = 4\pi mc/\mu_0 e \cong 17$ kA is proportional to the Alfven current $I_{Alfven} = I_A\beta\gamma$[9,10].

We nox examine the above equation by making the periodic wave ansatz that $\vec{a}'$ can be expanded in space harmonics,

$$\vec{a}' = \sum_{lmpq} |q > e^{-i(\beta+p)\zeta+im\theta} R_1^{|m|}(r')$$

$$\equiv \sum_{lmpq} A_{lmpq} |lmpq >$$

$$\equiv \sum_{mpq} A_{mpq}(r') |mpq >$$

$$\equiv \sum_{pq} A_{pq}(r',\theta) |pq > \tag{36}$$

where β is a factor generally complex, not to be confused with the electron velocity, $|pq> = |q> \exp\left[-i(\beta + p)\zeta\right]$, while the functions $R_1^{|m|}\exp(im\theta) \equiv |1m>$ form an orthonormal set of eigenfunctions of the operator $\nabla'^2 + \omega'^2 n^2$,

$$(\nabla'^2 + \omega'^2 n^2)|1m> = \epsilon^2 \gamma_{1m}^2 |1m> \tag{37}$$

being $\epsilon^2 \gamma_{1m}^2$ the relative eigenvalue and σ the width, generally complex, of the gaussian mode $|00>$ (see Eq. 55). The function $n(r')$ represents the refractive index profile of an optical fiber guiding modes whose transverse distribution coincides with that of the space harmonics $|1mpq>$. In the following we will choose a parabolic profile for $n(r')$, which is associated with a set of Gauss-Laguerre modes. Our choice was dictated by the possibility of using analytic expression for the integrals containing combinations of Gauss-Laguerre functions. The space harmonic index p can take positive and negative values. The polarization index q takes the values ± 1 only. $|1mpq>$ represents a circularly polarized wave with the field rotating clockwise $(q = -1)$ for increasing z at fixed time. A key assumption in the periodic wave expansion is to look for steady-state field distribution, i.e. independent of the initial conditions.

In the following we will consider modes $|1m>$ of the form

$$|1m> = e^{im\theta}R_1^{|m|}(r')$$

$$= \epsilon\sqrt{\frac{1!}{\pi(1 + |m|)!}} \; e^{-\frac{r^2}{2\sigma^2}} \; L_1^{|m|}\left(\frac{r^2}{\sigma^2}\right)\left(\frac{r}{\sigma}\right)^{|m|} e^{im\theta} \tag{38}$$

with $L_1^{|m|}$ the Laguerre polynomial and σ a parameter, generally complex, which can be chosen in such a way as to optimize the overlap of $|00>$ with the fundamental mode of propagation along the e-beam in absence of bending.

The above set of modes correponds to the index profile

$$n^2(r') = 1 - \left(\frac{\epsilon}{\sigma\omega}\right)^2 r'^2 \tag{39}$$

while the relative eigenvalues are given by[11]

$$\gamma_{1m}^2 = (\sigma\omega)^2 - 2(21 + |m| + 1) \tag{40}$$

For an e-beam current distributed with a Gaussian law,

$$J(\vec{x}_\perp) = \frac{I_{EB}}{2\pi\sigma_{EB}^2} \; e^{-r^2/2\sigma_{EB}^2} \tag{41}$$

where σ_{EB} represents the width of the e-beam, after some lengthy calculations we obtain

$$(\nabla_\perp^2 + k_w^2 \frac{\partial^2}{\partial \zeta^2} + \omega^2)\vec{a}' = - \frac{I_{EB}}{I_A} \frac{2}{\sigma_{EB}^2 \gamma_0} \{\vec{P}_0^{(0)}$$

$$+ \epsilon [\vec{P}_0^{(1)} - \frac{r\sqrt{x}}{\sigma_{EB}} (\vec{P}_0^{(0)} \sin(\zeta - \theta) - \hat{a}_w \vec{P}_1^{(0)} \ \hat{r})]$$

$$+ \epsilon^2 [\vec{P}_0^{(2)} - \frac{r\sqrt{x}}{\sigma_{EB}} \sin(\zeta - \theta)\vec{P}_0^{(1)}$$

$$+ \frac{x}{2} (\frac{r^2}{\sigma_{EB}^2}(1 - \cos 2(\zeta - \theta)) - 1) \ \vec{P}_0^{(0)}$$

$$- \frac{r\sqrt{x}}{\sigma_{EB}} \hat{a}_w \vec{P}_1^{(1)} \ \hat{r} + x\hat{a}_w \ P_1^{(0)} \cdot [\hat{r} \sin(\zeta - \theta) \ (\frac{r^2}{\sigma_{EB}^2} - 1)$$

$$+ \hat{\theta}\cos(\zeta - \theta)]]\} e^{- \frac{r^2}{2\sigma_{EB}^2}} \tag{42}$$

where $x \equiv \sigma^2/\sigma_{EB}^2$.

Now, replacing $\vec{a}'$ with the space harmonic expansion, we obtain a system of linear equations in A_{lmpq} the FEL system. The terms on the RHS of the above wave equation contain the exponential $e^{-r^2/2\sigma_{EB}^2}$ times some powers of r/σ_{EB} which can be expressed as products of the transverse modes times some matrix elements, namely

$$e^{-r^2/2\sigma_{EB}^2}|1m> = \sum_{l'} E_{11'}^m |1'm > \tag{43}$$

$$\frac{r^2}{\sigma_{EB}} e^{-r^2/2\sigma_{EB}^2}|1m> = \sum_{l'} E \ R_{11'}^m |1'm > \tag{44}$$

$$\frac{r^2}{\sigma_{EB}^2} e^{-r^2/2\sigma_{EB}^2}|1m> = \sum_{l'} E \ R \ R_{11'}^m |1'm > \tag{45}$$

where $E_{11'}^m$, $E \ R_{11'}^m$, $E \ R \ R_{11'}^m$ can be expressed in terms of hypergeometric functions[12].

The FEL system can be solved by iteration. We begin with suitable values of β and $A_{100,-1}$, while we put all the other terms equal to zero. In particular, we assume $A_{000,-1}=1$. Then, we calculate the other amplitudes A_{lmpq} by using the FEL system. Once calculated the other amplitudes A_{lmpq}

by using the FEL system and putting $A_{000,-1} = 1$. By iterating these operations a convergent sequence of values is obtained.

ZEROTH ORDER APPROXIMATION

If we retain only terms independent of the electron wiggling radius, the wave equation reduces to

$$(\nabla_-^2 + k_w^2 \frac{\partial^2}{\partial \zeta^2} + \omega^2)\vec{a}' = - \frac{I_{EB}}{I_A} \frac{2}{\sigma_{EB}^2 \gamma_o} \vec{P}(0) e^{-r^2/2\sigma_{EB}^2}$$

$$= - \frac{I_{EB}}{I_A} \frac{2}{\sigma_{EB}^2 \gamma_o} -\{4\alpha\hat{a} \frac{1}{\sqrt{2}} \sum_{pq} A_{p+q,q} (ia_p + b_p)|p>$$

$$- \gamma_o \sum_{pq} A_{pq}|pq>\} e^{-r^2/\sigma_{EB}^2} \tag{46}$$

where a_p and b_p are coefficients defined in Appendix. Near resonance a_p and b_p are large only for $p = 1$. For b_1 sufficiently large we can neglect a_1^p and γ_o^p, so that the FEL system reduces to

$$(\nabla_\perp^2 + \omega^2 n_{FEL}^2 - \beta^2 k_w^2) A_{0,-1} = 0 \tag{47}$$

$$(\nabla_\perp^2 + \omega^2 n_{FEL}^2 - (\beta + 2)^2 k_w^2)A_{21} = 0 \tag{48}$$

$$(\nabla_\perp^2 + \omega^2 - k_w^2 (\beta + p)^2)A_{pq} = 0, pq \neq 0,-1;2,1 \tag{49}$$

where

$$n_{FEL}^2 = 1 + Q \frac{b_1}{k_w^2} e^{-r^2/2\sigma_{EB}^2}$$

$$= 1 - \frac{Q}{(\omega + \nu - \beta k_w)^2} e^{-r^2/2\sigma_{EB}^2} \tag{50}$$

is the effective refractive index of the FEL amplifier while

$$Q = I_{EB}a_w^2 (1 + a_w^2)/I_A\sigma_{EB}^2\gamma_0^5 \tag{51}$$

For the Livermore laser $\lambda_w = 9.8$ cm, $a_w = 2.5$, $I_{EB} = 850A$, $\gamma_0 = 7$,

$\sigma_{EB}^2 = 10 \text{ cm}^2$, so that $Q = 1.348 \cdot 10^{-5} \text{ cm}^{-2}$.

If we put $A_{mpq} = 0$ with the exception of $A_{00,-1}$, the above system reduces to the wave equation generally used for studying optical guiding. Expanding $A_{00,-1}$ into space harmonics we transform the wave equation (47) into the homogeneous system

$$
\left[\frac{\gamma_{10}^2}{\sigma^2} + \omega^2 (1 - N_{11}^0) - \beta^2 k_w^2 - \frac{Q E_{11}^0 \omega^2}{(\omega + \nu - \beta k_w)^2} \right] A_{100,-1}
$$

$$
= \sum_{\substack{1' \\ 1' \neq 1}} \omega^2 \left| N_{11'}^0 + \frac{Q E_{11'}^0}{(\omega + \nu - \beta k_w)^2} \right] A_{1'00,-1} \tag{52}
$$

For solving the system (52) we choose first a suitable value of σ, for example using the roots of Eq. (59). Then, we calculate β by putting $A_{000,-1} = 1$ into the relative equation and using a trial set of $A_{100,-1}$, namely

$$
\frac{\gamma_{00}^2}{\sigma^2} + \omega^2 (1 - N_{00}^0) - \beta^2 k_w^2 - \frac{Q E_{00}^0 \omega^2}{(\omega + \nu - \beta k_w)^2}
$$

$$
= \sum_{1=1} \omega^2 \left[N_{01}^0 + \frac{Q E_{01}^0}{(\omega + \nu - \beta k_w)^2} \right] A_{1'00,-1} \tag{53}
$$

Now; letting $s = (k_w \beta - \nu - \omega)/\omega$, using the approximation $1 \, k_w^2 \beta^2 / \omega^2 \equiv -2(s + \nu/\omega)$ and relying on Eq. (40) we can recast Eq. (53) as

$$
s^3 + s^2 \left[-1 + \frac{2}{\sigma^2 \omega^2} + \frac{2\nu}{\omega} + \sum_{1=0}^{\infty} N_{01}^0 A_{100,-1} \right]
$$

$$
+ \frac{Q}{\omega^2} \sum_{1=0}^{\infty} E_{01}^0 A_{100,-1} = 1 \tag{54}
$$

Once found a root of (54), we solve the system (52) for $1 \neq 0$

$$
\left[\frac{\gamma_{11}^2}{\omega^2 \sigma^2} + 2 \left(s + \frac{\nu}{\omega} \right) \right] A_{100,-1} + \sum_{\substack{1' \\ 1' \neq 0}} N_{11'}^0 A_{1'00,-1}
$$

$$
+ \frac{1}{s^2} \frac{Q}{\omega^2} \sum_{\substack{1' \\ 1' \neq 0}} E_{11'}^0 A_{1'00,-1} = - N_{01}^0 - \frac{1}{s^2} \frac{Q}{\omega^2} E_{01}^0 \tag{55}
$$

Next, we recalculate s by using Eq. (54) with the new values of $A_{100,-1}$. By iterating this procedure, we generate a convergent sequence of values s and amplitudes $A_{100,-1}$.

For b_1 not very large we can calculate the mode profile and by following an approach developed for weakly guiding fibers[13,14]. In particular, we multiply the wave equation by $rA_{0,-1}$ and integrate from r=0 to r = ∞, thus obtaining

$$k_w^2 \beta^2 = \frac{\int_0^\infty \{\omega^2 n_{FEL}^2 A_{0,-1}^2 - (dA_{0,-1}/dr)^2\} rdr}{\int_0^\infty A_{0,-1}^2 \, rdr} \tag{56}$$

Approximating $A_{0,-1}$ by

$$A_{0,-1} = \frac{\epsilon}{\sqrt{\pi}} \exp\left(-\frac{r^2}{2\sigma^2}\right) \tag{57}$$

we have

$$1 - \frac{(k_w \beta)^2}{\omega^2} = \frac{Q}{1 + \sigma^2/2\sigma_{EB}^2} \frac{1}{(\omega + \nu - k_w \beta)^2} + \frac{1}{\omega^2 \sigma^2} \tag{58}$$

Now, imposing the condition $\partial\beta/\partial\sigma^2 = 0$ we have

$$s = \pm i \sqrt{2Q} \frac{x}{1 + 2x} \sigma_{EB} \tag{59}$$

so that, eliminating s we obtain a two branch cubic equation for $x \equiv \sigma^2/\sigma_{EB}^2$,

$$(\pm i 2^{3/2} \omega^2 \sigma_{EB}^3 \sqrt{Q} + 2\nu\omega\sigma_{EB}^2) x^3 + 4\omega\nu\sigma_{EB}^2 x^2 - 2x - 4 = 0 \tag{60}$$

SECOND ORDER APPROXIMATION

If we approximate the driving current by including terms up to the second order in ε the FEL system modifies as

$$(\nabla_\perp^2 + \omega^2 n_{FEL}^2 - \beta^2 k_w^2) A_{00,-1}$$

$$= -i \frac{Q}{2s^2} \sum_1 \left[i\epsilon \sum_{1'1''q} (A_{1'',1,2-q,q} X_{1''1'}^{1,-1} E_{1'1}^0 + A_{1'',-1,qq} X_{1''1'}^{-1,1} E_{1'1}^0 \right.$$

$$+ \frac{\epsilon^2 \alpha}{2} (\sum_{1'1''1'''q} A_{1'''00,-1} X_{1'''1''}^{0q} X_{1''1'}^{q,-q} E_{1'1}^0$$

55

$$+ x \sum_{1'1"} A_{1"00,-1} (-X^{1,-1}_{1"1'} + X^{-1,1}_{1"1'}) \, E \, R^0_{1'1}$$

$$+ 2x \sum_{1'} A_{1'00,-1} \, E \, R \, R^0_{1'1} - 2x \, A_{100,-1})] \, |10> \qquad (61)$$

the matrix elements being defined in the Appendix and having approximated b_1 with $-k^2/\omega^2 s^2$. Now, if we approximate the driving currents of $|1",1,2+q,q>$ and $|1",-1,qq>$ with terms of first order in ε we have

$$\left[\frac{\gamma^2_{1",1}}{\sigma^2} + \omega^2 (1 - N^1_{1"1"}) - (\beta + 2 + q)^2 \, k^2_w \right] A_{1",1,2+q,q}$$

$$= - i\varepsilon \frac{Q}{2s^2} \sum_{11'} A_{100,-1} (X^{01}_{11'} \, E^1_{1'1"} - x \, E \, R^1_{11"}) \qquad (62)$$

$$\left[\frac{\gamma^2_{1",-1}}{2} + \omega^2 (1 - N^{-1}_{1"1"}) - (\beta + q)^2 k^2_w \right] A_{1",-1,qq}$$

$$= - i\varepsilon \frac{Q}{2s^2} \sum_{11'} A_{100,-1} \, X^{0,-1}_{11'} \, E^{-1}_{1'1"} \qquad (63)$$

Solving for $A_{1",1,2+q,q}$ and $A_{1",-1,qq}$ we finally obtain

$$(\nabla^2_{\perp} + \omega^2 n^2_{FEL} - \beta^2 k^2_w) \, A_{00,-1} = - \varepsilon^2 \left(\frac{Q}{2s^2} \right)^2$$

$$\sum_{11'1"1'"1iv} A_{1iv00,-1} \left[\frac{(X^{01}_{1iv1'"} E^1_{1"'1"} - x \, E \, R^1_{1iv1"}) X^{1,-1}_{1"1'} E^0_{1'1}}{\frac{\gamma^2_{1",1}}{\sigma^2} + \omega^2(1 - N^1_{1"1"}) - (\beta + 2 + q)^2 k^2_w} \right.$$

$$\left. + \frac{X^{0,-1}_{1iv1'"} E^{-1}_{1'"1"} X^{1,-1}_{1"1'} E^0_{1'1}}{\frac{\gamma^2_{1",-1}}{\sigma^2} + \omega^2 (1 - N^{-1}_{1"1"}) - (\beta + q)^2 k^2_w} \right] 10>$$

$$+ -\varepsilon^2 \frac{Q}{2s^2} \frac{\alpha}{2} \sum_1 (\sum_{1'1"1'"q} A_{1'"00,-1} \, X^{0q}_{1'"1"} \, X^{q,-q}_{1"1'} \, E^0_{1'1}$$

$$+x \sum_{1'1"} A_{1"00,-1}(- X^{1,-1}_{1"1'}+X^{-1,1}_{1"1'}) E \, R^0_{1'1}+2x \sum_{1'} A_{1'00,-1} E \, R \, R^0_{1'1} -2A_{100,-1}) \, |10>$$

$$\equiv -\varepsilon^2 \frac{1}{4} \sum_s C_1(\beta)|10\rangle - \varepsilon^2 \frac{1}{2} \sum_s D_1(\beta)|10\rangle \tag{64}$$

Accordingly, Eqs. (53) and (55) are superseded by

$$\left[-\frac{\gamma_{11}^2}{\omega^2\sigma^2} + 2(s + \frac{\nu}{\omega})\right]A_{100,-1} + \varepsilon^2\left(\frac{C_1'}{4} + \frac{D_1'}{2}\right)$$

$$+ \sum_{1'\neq 0} N_{11'}^0 \, A_{1'00,-1} + \frac{1}{2}\frac{Q}{\omega^2} \sum_{1'\neq 0} E_{11'}^0 \, A_{1'00,-1}$$

$$= - N_{01}^0 - \frac{1}{2}\frac{Q}{\omega^2} E_{01}^0 - \varepsilon^2\left(\frac{C_1''}{4} + \frac{C_1''}{2}\right) \tag{65}$$

where C_1' and D_1' represent the parts of C_1 and D_1 which do not contain $A_{000,-1} = 1$, while C_1'' and D_1'' represent the remaining parts. In particular, for calculating C_1 and D_1 the propagation constant β appearing in the terms on the LHS of Eq. (63) can be approximated with the value obtained by solving Eqs. (58) and (59). In fact, β occurs in the denominators of the above terms in combination with q and 2, so that these terms do not depend critically on the value of β, in contrast with the denominator $\omega + \nu - k_w\beta$ appearing in Eq. (58).

Now, the system (64) can be solved iteratively by using the same procedure illustrated for (53) and (55).

On the other hand, we can also follow the same procedure used for calculating β and σ at the zeroth order in ε, by approximating $A_{00,-1}$ with a gaussian,

$$A_{00,-1} = \frac{\varepsilon}{\sqrt{\pi}} \frac{1 + \sigma^2/\sigma^{(2)}}{2} \exp\left(-\frac{r^2}{2\sigma^{(2)2}}\right) \tag{66}$$

where $\sigma^{(2)}$ is the new value of σ. The superscript 2 reminds us of the second order correction of the driving current. Next, multiplying the above equation by $rA_{00,-1}$ and integrating from 0 to ∞ we obtain the equation

$$1 - \frac{(k_w\beta)^2}{\omega^2} = \left(\frac{Q}{1 + \sigma^{(2)2}/2\sigma_{EB}^2} + \varepsilon^2 D\right)\frac{1}{(\omega + \nu - k_w\beta)^2}$$

$$+ \frac{C\varepsilon^2}{(\omega + \nu - k_w\infty)^4} + \frac{1}{\omega^2 \sigma^{(2)2}} \tag{67}$$

where C and D are suitable coefficients related to C_1 and D_1. Replacing β with s and imposing again the condition $\partial\beta/\partial\sigma^{(2)} = 0$ we obtain an algebraic equation for $\sigma^{(2)}$.

ALL ORDERS EXPRESSION OF THE DRIVING CURRENT IN THE RESONANT CASE

Near resonance a_p and b_p are large only for $p = 1$. Analogously, a_{pq} and b_{pq} blow up for $p = 1$, $q = \pm 1 (a_{11} \cong a_1, b_{11} \cong b_1)$, $p = 2$, $q = -1 (a_{2,-1} \cong -a_1, b_{2,-1} \cong -b_1)$ and $p = 0$, $q = 1 (a_{01} \cong -a_1, b_{01} \cong -b_1)$.

Accordingly, we can neglect in the expression of the driving current those integrals I which do not contain the factor $\zeta - \zeta'$ in their argument. This amounts to approximating $\vec{P}_0$ and $\vec{P}_1$ with

$$\frac{\vec{P}_0}{a_w} = \hat{a}_w (4\alpha - \nabla'^2) \int_0^\zeta e^{\tilde{X}-i\Delta} (\zeta - \zeta') \, \hat{a}_w \cdot \vec{a}'d\zeta' \tag{68}$$

$$\frac{\vec{P}_1}{a_w} = - \frac{\partial}{\partial\vec{x}'} \int_0^\zeta e^{\tilde{X}-i\Delta}(\zeta - \zeta') \, \hat{a}_w \cdot \vec{a}'d\zeta' \tag{69}$$

so that, the wave equation takes the form

$$(\nabla_\perp^2 + k_w^2 \frac{\partial^2}{\partial\zeta^2} + \omega^2) \, \vec{a}' = - \frac{Q\omega_r^2}{2\dot\alpha k_w^2}$$

$$\hat{a}_w \left[(4\alpha - \varepsilon^2\sigma^2\nabla^2) \, I_{\tilde{X}} - \varepsilon^2\sigma^2 \frac{\partial}{\partial\vec{x}_-} I_{\tilde{X}} \cdot \frac{\partial}{\partial\vec{x}_-}\right] e^{\tilde{X}(\zeta,0)} e^{-r^2/2\sigma_{EB}^2} \tag{70}$$

where

$$I_{\tilde{X}} \equiv \int_0^\zeta e^{\tilde{X}-i\Delta} (\zeta - \zeta') \, \hat{a}_w \cdot \vec{a}'d' \tag{71}$$

When the wiggling is neglected the wave equation reduces to the well studied cubic equation in the propagation constant, typical of the exponential growth regime [15,16,17].

$$(\nabla_\perp^2 + k_w^2 \frac{\partial^2}{\partial\zeta^2} + \omega^2)\vec{a}' = - \frac{2\omega_r^2}{k_w^2} \hat{a}_w \int_0^\zeta e^{-i\Delta} (\zeta - \zeta')\hat{a}_w \cdot \vec{a}'d' e^{-\frac{r^2}{2\sigma_{EB}^2}} \tag{72}$$

It is worth noting that the model used by Pantell et al.[8] for studying the electron wiggling influence on the optical guiding takes only into account the bending of the e-beam current. Although we arrived at the above equations by discussing the weights of the terms contributing to the driving

58

current of order less than 3 in the smallness parameter ε, it is straigt-forward to show that they hold true to all orders in ε under the assumption of neglecting the terms $a_{pqq'q''q'''\ldots q^n}$ with respect to $b_{pqq'q''q'''\ldots q^n}$ for every combination of indexes $pqq'q''\ldots q^n$.

APPENDIX - MATRIX ELEMENTS

As shown in Ref. 12 the matrix elements $X_{11'}^{mq}$, $N_{11'}^{m}$, $E_{11'}^{m}$, $E\,R_{11'}^{m}$, $E\,R\,R_{11'}^{m}$, occuring in the matrix representation of the wawe equation are given by

$$X_{11'}^{mq} = \eta\left(\sqrt{|m + q| + 1 + \frac{1 - \eta q}{2}}\,\delta_{11'} + \sqrt{1 + \frac{1 - \eta q}{2}}\,\delta_{1,1'+\eta q}\right) \tag{73}$$

$$N_{11'}^{m} = \delta_{11'} - \frac{1}{(\sigma\omega)^2}\sqrt{\frac{1!(|m| + 1')!}{1'!(|m| + 1)!}}\left[(21 + |m| + 1)\delta_{11'}\right.$$

$$\left. - (1 + 1)\delta_{1,1'-1} - (1 + |m|)\delta_{1,1'+1}\right] \tag{74}$$

$$E_{11'}^{m} = \frac{(1 + 1' + |m|)!}{\sqrt{1!1'!(1+|m|)!(1' +|m|)!}}\,\frac{(x/2)^{1+1'}}{(1 + x/2)^{1+1'+|m| + 1}}$$

$$F\left[-1, -1'; -1 -1' - |m|; 1 - \frac{4}{x^2}\right] \tag{75}$$

$$E\,R_{11'}^{m} = \frac{1}{\sqrt{x}}\sum_{1''q}q(X_{11''}^{mq}\,E_{1''1'}^{m} - E_{11''}^{m}\,X_{1''1'}^{mq}) \tag{76}$$

$$E\,R\,R_{11'}^{m} = \frac{1}{x}\sum_{1''}\sqrt{\frac{1!(|m| + 1'')!}{1''!(|m| + 1)!}}\left[(21 + |m| + 1)\,\delta_{11''}\right.$$

$$\left. - (1 + 1)\delta_{1,1''-1} - (1 + |m|)\delta_{1,1''+1}\right]E_{1''1'}^{m} \tag{77}$$

where $q = \pm 1$, $\eta = 1$ if $m \geq 0$, $m + q \geq 0$ and $\eta = -1$ otherwise, F is a polynomial coincident with the hypergeometric function and $x \equiv \sigma^2/\sigma_{EB}^2$ is the same quantity appearing in Eq. (59).

In addition,

$$a_p = \frac{i}{\beta + p - \frac{\omega + \nu}{k_w} - 1} \tag{78}$$

$$b_p = - \frac{1}{(\beta + p - \frac{\omega + \nu}{k_w} - 1)^2} \tag{79}$$

$$b_{pq} = b_p - b_{p+q} \tag{80}$$

$$b_{pqq'} = b_p - b_{p+q} - b_{p+q'} + b_{p+q+q'} \tag{81}$$

ACKNOWLEDGEMENTS

One of the authors (S.S.) wishes to thank the Lawrence Berkeley Laboratory for the kind hospitality offered during the completion of this work and the Italian National Council of Research for a NATO fellowship.

REFERENCES

1. C. M. Tang and P. Sprangle, in "Physics of Quantum Electronics", vol. 9, eds. S. F. Jacobs, G. T. Moore, H. S. Piloff, M. Sargent III, M. O. Scully and R. Spitzer, Addison-Wesley, Reading, MA (1982) p. 627

2. E. T. Scharlemann, A. M. Sessler and J. S. Wurtele, Proc. Int. Workshop on Coherent and Collective Properties in the Interaction of Relativistic Electrons and Electromagnetic Radiation, Como, Italy (1984); Phys. Rev. Lett. 54, 1925 (1985)

3. J. E. La Sala, D. G. G. Deacon and E. T. Scharlemann, N. I. M. A 250, 389 (1986)

4. G. T. Moore, Opt. Commun. 52, (1984) 46; 54 (1985) 121

5. A. Cutolo, S. V. Benson and J. M. Madey, Real time processing of picosecond pulses: application to FEL (preprint)

6. N. M. Kroll, in "Physics of Quantum Electronics", Addison Wesley, Reading, (1978) p. 115

7. P. Luchini and S. Solimeno, N. I. M., A 250 413 (1986)

8. R. H. Pantell, E. Fontana, J. Feinstein, Wave guiding by a wiggling FEL beam, talk given at the 7th International FEL Conference, Tahoe City, CA (1985)

9. J. D. Lawson, "The Physics of Charged Particle Beams", Oxford Univ. Press, Oxford (1978) p. 119

10. T. C. Marshall, "Free Electron Lasers", Macmillan, New York (1985)

11. S. Solimeno, B. Crosignani, P. Di Porto, "Guiding, Diffraction and Confinement of Optical Radiation", Chap. 8, Academic Press, Orlando (1986)

12. S. Solimeno, Yu-Juan Chen and A. M. Sessler, Lawrence Berkeley Laboratory Report (1987) (unpublished)

13. A. W. Snyder and J. D. Love, "Optical Waveguide Theory", Chapman and Hall, London (1983) p. 338

14. M. Xie and D. A. G. Deacon, <u>N. I. M.</u> A 250, 389 (1986)
15. I. B. Bernstein and J. L. Hirschfield, <u>Phys. Rev.</u> A 20; 1661 (1979)
16. E. Jerby and A. Gover, <u>IEEE J. Quantum Electron.</u> QE-21, 1041 (1985)
17. G. Dattoli, A. Marino, A. Renieri and F. Romanelli, <u>IEEE J. Quantum Electron.</u> QE-17, 1371 (1985)

THE FREE ELECTRON LASER: A BRIEF ANALYSIS OF THEORY AND EXPERIMENT

G. Dattoli and A. Torre

ENEA, Dip. TIB, U.S. Fisica Applicata - CRE Frascati

C.P. 65 - 00044 Frascati, Rome, Italy

INTRODUCTION

The intensive experimental and theoretical activity, originated by the successful operation of the first free electron laser (FEL) in Stanford[1] in 1977, has displayed the peculiarities of this new coherent source of radiation eventually dedicated to a number of both scientific and technological applications.

Presently the existing FELs cover a wide region of the electromagnetic spectrum and indeed their experimental feasibility has been proved from the visible to the millimeter region[2]. Furthermore existing projects tend to improve the working range down to the VUV[3] and even to X rays[4].

Even though the FEL is not yet a fully reliable facility some applications of this device have been undertaken in the wavelength region where the FEL can be considered complementary to the ordinary atomic or molecular lasers[5].

Strictly speaking the FEL is not a laser but it belongs to the family of travelling wave tubes (TWT) and it is one of its last born in a wider range of tunability.

In the field of coherent generation of radiation with an electron beam, the FEL brought the possibility of overcoming all the problems connected with miniaturization of the microwave tubes needed to operate at a shorter wavelength. The working principle of the FEL, as already stressed, is similar to that of TWT in which a TM wave propagates in a waveguide and modulates a copropagating electron beam (see Fig. 1). This modulation is due to the fact that the electric field of the TM wave is longitudinal and thus parallel to the electron velocity. Therefore we get that:

$$\frac{d\mathcal{E}}{dt} = - e \, \vec{E} \cdot \vec{V} \neq 0 \tag{1}$$

This energy modulation transforms into a density modulation at the same wavelength of the input radiation. The final result is a coherent emission from each packet and thus an amplification of the input wave. It is well known that these devices can work up to the centimeter region, a shorter region would require a waveguide with such small dimensions practically impossible to be realized. On the other hand in the FEL the electrons are constrained to transverse motion so that their energy modulation is provided by a TE wave which propagates in an optical cavity. The element transforming the longitudinal nonradiating electron motion into a transverse one is the undulator, i.e., a magnetic device with alternating N-S poles arranged as in Fig. 2.

In this case too the energy modulation transforms into a density modulation and the emission-amplification mechanism takes place according to what has been previously illustrated.

We will see in more detail in the next section that the operating wavelength λ is a function of the undulator parameters and of the electron energy, in particular:

$$\lambda = \frac{\lambda_u}{2\gamma^2} (1 + K^2) \tag{2}$$

where λ_u is the magnet period, K is the undulator parameter, linked to its on-axis field (see below), and γ is the electron relativistic factor.

The above relation states that using an electron beam with the appropriate energy it is possible to get radiation covering practically all the e.m. spectrum. However as we will see in the next section there are problems connected with e-beam quality (current, energy spread and emittances), which pose severe limitations to the operation at a shorter wavelength.

In this note we will give a general overview of the FEL physics without entering into any particular detail. The paper is intended as an elementary introduction to this new type of coherent source and a deeper insight can be found in Refs. 6, 7.

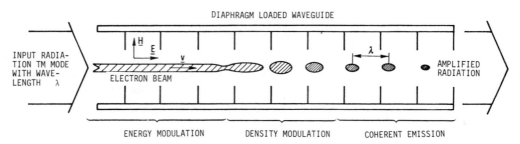

Fig. 1. A schematic representation of a TWT coherent emission process.

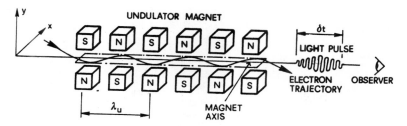

Fig. 2. Undulator magnet geometry.

SPONTANEOUS EMISSION AND GAIN IN FEL

In atomic and molecular lasers the spontaneous emission is provided by a transition between quantum stats and thus the wavelength depends on the relative energy gap. On the other hand the mechanism underlying the FEL emission is classical and is due to the electron acceleration during its motion in the undulator. To state this point more quantitatively let us notice that an electron moving in a magnetic transverse field of the type:

$$B = B_0 \cos (2\pi z/\lambda_u)$$ (3)

undergoes a Lorentz force which in turn causes a transverse velocity given by:

$$V_T = - \frac{cK\sqrt{2}}{\gamma} \sin (2\pi z/\lambda_u)$$ (4)

where

$$K = \frac{eB_0\lambda_u}{2\sqrt{2\pi}m_0c^2}$$

and B_0 is the on-axis magnetic field. The average longitudinal velocity will be therefore (where the symbol $< >$ denotes an average on an undulator period):

$$V_0^2 = V^2 - <V_T^2> = c^2 (1 - \frac{1}{2\gamma^2}(1+K^2))$$ (5)

In the transverse plane the electron is oscillating at a frequency

$$\omega_u = \frac{2\pi\lambda_u}{c}$$ (6)

which in the electron rest frame (namely, in the frame in which the longitudinal velocity is zero) transforms to:

$$\omega' = \frac{\omega_u}{\sqrt{1-(V_0/c)^2}} \tag{7}$$

In this frame the electron radiates isotropically at frequency ω'. Going back to the laboratory frame we will also have a Doppler shift of the emitted radiation so that we finally get:

$$\omega_0 = \sqrt{\frac{1+V_0/c}{1-V_0/c}}\ \omega' \cong \frac{4\pi c}{\lambda_u}\ \frac{\gamma^2}{1+K^2} \tag{8}$$

The width of the emitted lines is fixed by the time duration Δt of the light pulse due to each electron which is linked to the difference between the electron and radiation transit times in the undulator, i.e.,

$$\Delta t = \frac{L}{V_0} - \frac{L}{c} \cong (1 - \frac{V_0}{c})\ \frac{N\lambda_u}{c} \cong \frac{2\pi N}{\omega_0} \tag{9}$$

where L and N are the undulator length and number of periods respectively ($L=N\lambda_u$). Finally since $\Delta\omega\Delta t \cong \pi$, we get the relative width of the emission line:

$$(\frac{\Delta\omega}{\omega})_0 \cong \frac{1}{2N} \tag{10}$$

Therefore for typical undulators with N=50 one gets a relative bandwidth of 1%. Such a small width is the distinctive feature of the undulator radiation with respect to that of synchrotron emission which is considerably wider due to a very short light pulse (see Fig. 3, where the synchrotron and undulator spectra are reported for comparative purposes).

The undulator spectrum shape is also given by:

$$\frac{dP}{d\nu} \propto (\frac{\sin\nu/2}{\nu/2})^2$$

$$\nu = 2\pi N\ \frac{\omega_0 - \omega}{\omega_0} \tag{11}$$

Let us now assume that the emission process in the undulator takes place in the presence of a radiation field, with frequency near that of spontaneous emission given by Eq.(11) copropagating with the electrons. In this case a variation of the intensity of the input radiation field can be observed and this process can be referred to as stimulated emission. According to the energy conservation the intensity variation of the field will be proportional to the energy variation of the e-beam, which can be written as follows:

66

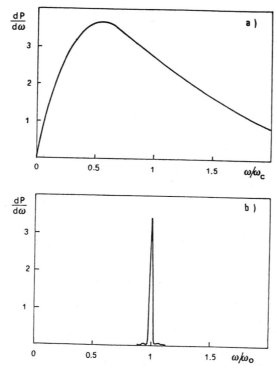

Fig. 3. Comparison between synchrotron (a) and undulator (b) emission spectra.

$$\dot{\gamma} = \frac{eE_L V_T}{m_0 c^2} \tag{12}$$

where V_T is given in Eq.(4), E_L is the electric field of the wave:

$$E_L = E_0 \cos (\omega t - kz + \Phi_L) \tag{13}$$

and Φ_L is the relative phase between field and electron.

Inserting equations (13) and (4) in (12), we get:

$$\dot{\gamma} = - \frac{eE_0 K}{m_0 c \gamma \sqrt{2}} \sin\Psi \tag{14}$$

$$\Psi = (k + \frac{2\pi}{\lambda_u}) z - \omega t - \Phi_L \tag{15}$$

From Eq. (5) ($\dot{z} = V_0$) it is easy to see that:

$$\ddot{z} = \frac{c\dot{\gamma}}{\gamma^3} (1 + K^2) \tag{16}$$

Substituting Eq. (14) in Eq. (16) , by means of Eq. (15), we obtain (see ref. 6 for further comments):

$$\ddot{\psi} = -\Omega^2 \sin\psi \tag{17}$$

where

$$\Omega^2 = \frac{e^2 E_0 B_0}{(m_0 c\gamma)^2} \tag{18}$$

The result (17) states that the basic FEL dynamics can be understood by means of a pendulum-like equation. Equation (17) can also be written in the form

$$\dot{\psi}^2 - \dot{\psi}^2 (0) = 2\Omega^2 (\cos\psi - \cos\psi (0)) \tag{19}$$

where

$$\psi (0) = - (\omega t_0 + \Phi_L) \quad ; \quad \dot{\psi}(0) = \frac{c}{L} v \tag{20}$$

The analytical solution of (19) can be obtained in terms of elliptic$^{(*)}$ functions but since we are interested in the small signal regime$^{(*)}$ we can use a perturbative solution which immediately yields the FEL gain. Defining indeed:

$$G = -m_0 c^2 \frac{\Delta\gamma}{W_0} \qquad W_0 = \frac{1}{8\pi} E_0^2 V \qquad (V = \text{mode volume}) \tag{21}$$

at the lowest order in Ω after some algebra we get^6

$$G = - g_0 \frac{d}{dv} (\frac{\sin v/2}{v/2})^2 \tag{22}$$

where g_0 is the gain coefficient explicity defined by

$$g_0 = \frac{2\pi^2}{\gamma} \frac{\lambda_0 L}{\Sigma_E} \frac{I}{I_0} F \frac{K^2}{1+K^2} (\frac{\Delta\omega}{\omega})_0^{-2} \tag{23}$$

Where I and I_0 are respectively the e-beam and Alfven current ($I_0 \simeq 17kA$), Σ_E is the e-beam cross section and F is a "filling-factor" which takes into account that only the electron inside the laser mode section Σ_L contribute to the gain, F can be written in the form:

---

* i.e., when the laser electric field is weak enough such that the coupling strength $\Omega \ll c/L$.

$$F = \begin{cases} 1 & \text{if } \Sigma_E > \Sigma_L \\ \Sigma_E / \Sigma_L & \text{if } \Sigma_E < \Sigma_L \end{cases}$$

In this case the gain is proportional to the derivative of what we have called FEL spontaneous emission (for the shape of the gain curve see Fig. 4).

Before adding further considerations let us define the FEL efficiency η as the ratio between the output "laser" power (P_L) and the electron beam power (P_E)(*)

$$\eta = \frac{P_L}{P_E} \tag{24}$$

We use a simple qualitative argument to derive the value of η. It is clear that when the electrons lose energy they do not satisfy the resonance condition anymore; after some time their energy variation can be so strong as to bring ν outside the positive part of the gain curve. The maximum variation of ν which is compatible with positive gain is

$$\Delta\nu \leq 2\pi \tag{25}$$

which implies a maximum relative energy variation

$$\frac{\Delta\mathscr{E}}{\mathscr{E}} = \frac{\Delta\nu}{\nu} \sim \frac{1}{2N} \tag{26}$$

Finally, $\eta \sim \dfrac{1}{2N}$ \hfill (27)

We have so far discussed the gain in the hypothesis that the e-beam is monoenergetic and with zero divergence. However a correct evaluation of the gain requires the inclusion of the effects due to the non-monochromaticity and nonzero divergence of a real e-beam. These contributions act as a kind of inhomogeneous broadening causing both a reduction of peak gain and broadening of the linewidth. To give an idea of how these effects should be included, let us notice that the central frequency emission is given by

$$\omega_0 = \frac{4\pi c\gamma^2}{\lambda_u} \left(\frac{1}{1+K^2+\gamma^2\theta^2} \right) \tag{28}$$

which differs from that given in Eq. (8) because the contribution from off-axis radiating electrons has also been taken into account (θ is the angle of bending of the average electron trajectory).

Eventual shift of the central frequency may be due to energy, angle

---

* Recall that $P_E(MW) = \mathscr{E}(MeV) \cdot I(A)$ where $\mathscr{E}$ and I are the energy and current of the electron beam expressed in MeV and Ampere respectively.

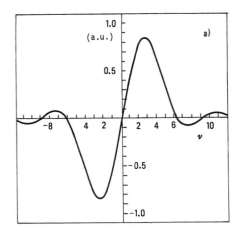

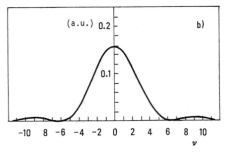

Fig. 4. (a) Small signal gain; (b) spontaneous spectrum.

and magnetic variation from the nominal values. The relative shift can be immediately evaluated from Eq. (28) and reads:

$$\frac{\delta\omega}{\omega} = 2\varepsilon - \frac{1}{1+K^2} \left[(\gamma\theta)^2 + 2K^2 \frac{\delta B}{B} \right]$$ (29)

where ε is the relative energy shift

$$\varepsilon = \frac{\delta\gamma}{\gamma}$$

and $\delta B/B$ accounts for the transverse magnet inhomogeneities and depends on the undulator geometry[6].

Assuming a gaussian distribution of the electron beam in energy, angle and transverse dimensions and defining the r.m.s. inhomogeneous width as:

$$(\Delta\omega/\omega)_i = (<(\delta\omega/\omega)^2> - <\delta\omega/\omega>^2)^{1/2}$$ (30)

where the symbol $< >$ means average on the electron distribution, we get the following total inhomogeneous bandwidth:

$$\left(\frac{\Delta\omega}{\omega}\right)_T = \left[\left(\frac{\Delta\omega}{\omega}\right)_0^2 + \left(\frac{\Delta\omega}{\omega}\right)_\varepsilon^2 + \left(\frac{\Delta\omega}{\omega}\right)_x^2 + \left(\frac{\Delta\omega}{\omega}\right)_y^2\right]^{1/2} \tag{31}$$

$$\left(\frac{\Delta\omega}{\omega}\right)_\varepsilon = 2\sigma_\varepsilon \tag{32a.}$$

$$\left(\frac{\Delta\omega}{\omega}\right)_{x,y} = \sqrt{2|h_{x,y}|}\ \frac{\gamma K}{1+K^2}\ \frac{E_{x,y}}{\lambda_u} \tag{32b.}$$

σ_ε is the r.m.s. of the electron beam energy distribution, $E_{x,y}$ is the transverse emittance and $h_{x,y}$ are parameters depending on the geometry of the the undulator magnet (for further comments see ref. 6).

Introducing now the parameters

$$\mu_{\varepsilon,x,y} = \left(\frac{\Delta\omega}{\omega}\right)_{\varepsilon,x,y}\left(\frac{\Delta\omega}{\omega}\right)_0^{-1} \tag{33}$$

which are the ratio between inhomogeneous and homogeneous broadening we can say that when $\mu_{\varepsilon,x,y} \ll 1$ the FEL is operating in homogeneous regime an the inhomogeneous contributions can be neglected, otherwise severe limitations to the gain may arise as shown in Fig. 5.

A further difficulty may be due to FEL operating with radiofrequency devices. The e-beam furnished by an r.f. accelerator has a structure displayed in Fig. 6. It has been shown that this kind of structure induces an analogous structure on the laser beam which may be visualized as a series of pulses. To have synchronism between electron and optical bunches we must require that the electron bunch-bunch time distance T be equal to a round trip time. However owing to the different speeds between the two bunches the optical one will be ahead of a quantity

$$\Delta = N\lambda \tag{34}$$

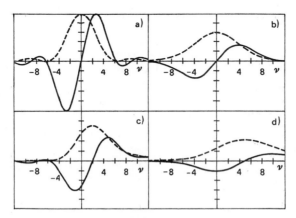

Fig. 5. Spectrum (dashed line) and gain (solid line) ($h_x > 0, h_y > 0$) :
(a) $\mu_\varepsilon = \mu_x = \mu_y = 0$ (homogeneous regime); (b) $\mu_\varepsilon = 1, \mu_x = \mu_y = 0$; (c) $\mu_\varepsilon = 0, \mu_x = 1, \mu_y = 0$; (d) $\mu_\varepsilon = \mu_x = \mu_y = 1$.

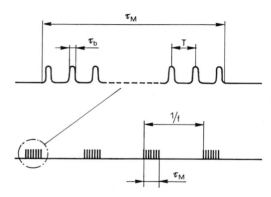

Fig. 6. R.f. electron beam temporal structure: τ_b microbunch time duration; T bunch-bunch time distance; τ_M macrobunch time duration; f accelerator repetition rate.

after one undulator passage. This slippage has two consequences on the gain:

(a) the electrons and photons do not overlap for all the time of interaction;

(b) since the front part of the optical pulse sees a decreasing e-beam distribution its backward part will experience larger gain, therefore the optical bunch centroid will be slowed down.

As a consequence, the cavity length should be the nominal length, fixed by the electron bunch-bunch distance, minus a certain amount δL necessary to ensure the synchronism between input electron and photon bunches.

To summarize the gain is a rather complicated function including the effect of inhomogeneities, the effect of slippage and of the cavity mismatch with respect to the nominal round trip length. The dependence of the gain vs. δL is shown in Fig. 7 ($\theta = -\delta L/\Delta g_0$), while the dependence on the slippage can be represented by the simple relationship[8]

$$G\,(\mu_c) \propto \frac{1}{1+\mu_c/3} \tag{35}$$

where $\mu_c = \Delta/\sigma_z$ (σ_z being the r.m.s. of the electron bunch longitudinal distribution) is a parameter which gives a measure of the relative slippage between electron and optical pulses. Also the efficiency is affected by the lethargy machanism and the dependence of this quantity on δL is shown in Fig. 7. The maximum of the efficiency is however around 1/2N.

It is clear that one of the crucial parameters in the FEL is the efficiency. An improvement of η may give larger output power. Different schemes have been proposed to get larger values of η making undulators called "tapered". We have seen that the poor FEL efficiency is due to the energy loss of the electron and the consequent lack of resonance. These undulators are designed in such a way to fulfill the resonance condition

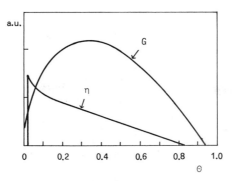

Fig. 7. Gain and efficiency vs. θ.

during all the interaction time by suitably compensating the energy loss by a variation of the undulator period or the peak magnetic field. Schemes of this type have improved the efficiency even of an order of magnitude[9].

The overall efficiency can also be enhanced by recovering the electron beam energy as in the UCSB-FEL[10] experiment. In this case we have:

$$\eta_T = \frac{\eta}{\eta + (1 - \eta_R)} \qquad (36)$$

where η_R is the electron energy recovery efficiency (values quoted in the literature for η_R are around 98%).

This last discussion may lead to an erroneous conclusion for FELs operating with storage rings, where the electron beam is continuosly re-injected in the interaction region and not lost after one passage as in low energy accelerators. A very simple, qualitative argument, but substantially correct, proves that also for storage ring FELs the efficiency is of the order of 1/2N. We have seen that the FEL interaction acts as a kind of noise on the electron beam, thus introducing an energy spread which will cause an inhomogeneous bradening which in turn will reduce the gain. When the induced relative energy spread is of the order of 1/2N the laser process will stop, then we should wait a time of the order of the damping time[(*)] τ_s to have a new laser pulse.

Finally, the laser average power will be related to the damping time. More quantitatively:

$$P_L = \frac{\mathcal{N}}{\tau_s} \, \mathcal{E} \cdot \frac{\Delta \mathcal{E}}{\mathcal{E}} \qquad (37)$$

where $\mathcal{N}$ is the number of particles in the beam. Since

---

* The damping time of a storage ring is the characteristic time in which the off-energy particles reduce to synchronous particles, i.e., the particles with the machine nominal energy.

$$\tau_s = \frac{T \mathcal{E}}{U_0} \tag{38}$$

where U_0 is the energy radiated per turn in the bending magnet of the storage ring and T is the revolution period, we get that:

$$P_L \cong \frac{1}{2N} P_s \tag{39}$$

and $P_s = \mathcal{N} \mathcal{E}/\tau_s$ is the synchroton radiation power. In conclusion the storage ring FEL power is related to the synchrotron power via the usual efficiency factor[11].

CONCLUSIONS

In the previous sections we have touched on the essential features of the FEL physics. We have not presented an exhaustive account of any of the discussed topics but we have tried to give a feeling of which are the problems underlying the new laser process.

FEL is only ten years old, the modern era of this laser concept started in 1976 when at the University of Stanford J. Madey and coworkers used the FEL to amplify the 10.6 μm output of a CO_2 laser. From that time to now many FEL sources have been operated and a scenario characterizing their properties is given in Table I. Even though ten years have elapsed from the first FEL experiments it is not easy to make a balance of activity and may be the next future will be able to say how much the FEL is interesting for applications.

Many of the most promising uses of FEL can be envisaged in those wavelength regions where efficient, tunable, high power conventional lasers do not operate. As suggested in Fig. 8, particular interest should be given to sources in the infrared or far-infrared where the FEL is expected to furnish a considerable amount of power. This region is of primary interest for solid state physics and at shorter infrared wavelengths FEL may be an interesting source for chemical and molecular spectroscopy studies.

Medicine is a further area which has shown a noticeable interest for FEL. It has been indeed suggested[12] that FEL may be an actractive alternative to the use of the CO_2 or NdYAG laser for microsurgery and phototherapy.

Table I. FEL SCENARIO

EXPERIMENTS	STATUS	$\lambda(\mu m)$	ACCELERATOR	UNDULATOR MAGNET
LBL-LLNL (Livermore)	High gain amplifier of spontaneous emission	8×10^3	Induction Linac	Electromagnet
UCSB				
Single stage	Oscillator	400	Pelletron	Permanent magnet
Two stage	Project	1-10		Laser wave
AT&T Bell	Spontaneous emission	100-40	Microtron	Electromagnet
ENEA (Frascati)	Oscillator	10	Microtron	Electromagnet
		10-30		Permanent magnet
YEREVAN	Oscillator	30-40	Microtron	Electromagnet
MSNW-BAC	Amplifier	10.6	Linac	Permanent magnet
LANL (Los Alamos)	Oscillator	9-11	Linac	Permanent magnet
UK Project	Spontaneous emission	2-20	Linac	Permanent magnet
Stanford	Oscillator	3.3	Superconducting Linac	Superconducting
Stanford MK III	Oscillator	2.6-3.1	Linac	Electromagnet
TRW-Stanford	Oscillator	1.06	Superconducting Linac	Permanent magnet
Orsay	Oscillator	0.646	Storage Ring	Permanent magnet
Novosibirsky	Spontaneous emission	0.63	Storage Ring	Permanent magnet
INFN (Frascati)	Amplifier	0.514	Storage Ring	Electromagnet
Brookhaven	Spontaneous emission	Visible	Storage Ring	Permanent magnet
SOR-RING (Japan)	Project	Visible	Storage Ring	Permanent magnet

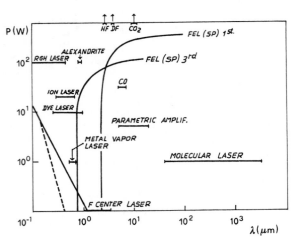

Fig. 8. Comparative chart between FEL and conventional coherent sources.
Average power vs. λ. Curves FEL (sp): single passage FEL average
power vs. λ 1st and 3rd harmonic respectively, maximum e.b. power
20 MW, K=1, λ_u =5 cm, N=50, L=6 m, τ_M =12 μsec, τ_b =30 sec. Straight
curves: FEL storage ring average power vs. λ without (continuous)
and with (dashed) Touschek effect respectively. (I=100 A).

REFERENCES

1. D. A. G. Deacon, L. R. Elias, J. M. J. Madey, G. J. Ramian, H. A.
   Schwettman,T. J. Smith, First Operation of a Free Electron Laser
   Phys. Rev. Lett., 38:892 (1977)

2. See e.g. 8th Conference on Free Electron Lasers, Glasgow 1986, Nucl.
   Instrum. Methods., A259 (1987).

3. See e.g. B. E. Newnam, Prospects for Free Electron Laser Operation
   in the Extreme Ultraviolet. Talk delivered at XV International
   Quantum Electronics Conference, 26 April - 1 May 1987. Baltimore
   USA

4. J. B. Murphy, C. Pellegrini, Free Electron Lasers for the XUV Spectral
   Region, Nucl. Instrum. Methods, A237:159 (1985)

5. See e.g., Applications of Free Electron Lasers, D. A. G. Deacon and
   A. De Angelis, eds., Nucl. Instrum. Methods, A 239 (1985)

6. G. Dattoli and A. Renieri, "Experimental and Theoretical Aspects of
   the Free Electron Laser", Laser Handbook, Vol. IV, M.L. Stitch and
   M. S. Bass, eds., North Holland, Amsterdam (1985)

7. W. B. Colson, A. M. Sessler, Free Electron Laser, Annu. Rev. Nucl. Part.
   Part. Sci., 35:25 (1985)

8. G. Dattoli, T. Letardi, J. M. J. Madey, A. Renieri, Lawson-Penner Limit
   and Single Passage Free Electron Lasers Performances, IEEE J. Quantum
   Electron., QE-20:637 (1984)

9. T. J. Orzechowsky, B. Anderson, W. M. Fawley, D. Prosnitz, E. T.
   Scharlemann and S. Yarema, Microwave Radiation from a High-gain Free
   Electron Laser Amplifier, Phys. Rev. Lett., 54:889 (1985)

10. L. R. Elias, J. Hu and G. Ramian, The UCSB Electrostatic Accelerator
    Free Electron Laser: First Operation, Nucl. Instrum. Methods A237:

203 (1985)

11. A. Renieri, Storage Ring Operation of the Free Electron Laser: the amplifier, Nuovo Cimento, 53B:160 (1979)

12. R. Ramponi and O. Svelto, Potential Applications of Free Electron Lasers in Biomedicine, Nucl. Instrum. Methods A239:386 (1985)

    L. J. Cerullo, Laser Applications in Neurosurgery, Nucl. Instrum. Methods, A239:385 (1985)

    V. A. Fasano, G. F. Lombard, R. Urciuoli, F. Benech and R. M. Ponzio, New Technologies in Neurosurgery: Effects on the Conventional Techniques and Anaesthesiological Considerations, Nucl. Instrum. Methods, A239:414(1985)

INTEGRATED HIGH-POWER-CO$_2$-LASER WITH COAXIAL TURBO-BLOWER

AND RF-EXCITATION

R. Beck

Lasertechnik GmbH

Heusenstamm, Fed. Rep. of Germany

Modern CO$_2$-lasers use the gas-transport principle to remove the waste heat from the discharge region. The gas-exchange rate which is necessary for lasers in the kW-range amounts to appr. 1 m^3/s. To generate volume flows in this order of magnitude usually axial fans or Roots-blowers are employed.

Roots-blowers are displacement pumps. They are able to sustain large pressure differences which are encountered in fast-axial-flow lasers of conventional design where because of the separate mounting of discharge tubes, heat exchanger, and pump extensive gas ducting is necessary. Contrary to Roots-blowers turbo-blowers (radial fans) are of a very simple design. They generate large gas flows provided the pressure drop can be limited to very low values. Fig. 1 shows the typical behavior of Roots- and turbo-blower in comparison. It can be seen that with the Roots-blower the volume flow changes very little with varying pressure difference, whereas the volume flow generated by the turbo-blower depends strongly on the pressure difference, i.e. the flow resistance.

Turbo-blowers can be used with advantage in fast-axial-flow lasers provided the recirculation losses are minimized. We have developed such a laser in the past four years by integration of the recirculation system into the laser chamber with a coaxial design.

The principle of this system is explained with reference to fig. 2. Discharge tube, turbo-blower, and heat exchangers are housed in a common gas-flow chamber (laser chamber). All these elements are coaxial to the resonator axis. The turbo-impeller disc is covered with a watercooled plate (cover disc). The heat exchangers are inserted into this plate so that the coolant (water) can circulate in the following way: from the lower· set of heat exchangers (only one is shown in fig. 2) through the cover disc to the upper set of heat exchangers. The heat exchangers are made from copper tubes to which a large number of copper wires are soldered. The cover disc is fitted with a diffusor ring of small blades which gives a pressure recovery of 1 mbar. The gas flow is guided to the heat exchangers through

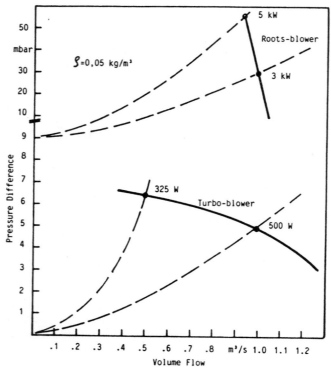

Fig. 1. Typical behavior of Roots- and turbo-blower in comparison. The indicated density is representative for a typical laser-gas-mixture. The operating points for both types are shown for two different flow resistances with the necessary drive powers.

Dimensions and weights (without motor)	Speed/rpm
* Roots-blower 1x0.5x0.5 m^3, 400 kg	3000
* Turbo-blower 30cm Ø x 5 cm, < 10 kg	20000

an annular channel which is formed between the cover disc and the inner wall of the gas-flow chamber. The gas passes the heat exchangers with relatively low velocity because of the large flow cross section in this area of the flow chamber. The gas is cooled back to 25° C very effectively and introduced again into the discharge tube. The turbo-blower operates with a speed of 20,000 rpm and generates a volume flow of 0.15 m^3/s. The pressure drop of the system is not larger than 4 mbar.

The motor which drives the blower is fitted with gas-bearings. These are fed with the laser gas at a pressure of 6 bar. Because of the low friction losses of the bearing and the very low recirculation losses the necessary drive power for the motor can be limited to 300 - 400 W. This is a factor of 10 less than the power necessary for a Roots-blower. An additional advantage of the gas-bearing is the elimination of a possible contamination of the laser gas by lubrificants.

The hollow shaft of the drive motor is fitted with an impeller disc on each end. Consequently on each side of the motor a gas-recirculation chamber is attached. The resulting twin-module contains two discharge sections and

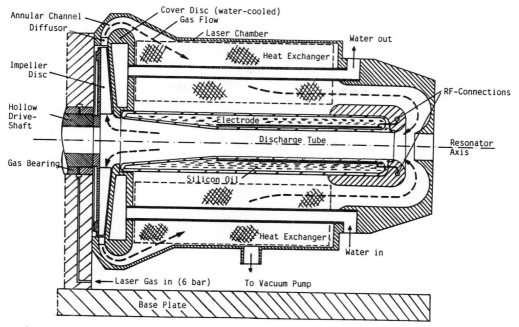

Fig. 2. Principle of integrated CO_2-laser with coaxial turbo-blower.

represents the basic unit which can be stacked together the output power to the desired level.

Two of these modules are combined in laser of 1.2 kW rated output power. The discharge is excited by capacitively coupled RF-power (13.56 MHz). The advantages of this method of excitation are based on the very stable and uniform discharge which is obtained up to very large specific input powers. Our discharge system employs transverse capacitive electrodes embedded in flowing silicon oil. The volume of one discharge section is 57 cm^3, the specific input power amounts to 30 W/cm^3.

POLARIZATION AND NONLINEAR PROPAGATION IN SINGLE-MODE FIBRES

B. Crosignani** and B. Daino

Fondazione Ugo Bordoni

Via Baldassarre Castiglione, 59, 00142 Roma, Italy

INTRODUCTION

A monochromatic electromagnetic field inside a straight real lossless single-mode fiber (see Fig. 1) can be written, in full generality, as a superposition of two linearly polarized orthogonal modes in the form

$$\underline{E} = \hat{x}\varepsilon(\underline{r})A_x(z)\exp(i(\omega t - \beta_x z)) + \hat{y}\varepsilon(\underline{r})A_y(z)\exp(i(\omega t - \beta_y z))$$

$$= \hat{x}\varepsilon(\underline{r})a_x(z)\exp(i\omega t) + \hat{y}\varepsilon(\underline{r})a_y(z)\exp(i\omega t) \qquad (1)$$

where $\varepsilon(\underline{r})$ is the spatial configuration of the mode.

The state of polarization (SOP) of the field, at the section z, can be conveniently expressed by the Jones vector

$$\underline{J}(z) = \begin{vmatrix} A_x(z) \\ A_y(z) \end{vmatrix} \qquad (2)$$

This allows, given the state of polarization at the input section of the fiber $\underline{J}(0)$, to express the output SOP as

$$\underline{J}(L) = \underline{\underline{M}} \cdot \underline{J}(0), \qquad (3)$$

$\underline{\underline{M}}$ being a unitary matrix of the form

$$\underline{\underline{M}} = \begin{vmatrix} u_1 & u_2 \\ -u_2{}^* & u_1{}^* \end{vmatrix} \qquad (4)$$

** also with Dipartimento di Fisica, Università de L'Aquila, (Italy).

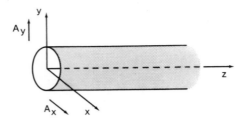

Fig. 1. Single-mode optical fiber.

where, because of energy conservation,

$$|u_1|^2 + |u_2|^2 = 1$$

Here, $u_{1,2}$ are complicated functions of the wavelength and their actual values depend on the particular properties of the fiber along its length.

To see in more detail how the SOP evolves during the propagation along the fiber, we can resort to its representation in terms of the Stokes parameters. As it is well known, these are defined by the relations[1]

$$
\begin{aligned}
S_0 &= a_x a_x{}^* + a_y a_y{}^* \\
S_1 &= a_x a_x{}^* - a_y a_y{}^* \\
S_2 &= a_x{}^* a_y + a_x a_y{}^* \\
iS_3 &= a_x{}^* a_y - a_x a_y{}^*
\end{aligned}
\tag{5}
$$

Normalizing S_1, S_2 and S_3 with respect to S_0 (which gives the total intensity), each SOP corresponds to a point on the surface of a unitary sphere (the Poincaré sphere, see Fig. 2) in a cartesian space, whose co-ordinates are $s_1 = S_1/S_0, s_2 = S_2/S_0$ and $s_3 = S_3/S_0$. We can thus associate, to each SOP, a unitary vector in this space, defined by

$$\underline{s} = s_1 \hat{s}_1 + s_2 \hat{s}_2 + s_3 \hat{s}_3 \tag{6}$$

Although adding nothing new with respect to other methods, as for example the Jones calculus, in studying the propagation properties of po-larization, the Poincaré sphere representation has some distinct features which makes it particularly useful in some circumstances. It is worthwhile to remember that the propagation of the SOP in any linear optical medium is described by

$$d\underline{S}/dz = \underline{\Omega} \times \underline{S} \tag{7}$$

where $\underline{\Omega}$ is a vector describing the infinitesimal, rigid rotation of the Poincaré sphere in traversing the optical layer dz. It has been shown that $\underline{\Omega}$ can be obtained as the vectorial sum of the different vectors describing the local birefringence of the medium (linear birefringence, circular).

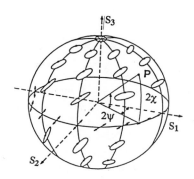

Fig. 2. The Poincaré sphere.

The change of the SOP after a length L of a medium possessing bire-fringence $\underline{\Omega}(z)$, is obtained as the integral of the elementary rotations (7), and corresponds to a finite, rigid rotation of the sphere. The intersections of the axis of rotation with the sphere individuate the two eigenstates which characterize the medium (after the length L and at the given wave-length), i.e. the two SOP's which remain identical after propagation.

It may be also shown that, in a small frequency interval $\Delta\omega$ around ω, the change in the output SOP after any whatsoever linear optical system, due to a change in the frequency of the wave, gives a small rotation of the sphere, which, at the first order, is proportional to the change in frequency and has a constant axis of rotation. The intersections of this axis with the sphere individuate the so-called "principal states of polar-ization". It is referring to this states that it is possible to define the polarization delay and the polarization dispersion[2].

CONTROL OF THE STATE OF POLARIZATION OF THE FIELD PROPAGATING
IN A SINGLE-MODE FIBER

The polychromatic electric field inside a straight real single-mode optical fiber can be written as

$$\underline{E}(\underline{r},z,t) = \varepsilon(\underline{r})\hat{x}\exp(i\omega_0 t - i\beta_x(\omega_0)z)A_x(z,t) +$$

$$\varepsilon(\underline{r})\hat{y}\exp(i\omega_0 t - i\beta_y(\omega_0)z)A_y(z,t), \qquad (\underline{r}=x,y) \qquad (8)$$

where $\varepsilon\hat{x}, \varepsilon\hat{y}$ are linearly-polarized eigenmodes, β_x and β_y the relative propagation constants, ω_0 the mid-frequency of the field and A_x, A_y the modal amplitudes.

In the linear regime, that is whenever nonlinear optical effects can be neglected, the A's can be shown to obey the set of coupled differential equations

$$(\partial/\partial z + 1/V_x \, \partial/\partial t)A_x = iK(z)e^{i\delta\beta z}A_y$$

$$(\partial/\partial z + 1/V_y \, \partial/\partial t)A_y = iK(z)e^{-i\delta\beta z}A_x \qquad (9)$$

where V_x and V_y are the group velocities of the two linearly polarized modes

$$V_x = (d\beta_x/d\omega)_{\omega_0}^{-1}, \quad V_y = (d\beta_y/d\omega)_{\omega_0}^{-1} \tag{10}$$

$\delta\beta = \beta_x(\omega_0) - \beta_y(\omega_0)$, $K(z)$ is random real coupling coefficient related to the local deviation of the refractive index from that pertaining to the ideal fiber and we have neglected for the sake simplicity chromatic dispersion. An ordinary fiber is not particularly birefringent, that is $\beta_x \sim \beta_y$ (in practice $\delta\beta = 10^{-6} - 10^{-8}k$, where $k = \omega_0/c$). Thus, the oscillating factors $\exp(\pm i\delta\beta z)$ vary on a distance large compared with the typical variation scale of $K(z)$ and the coupling mechanism is particularly efficient. Viceversa, if one is able to produce fibers with a high birefringence (typically $\delta\beta \sim 10^{-3} - 10^{-5}k$), then the oscillating factors $\exp(\pm i\delta\beta z)$ vary on a scale small compared with that of $K(z)$ so that coupling tends to average to zero and to become inefficient. If this happens to be the case, the power initially launched on one of the two polarization states does not leak into the other one and linear polarization is maintained (polarization-preserving fibers).

It is obvious that large birefringence introduces two well defined birefringent axes along one of which light has to be launched in order to preserve its linear polarization. The ideal polarization preserving fiber would be a perfectly circular one ($\beta_x = \beta_y$) with $K(z) = 0$, which would maintain any state of input polarization: in real life, the introduction of high-birefringence is the price one has to pay in order to suppress the mechanism of mode-coupling. There is, however, another possibility of neutralizing the effect of deviation from ideality leading to coupling, which basically consists in superposing a specific deterministic perturbation taking the form of a strong twist to the real straight fiber[3].

In the reference frame which rotates, for example counter-clock-wise, with the twist, the field can be written as

$$\underline{E}(\rho,z,t) = \varepsilon(\underline{\rho})\hat{\xi} \ \exp(-i\beta_x(\omega_0) + i\omega_0 t) \ A_\xi(z,t) \ +$$
$$\varepsilon(\underline{\rho})\hat{\eta} \ \exp(-i\beta_y(\omega_0) + i\omega_0 t) \ A_\eta(z,t), \quad \underline{\rho} = (\xi,\eta) \tag{11}$$

where $\hat{\xi}$ and $\hat{\eta}$ are the local birefringence axes and $\varepsilon(\underline{\rho})\hat{\xi}$ and $\varepsilon(\underline{\rho})\hat{\eta}$ the linearly polarized local eigenmodes. The evolution of the field inside the fiber is now described by the system of coupled equations

$$(\partial/\partial z + 1/V_x \ \partial/\partial t) \ A_\xi = (iK(z) + T) \ e^{i\delta\beta z} A_\eta$$
$$(\partial/\partial z + 1/V_y \ \partial/\partial t) \ A_\eta = (iK(z) - T) \ e^{-i\delta\beta z} A_\xi \tag{12}$$

where $T = (i - g')\tau$ and τ is the twist rate, $g'\tau \approx 0.08$ being the elasto-optically induced optical activity.

It is convenient at this point to introduce the local right and left circularly polarized eigenmodes $\varepsilon(\rho)\hat{e}^+$ and $\varepsilon(\rho)\hat{e}^-$, where

$$\hat{e}^+ = 1/\sqrt{2} \ (\hat{\xi} + i\hat{\eta}), \quad \hat{e}^- = 1 \ \sqrt{2} \ (\hat{\xi} - i\hat{\eta}), \tag{13}$$

and rewrite the field $E(\rho,z,t,)$ in the form

$$E(\rho,z,t) = \varepsilon(\underline{\rho})\hat{e}^+ e^{-i\beta^+ z + i\omega_0 t} A^+(z,t) +$$
$$\varepsilon(\underline{\rho})\hat{e}^- e^{-i\beta^- + i\omega_0 t} A^-(z,t) \qquad (14)$$

where the propagation constants β^+ and β^- are "a priori" unknown. By comparing Eqs. (11), (13) and (14) one has

$$A_\xi = (1/\sqrt{2}) e^{i\beta_x z} (e^{-i\beta^+ z} A^+ + e^{-\beta^- z} A^-)$$
$$A_\eta = (i/\sqrt{2}) e^{i\beta_y z} (e^{-i\beta^+ z} A^+ - e^{-i\beta^- z} A^-) \qquad (15)$$

which can be substitutes into the set of eqs. (12). By doing this, it is possible to show that, in the limit of large twist rate,

$$\beta^+ = (\beta_x + \beta_y)/2 - T \equiv \bar{\beta} - T$$
$$\beta^- = (\beta_x + \beta_y)/2 + T \equiv \bar{\beta} + T , \qquad (16)$$

while the amplitudes A^+ and A^- obey the set of uncoupled equations

$$(\partial/\partial z + 1/V \, \partial/\partial t) A^+ = 0$$
$$(\partial/\partial z + 1/V \, \partial/\partial t) A^- = 0 , \qquad (17)$$

with

$$1/V = (\tfrac{1}{2})(1/V_x + 1/V_y). \qquad (18)$$

According to Eqs. (17), the amplitudes A^+ and A^- of the two circularly polarized states travel with the same group velocity V and the fiber imperfections, associated with $K(z)$, have no influence on their evolution.

In the "laboratory" frame rotating at a rate τ in the clock-wise-direction,

$$x = \cos(\tau z)\xi - \sin(\tau z)\eta$$
$$y = \sin(\tau z)\xi + \cos(\tau z)\eta , \qquad (19)$$

the field reads

$$\underline{E}(\underline{r},z,t) = \varepsilon(\underline{r})\hat{p}^+ e^{-i\bar{\beta}z - ig'\tau z + i\omega_0 t} A^+(0, t-z/V) +$$
$$\varepsilon(\underline{r})\hat{p}^- e^{-i\bar{\beta}z + ig'\tau z + i\omega_0 t} A^-(0, t-z/V), \qquad (20)$$

where

$$\hat{p}^+ = (\hat{x} + i\hat{y})/\sqrt{2} , \quad \hat{p}^- = (\hat{x} - i\hat{y})/\sqrt{2} \qquad (21)$$

Equation (20) shows that a highly-twisted fiber will preserve any circularly polarized state launched at $z=0$ (right or left). Besides, a spun fiber, a highly twisted fiber in which the spinning is introduced during the drawing process in order to minimize the elasto-optically induced optical activity $(g' \to 0)$, behaves as a perfectly circular isotropic waveguide and is in principle capable of maintaining any input state of polarization.

NONLINEAR EFFECTS ASSOCIATED WITH SELF-MODULATION OF THE REFRACTIVE INDEX

Among the various nonlinear optical processes affecting propagation in a single-mode fiber, which actually enhances them because of the long interaction length is able to provide, we will describe the one associated with self-modulation of the refractive index (optical Kerr effect), consisting of a nonlinear contribution to its linear part proportional to the intensity of the field itself.

Our choice is due to the fact that this effect can become relevant at relatively low power levels, small in particular if compared with those necessary to evidentiate over short fiber lengths other nonlinear phenomena as stimulated Raman and Brillouin scattering, a circumstances which, together with the almost instantaneous response of the medium to the intensity variations, makes it possible to conceive new types of optical devices based on its explotation. Although its role can be also very important over long fiber lengths for balancing the influence of chromatic dispersion and thus providing the possibility of transmitting distortionless pulses in the form of envelope solitons, we consider in this section the situation in which chromatic dispersion can be neglected, this being actually the case for short fiber samples.

In an unbounded medium, optical Kerr effect is described by the relation

$$n = n_1 + n_2 \, |E|^2 \tag{22}$$

where n_1 is the linear refractive index, E the propagating field (assumed to be linearly polarized) and n_2 the nonlinear refractive index coefficient (typically, for silica, $n_2 \sim 10^{-22} (m/V)^2$). The relation (22) becomes more involved in a single mode fiber where, due to the intrinsic birefringence of the propagation medium, it acquires a tensorial character. The general theory of nonlinear propagation in a single mode fiber in the presence of optical Kerr effect having been developed elsewhere, we limit ourselves to write down the nonlinear system of equations describing the evolution of the amplitudes of the electric field eigenmodes.

In the case of a straight untwisted fiber it reads[4]

$$(\partial/\partial z + 1/V_x \, \partial/\partial t) \, A_x = iK(z) \, e^{i\delta\beta z} \, A_y - iR(|A_x|^2 + (2/3)|A_y|^2)A_x$$

$$-(i/3)R \, e^{2i\delta\beta z} \, A_y^2 A_x^*$$

$$(\partial/\partial z + 1/V_y \, \partial/\partial t) \, A_y = iK(z) \, e^{-i\delta\beta z} \, A_x - iR(|A_y|^2 + (2/3)|A_x|^2)A_y$$

$$-(i/3)R \, e^{-2i\delta\beta z} \, A_x^2 A_y \tag{23}$$

where, if the transverse field configuration is normalized to one

$$\iint_{-\infty}^{+\infty} \varepsilon^2(\underline{r}) \, dxdy = 1,$$

(24)

the coefficient R takes the simple form

$$R = k \, n_2 \iint_{-\infty}^{+\infty} \varepsilon^4(r) \, dxdy.$$

(25)

In the case of a highly-twisted fiber one has, in terms of the amplitudes A^+ and A^-,

$$(\partial/\partial z + 1/V \, \partial/\partial t) \, A^+ = -i(2/3)R(|A^+|^2 + 2|A^-|^2)A^+$$

$$(\partial/\partial z - 1/V \, \partial/\partial t) \, A^- = -i(2/3)R(|A^-|^2 + 2|A^+|^2)A^-$$

(26)

It is worth noting that, even coupling is not present (K = 0) or made ineffective by high birefringence, the nonlinear effect is able to induce coupling between the two polarization states. In the case of a high-bire-fringent or highly-twisted fiber, this coupling does not result in any energy exchange between the two modes

$$|A_i(z,t)| = |A_i(0,t - z/V_i)| \quad , \quad i=x,y,$$

(27)

$$|A^i(z,t)| = |A^i \, 0,t - z/V)| \quad , \quad i=+,-$$

(28)

but only in a phase-modulation of each mode consisting in a self-modulation and a phase-variation associated with the intensity variation of the field traveling on the orthogonally polarized state, that is

$$A_x(z,t) = A_x(0,t-z/V_x)\exp(-iR|A_x(0,t-z/V_x)|^2 z) \times$$

$$\exp(-(\frac{2}{3})iR\int_0^z |A_y\{0,t -z'/V_y + (z'-z)/V_x\}|^2 dz')$$

(29)

and

$$A^+(z,t) = A^+(0,t - z/V)\exp(-i(\frac{2}{3})R|A^+(0,t -z/V)|^2 z) \times$$

$$\exp(-(4/3)iR|A^-(0,t-z/V)|^2 z),$$

(30)

similar expressions holding true for A_y and A^-.

Conversely, also if no coupling is present (K = 0), the set of Eqs. (23) implies, for a low-birefringence fiber ($\delta\beta \doteq 0$), an energy exchange between the two linearly polarized eigenmodes. We shall deal more exten-sively with this case in next section, in connection with polarization instability.

EVOLUTION OF THE STATE OF POLARIZATION
ALONG NONLINEAR SINGLE-MODE BIREFRINGENT FIBER

Let us consider again the set of coupled equations (23) with K(z)=0. The evolution of the SOP is most conveniently described in the monochromatic case, in terms of the Stokes vector $\underline{S}$ ant the Poincaré sphere introduced in the previous sections. In effect, after introducing the linear and nonlinear "angular velocities" $\underline{\Omega}_L$ and $\underline{\Omega}_{NL}$ respectively given by

$$\underline{\Omega}_L = (\delta\beta,0,0) \qquad , \qquad \underline{\Omega}_{NL} = (0,0,\, -(2/3)RS_3) \qquad (31)$$

the set of Eqs. (23) takes, in the stationary case ($\partial/\partial t = 0$), the form pertaining to a (rigid plus nonrigid) rotation,

$$d\underline{S}/dz = (\underline{\Omega}_L + \underline{\Omega}_{NL}) \times \underline{S}, \qquad (32)$$

whose esplicit expression reads[5]

$$
\begin{aligned}
dS_1/dz &= (2/3)\,RS_3 S_2 \\
dS_2/dz &= -(2/3)\,(RS_3 S_1 - \delta\beta\, S_3) \\
dS_3/dz &= \delta\beta\, S_2.
\end{aligned}
\qquad (33)
$$

By considering, for the sake of simplicity, a particular case, corresponding to choose

$$\delta\beta = RS_o/3 = R(S_1^2 + S_2^2 + S_3^2)/3 \quad , \qquad (34)$$

the evolution of the state of polarization of a field almost circularly polarized at the fiber input z=0 (initial point on the Poincaré sphere close to the S_3-axis) corresponds to the trajectories shown in Fig. 3. It is evident that small differences in the initial condition can result, after some traveled distance z (curvilinear abscissa on the trajectory), in widely separated polarization states (polarization instability). This fact is most clearly seen in Fig. 4, which shows S_3 at some value of z as a function of S_1 at z = 0. There is a range of value of $S_1(0)$ around $S_1(0)$=(0) such that the change $\Delta S_3(z)$ of the output parameter S_3 is proportional to $\Delta S_1(0)$. For certain values of $\delta\beta L$, the slope is almost vertical and a scheme can be for example easily conceived in which the fiber operates as an optically activated polarization switch.

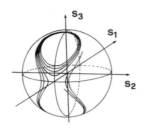

Fig. 3. Evolution of the SOP for a set of initial conditions close to the left circular polarization.

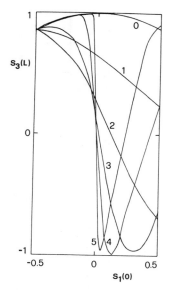

Fig. 4. Fiber output parameter S_3 as a function of the input value of S_1 for different fiber lengths. The numbers close to each curve give $\Delta\beta L$.

INTERPLAY BETWEEN SELF-PHASE MODULATION AND CHROMATIC DISPERSION:
ENVELOPE SOLITONS

By referring to a polarization-preserving fiber supporting only one state of polarization, the equation of evolution of the amplitude ϕ of any of the two modes reads, if chromatic dispersion is taken into account,

$$(\partial/\partial z + 1/V \; \partial/\partial t - i/2A \; \partial^2/\partial t^2) \; \phi = -iR|\phi|^2\phi \qquad (35)$$

where

$$A = (d^2\beta/d\omega^2)_{\omega_o}^{-1} \quad , \qquad (36)$$

the so called group-velocity dispersion, is responsible for chromatic dispersion.

It is not difficult to verify that eq.(35) admits of the exact analytical solution

$$\phi \; (z,t) = e^{i\chi z}\phi_o \; \text{sech} \; ((t-z/V)|\tau_o) \; , \qquad (37)$$

where sech = 1/cosh, provided that

$$2A \; \tau_o^2\chi \; = 1$$

$$A \; \tau_o^2R \; |\phi_o|^2 = -1 \qquad (38)$$

A solution of the kind given in Eq.(37) is called an envelope soliton

and physically corresponds to an exact balance between pulse broadening
due to chromatic dispersion and pulse compression associated, in the regime
of negative group-velocity dispersion A $<0(\lambda \geq 1.3$ μm for silica), with self-
phase modulation.

After introducing the dimensionless variables

$$\xi = z/|A|\tau_o^2$$

$$s = (t-z/V)/\tau_o$$

$$u = \tau_o(R|A|^{1/2})\phi \quad , \tag{39}$$

Eq. (35) can be rewritten in the dimensionless form

$$\partial u/\partial \xi - (i/2)\partial^2 u/\partial \xi^2 = -i|u|^2 u \quad , \tag{40}$$

sometimes referred to as nonlinear Schrodinger equation. It is possible
to show that, besides the fundamental soliton described by Eq. (37), there
are a number of higher-order solitons which develop structure periodically in z corresponding to input pulse of hyperbolic secant shape and amplitudes integral multiples of the amplitude of the fundamental soliton. They
are excited by input pulses of the kind[6]

$$u(\xi = 0,s) = M \operatorname{sech}(s) \tag{41}$$

the integer M being the order of soliton (M = 1 corresponding to the fundamental soliton).

It is possible to give analytic expression only to the two first
solitons which read

$$u_1(\xi,s) = e^{-i\xi/2} \operatorname{sech}(s) \tag{42}$$

and

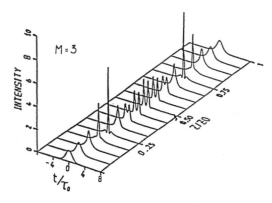

Fig. 5. The M = 3 soliton.

$$u_2(\xi,s) = 4\ e^{-i\xi/2}(\cosh(3s) + 3\ e^{-4i\xi}\cosh(s))\ x$$

$$(\cosh(4s) + 4\cosh(2s) + 3\cos(4\xi))^{-1}\ . \tag{43}$$

All solitons (with the exception of u_1) possess a common period $\xi_0 = \pi/2$ or, in the z variable

$$z_0 = (\pi/2)|A|\tau_0^2\ . \tag{44}$$

In fig. 5 is shown a perspective plot of the temporal pulse shape at various points along the fiber for the M = 3 soliton. From this behavior, one can induce that an optimal pulse narrowing will occur at certain fiber lengths.

REFERENCES

1. M. Born and E. Wolf, "Principles of Optics", 4th ed., Pergamon, Oxford (1970)

2. R. Ulrich and A. Simon, Polarization optics of twisted single-mode fibers , Appl. Opt. 18, 2241 (1979)

3. D. N. Payne, A. J. Barlow and J. J. Ramskov-Hansen, Development of low-and high-birefringence optical fibers , IEEE J. Quantum Elect. QE-18, 447 (1982)

4. B. Crosignani, B. Daino and P. Di Porto, Depolarization of light due to the optical Kerr effect in low-birefringence single-mode fibers, J. Opt. Soc. Am., B3, 1120 (1986)

5. B. Daino, G. Gregori and S. Wabnitz, New all-optical devices based and third-order nonlinearity of birefringent fibers", Opt. Lett. 11, 42 (1986)

6. N. J. Doran and K. J. Blow, Solitons in optical communications, IEEE J. Quantum Electr., QE-19, 1883 (1983)

COUPLED - CAVITY EFFECTS IN SEMICONDUCTOR LASERS

H. D. Liu

Dept. of Physics, Peking University

Beijing, China

INTRODUCTION

The semiconductor laser involves the use of semiconducting material as the gain medium. However, there are several pumping approaches: e.g. optical pumping, similar to solid state lasers; electron beam pumping and injection pumping, the most efficient and easy way. In the latter case, we have a diode laser or injection laser.

We will concentrate on the diode lasers throughout this paper, but the theories presented here can be applied to other type of semiconductor lasers. Thus, it is not necessary to confine the title of this talk to a specific pumping technique.

Semiconductor diode lasers have been and can be used as the light sources in many fields:

- Optical fibre communications: the lasing wavelength depends on the transmission windows of the fibres: Silicon fibre, at 0.85 µm, 1.3 µm, and 1.55 µm, and Polymer fibre, at 0.65µm. Perhaps in the future the super-low loss optical fibres and lasing wavelength will lie at 2-5µm.
- Optical fibre sensing, at the wavelength of 0.85µm or shorter.
- Optical storage, reading and writing, including audio and video disks (better known as compact disk and laser disk); CD-ROM; write once optical disk or DRAW; and EFRAW. These devices require diode lasers operating at 0.78 µm or shorter.
- Laser printing: the lasing wavelength has to match the spectral sensitivity of the photosensitive materials: usually $\lambda < 0.8$µm.
- Industrial and other applications: automation, monitoring, bar-code reading, etc.
- Laser displays, which require a lasing wavelength in the visible range.

As we can see, semiconductor lasers are one of the most popular kind of laser devices, because of their outstanding features:

- very compact or very tiny
- high efficiency
- low working voltage and current
- can be directly modulated at high frequency, $>10^9$ Hz
- can be very inexpensive, because these lasers are relatively easy to put into mass production. Some day, semiconductor lasers will be seen as one of the conventional electronic devices, like a transistor. When you work in a lab and need the semiconductor laser, you may take a few and put them in your pocket. If a laser drops on to the floor, you don't feel like bending over the floor trying to find it; just pick up another one from your pocket, that's all.

Of course, today these lasers still have many shortcoming or disadvantages; in our impression, as far as optical quality is concerned, (such as coherence or linewidth, optical power output, and so on), semiconductor lasers cannot presently compete with other laser devices. But the situation is rapidly changing now, and new functions and better performances are expected. Let's look at the research and development efforts that have occurred in the past quarter of a century:

- in the 1960's, semiconductor lasers developed from the homojunction to the heterojunction structure (Fig. 1). The problems were in the growth of materials, many layers, very thin, but requiring very good crystalline quality. For example, the first attempt was to make use of GaAsP/GaAs, but it failed because of the lattice mismatch.
- In the 1970's, people concentrated on the stripe geometry along the lateral direction, mainly for achieving the stable transverse fundamental mode operation that is required for efficient and stable coupling of the laser light into multimode optical fibres. Of course, in other applications as well people need a single laser spot instead of a higher order mode. The stripe laser geometry can be illustrated in Fig. 2.

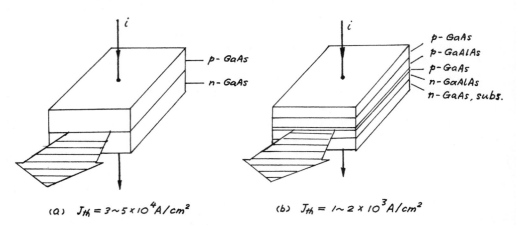

(a) $J_{th} = 3 \sim 5 \times 10^4 A/cm^2$ (b) $J_{th} = 1 \sim 2 \times 10^3 A/cm^2$

Fig. 1. Semiconductor lasers: (a) Homojunction structure; (b) Heterojunction structure.

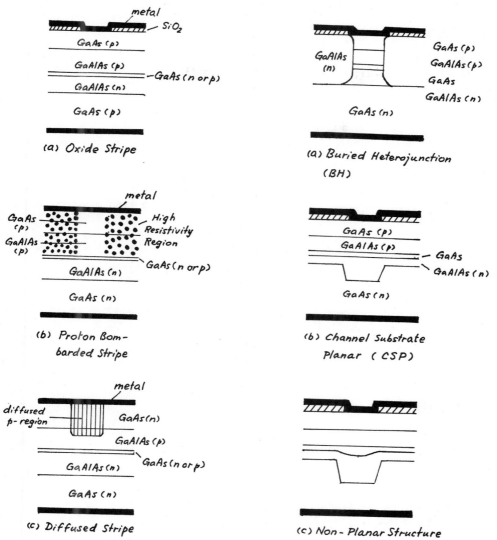

Fig. 2. Gain-guided stripe lasers.   Fig. 3. Index-guided stripe lasers.

- in the 1980's, people need better laser performance and new functions, and so work is proceeding on coupled-cavity structures.

In the early developing stage of GaAs homojunction lasers, soupledcavity structures had been proposed and tested for a few kinds of laser switching devices, e.g., bistable laser devices and laser half adders. However, people had not attached importance to them, because at that time the main problem was the threshold current density.

Recently, coupled-cavity effects and the related structures of heterojunction lasers have become the subject of much study, because of their potential for single frequency operation even at very high frequency modulation, for high optical power output with remarkably lower beam divergence, and for optical pulse shaping and optical logic as well.

In this paper, my intention is to give a tutorial treatment of coupled cavity effects in two fundamental coupling geometries of semiconductor lasers: longitudinal coupling and lateral coupling structures.

SINGLE FREQUENCY OPERATION AND COUPLED-CAVITY LASERS

In the early 1980's, with the advent of low-loss single mode fibres at 1.55μm, it became obvious that the development of a single-frequency semiconductor laser with narrow linewidth operating at this wavelength would be very important in very high-data-rate long-distance transmission. Here,

(1) single frequency means a true single longitudinal mode at high frequency modulation; however, many semiconductor lasers show single longitudinal mode operation in the static state, or a low frequency modulation, and show multimode operation or mode hopping at high frequency modulation.

(2) Even single frequency lasers suffer a frequency shift during their pulse duration, which is known as frequency chirping. This effect bro dens the linewidth of these lasers, and it results from carrier density variation within the cavity which turns to change the refractive index of the cavity.

(3) Frequency stability against external reflections poses yet another problem.

The recent interest in coherent transmission systems has propelled semiconductor laser research into another horizon. These applications demand a semiconductor laser that is:

(i) - true single frequency with a linewidth from 1MHz to <10KHz;
(ii) - insensible to external reflections, and
(iii) - electrically tunable.

Mode	$\Delta\nu$
FSK	≤ 1 MHz
PSK	≤ 10 KHz

The above requirements come from optical communications. Let's look at optical storage technology, and see what its requirements are for a laser source.

The storage capacity of optical disk is $\sim 10^7$ bits/cm^2, two orders of magnitude larger than the magnetic disk, and more than three orders of magnitude bigger than the floppy disk. But even so, we still need bigger storage capacity for many purposes. One way for increasing the storage capacity of the optical disk is to make it three dimensional: adding the frequency dimension to the conventional optical disk, e.g., making use of the so called "two photon gated spectra hole burning" effect. The material used in the experiment is BaClF. Sm^{2+}; its absorption band has a linewidth

of about 50 GHz. If the laser source's linewidth is $\sim$ 5 MHz, then we get
hole burning in the absorption spectral band with the hole width similar
to the laser linewidth. In this case, we set the longitudinal mode separation to be 50 MHz; then at one point on the optical disk we can write
N 5 GHz/50 MHz $\gtrsim 10^3$ bits. That means now the storage capacity increases
three order of magnitude, $\sim 10^{10}$ bits/cm^2. That would be wonderful: again
in the optical storage, we need the laser source to have single frequency
operation with narrow linewidth, and be electrically tunable. Of course,
the lasing wavelength will be quite different from the case of optical
communication. An argon ion laser at 5145 Å and a dye laser at $\sim$6900 Å
were used in the above experiment. In general, we require a single frequency
semiconductor laser lasing at shorter wavelengths, or even in the visible
region.

So far, two main kinds of single frequency semiconductor lasers have
been developed: distributed feedback (DFB) lasers, and coupled lasers, the
latter including cleaved-coupled-cavity (C^3) lasers, and external cavity
lasers.

Let us now discuss how well these lasers address the three problems
outlined above:

(1) the linewidth power product for these different lasers is shown
    in Table 1.
(2) Sensitivity to external reflection: this is a kind of coupled-
    cavity effect, but not intentionally controlled. For example,
    see Fig. 4.
(3) Frequency chirping: this can be measured by means of time-resolved
    spectroscopy (Modulation $\geq$ 1 G bits/s). (See Fig. 5).

Table 1.

	DFB	C^3	external cavity
linewidth power	30 MHz · mW	3 MHz · mW	18 kHz · mW
$\Delta\nu_{min}$	4 MHz	250 MHz	5 kHz

Thus the chirping problem remains unsolved. However, we believe that
by using the coupled-cavity structure, frequency chirping can be compensated
to some extent by controlling the current and injection rate separately.
Several configurations of coupled-cavity lasers for single frequency operation are shown in Fig. 6.

Compared with the above coupled-cavity lasers, DFB lasers are much
more sophisticated to fabricate, especially in the shorter wavelength
region.

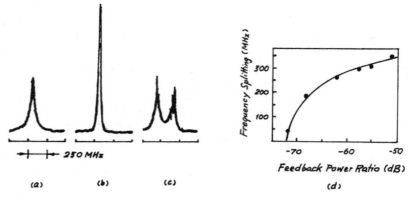

(a)   (b)   (c)   (d)

Fig. 4. Fabry-Perot spectra and frequency splitting of a 1.5 μm DFB laser in various feedback conditions: (a) laser spectrum with no feedback; (b) feedback power ratio 6.4x10⁻⁷, feedback phase adjusted for minimum linewidth; (c) same feedback power ratio as (b) with feedback phase adjusted for maximum line splitting; (d) plot of frequency separation of split modes vs feedback power ratio.

COUPLED-CAVITY EFFECTS ON MODE DISCRIMINATION IN SEMICONDUCTOR LASERS

Let us take the C^3 laser as an example. Fig. 7a gives a schematic of a C^3 laser, in which two standard Fabry-Perot (FP) cavities of length 1_1 and 1_2, respectively, are in-line coupled longitudinally via a coupling gap of width d.

The coupling gap is infact as a resonant cavity, as we will see later on. The gap is formed by the single cleavage technique, but also can be made by other techniques, e.g., wet chemical etching or reactive ion etching.

As is known, the FP mode spacing for a single FB cavity laser is given by

$$\Delta\lambda \sim \lambda_0^2/2n_{eff}1 \tag{1}$$

n_{eff} and 1 are the effective refractive index and the cavity length, respectively. For cavity 1 and 2, $\Delta\lambda$ is different because of 1 and n_{eff}.

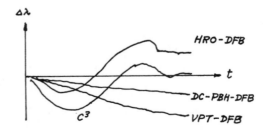

Fig. 5. Frequency chirping for various types of lasers.

100

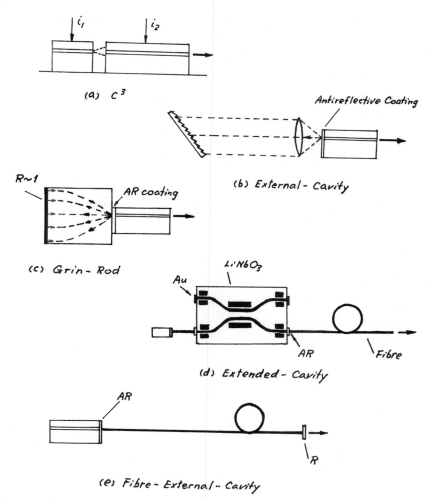

Fig. 6. Several configurations of coupled-cavity lasers for single frequency operation.

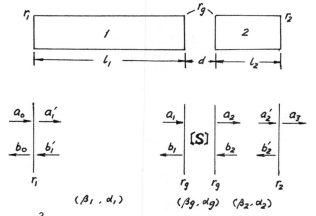

Fig. 7. C^3 laser and the linear wave scattering model.

Since the two cavities are coupled, those FP modes from each cavity that coincide spectrally will interfere constructively and become the enforced FP mode of the whole system, whereas the other modes interfere destructively and are suppressed. The spectral spacing of these enforced FP modes, Λ, can be estimated as

$$\Lambda \sim |\Delta\lambda_1 \Delta\lambda_2 / (\Delta\lambda_1 - \Delta\lambda_2)|$$

where $\Delta\lambda_2 \cong \Delta\lambda_2$ is assumed if $\Delta\lambda_2 >> \Delta\lambda_1$, $\Lambda \sim \Delta\lambda_2$. It is then obvious that the coupling effect results in better mode discrimination because Λ is now significantly larger than $\Delta\lambda$ or $\Delta\lambda_2$, while the gain profile remains essentially the same.

Scattering matrix

The coupled cavity effect in the C^3 laser can be further analyzed with a linear wave scattering theory. As shown in Fig. 7b, the coupling between two end cavities is characterized by a general scattering matrix (S)

$$\begin{Bmatrix} b_1 \\ a_2 \end{Bmatrix} = \begin{Bmatrix} S_{11} & S_{12} \\ S_{21} & S_{22} \end{Bmatrix} \begin{Bmatrix} a_1 \\ b_2 \end{Bmatrix} \tag{2}$$

in which a_1, a_2, b_1, b_2 are the field amplitudes and S_{ij} are the scattering parameters, determined by the amplitude reflection and transmission factors of the coupling gap

$$\begin{cases} S_{12} = S_{21} = \dfrac{t(1-r_g^2)}{1-r_g^2 t'} \\[3mm] S_{11} = S_{22} = \dfrac{r_g t'(1-r_g^2)}{1-r_g^2 t'} \end{cases} \tag{3}$$

where $t=\exp(i-i\ B_g\ d)$, $t'=\exp(-i \cdot 2B_g\ d)$, are the single-pass and round-trip amplitude transmission factors in the coupling gap; these factors include all losses for emission from a fundamental transverse waveguide mode and back into the same mode in the other or same cavity, respectively, and r_g is the amplitude reflection factor at the interface of the coupling gap.

Threshold Condition

Now, we introduce the concept of an effective mirror with reflectivity $R(\lambda)$

$$R = (\lambda) = \frac{b_1}{a_2} = S_{11} + \frac{S_{12} S_{21} r_2 t_2^2}{1-S_{22} r_2 t_2^2} \tag{4}$$

which combines the effects of cavity 2 and the coupling gap. In other words, in $R(\lambda)$ everything outside of cavity 1 has been included.

From Fig. 7b we define the transfer function for the whole system

$$H(\lambda) = \frac{a_3}{a_0} = \frac{t_1 t_2 \sqrt{(1-r_1^2)(1-r_2^2)}}{1-(S_{12}^2-S_{11}^2)r_1 r_2 t_1^2 t_2^2 - S_{11}(r_1 t_1^2 + r_2 t_2^2)} \tag{5}$$

where $t_1 = \exp(-iB_2 1_2)$, $t_2 = \exp(i-B_2 1_2)$ are the transmission factors in cavity 1 and 2, respectively. We focus on the zeros (B_{th1}, B_{th2}) of the denominator of $H(\lambda)$,

$$1-(S_{12}-S_{11})r_1 r_2 t_1^2 t_2^2 - S_{11}(r_1 t_1^2 + r_2 t_2^2) = \emptyset \tag{6}$$

these routs give the threshold gain $\alpha_{th\ i}$ and wavelength $\lambda_{th\ i}$ for the possible modes of the coupled-cavity system. Eq. (6) can be written as

$$R(\lambda) \cdot r_1 \cdot t_1^2 = 1 \tag{7}$$

This is the threshold condition for a FP cavity laser and we see now the C^3 laser is really equivalent to a single FP cavity laser, but with one of the cavity mirrors replaced by a "dispersive mirror" $R(\lambda)$. From Eq. (7), we have

$$\begin{cases} \alpha_{th\ i} = \dfrac{1}{1_1} \ln \dfrac{1}{|r_1 R(\lambda)|} \\[2em] 2\beta_1 1_1 - \phi_R = 2\pi m_1, \quad m_1 = 0, \pm 1, \ldots \end{cases} \tag{8}$$

in which ϕ_R is the phase angle of $R(\lambda)$. Now it is clear that the threshold condition is dependent on wavelength, and here results in mode discrimination.

Mode Discrimination Effect

We have seen that the coupling effect can be described by a scattering matrix, and the only parameter to be determined is t or t' or α_g. Assuming the beam in the coupling gap, is a Gaussian beam, it is easily estimated as

$$t' = \frac{1}{\sqrt{1+(\dfrac{\lambda 1g}{\pi w_0^2})^2}} e^{-i2\beta_g d}$$

where w_0 is the beam waist, and the refractive index equals unity in the gap. Then, (S) is determined. Fig. 8 shows $|S_{12}/S_{11}| \sim (d/\lambda)$.

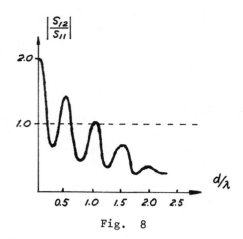

Fig. 8

We can see at $d=2N\cdot\frac{\lambda}{4}$ that $|S_{12}/S_{11}|$ reaches its maximum and at $d=(2N+1)\frac{\lambda}{4}$, $|S_{12}/S_{11}|$ decreases to its minimum. d is a very sensitive parameter; when d/λ increases, $|S_{12}/S_{11}|$ tends to level out, and so the mode discrimination effect will disappear.

Case (1). For the case of $1_1 \gg 1_2$, assuming $\alpha_2 1_2 \approx 0$, we have approximately

$$\left|S_{22} r_2 t_2^2\right| \ll 1, \quad \left|\frac{S_{12}}{S_{11}} r_2 t_2^2\right| \ll 1,$$

$$R(\lambda) \cong |S_{11}| \left|1+\left|\frac{S_{12} r_2}{S_{11}}\right| e^{\alpha_2 1_2} \cos(2\beta_2 1_2 + \xi)\right| \tag{10}$$

in which ξ is the phase angle of $\frac{S_{12} r_2}{S_{11}}$. Substituting Eq. (10) into Eq. (8), we have

$$\alpha_{th\,1} \cong \frac{1}{1_1} \ln\frac{1}{|S_{11} r_1|} - \frac{1}{1_1} \left|\frac{S_{12}^2 r_2}{S_{11}}\right| e^{\alpha_2 1_2} \cos(2\beta_2 1_2 + \xi) \tag{11}$$

The results can be shown schematically in Fig. 9. The period of α_{th1} is determined by cavity 2 and is approximately $\Delta\lambda_{2FP}$, the FP mode spacing of cavity 2. Two neighboring modes will have different thresholds. From the figure, we can estimate according to Eq. (11):

$$\Delta\alpha_{th1} \cong \frac{1}{1_1^3} \frac{2n_2^2 1_2^2}{n_1^2} \left|\frac{S_{12}^2 r_2}{S_{11}}\right| e^{\alpha_2 1_2}$$

Let's set $\Delta\alpha = \Delta\alpha_1 + \Delta_g$, the value needed for mode discrimination; then, for the enforced mode, we have $\Delta\alpha_{th1} = 0$; assuming carrier density dependence of the gain, this can be written as:

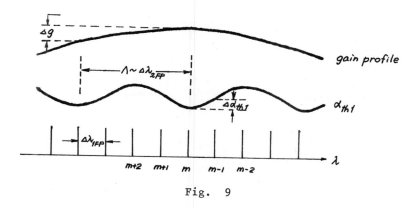

Fig. 9

$$g = a \, \Gamma \, (N-N_0) - b \, \Gamma \, (\bar{\lambda}-\lambda)^2.$$

Then we must have $\Delta g = b \Gamma \, \Delta\lambda_{2FP}^2 \geq \Delta\alpha$, $\Delta\lambda_{2FP} = \dfrac{\lambda^2}{2n_2 l_2}$, and therefore

$$l_2 \leq \frac{\lambda^2}{2n_2} \sqrt{\frac{b\Gamma}{\Delta\alpha}}.$$

For the FP modes of cavity 1, $\Delta g \simeq 0$, and thus requires

$$\Delta\alpha_{th1} \geq \Delta\alpha$$

and therefore

$$l_1 \leq \sqrt[3]{\frac{1}{\Delta\alpha}(2\pi^2 \, l_2^2 \left| \frac{S_{12}^2 \, r_2}{S_{11}} \right| e^{\alpha_2 l_2})}.$$

From the above discussion, we can see that the parameters l_1, l_2 and d have to be properly chosen in order to achieve single mode operation.

Case (2): $l_1 // l_2$. In this case, Eq. (6) can be transformed to

$$\frac{\exp\left[i 2\beta_1 \, l_1 - \alpha_1 l_1\right]}{r_1 S_{11}} - 1 \quad \frac{\exp\left[i 2\beta_2 l_2 - \alpha_2 l_2\right]}{r_2 S_{22}} - 1 = \left(\frac{S_{12}}{S_{11}}\right)^2. \tag{12}$$

Solving Eq. (12) with the help of a computer, we obtain the result given in Fig. 10. The mode discrimination is much stronger in the case of $d=\lambda$ than that in case of $d=\dfrac{5}{4}\lambda$.

SEMICONDUCTOR LASERS ARRAYS

Many techniques have been developed for fabricating injection lasers that can realiably emit 20 to 50 mW or more, as depicted in Fig. 11. However, there are inherent limits to the power obtainable from a single

105

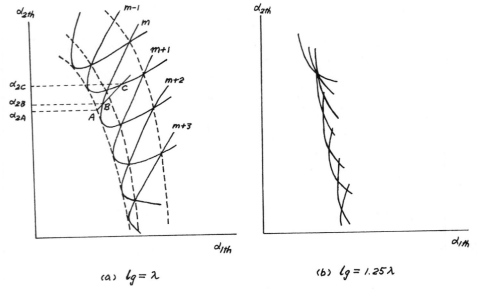

(a) $lg = \lambda$ (b) $lg = 1.25\lambda$

Fig. 10

cavity laser. Above a certain power, no matter what its construction, the tiny device will be destroyed by facet-mirror damage or by heating from the electrical power dissipated within it. On the other hand, if you really have a high power semiconductor laser, you can imagine many important applications for it. This is an important motivation for developing the Laser Array. In this case, we hope to combine the optical power coming from each elemenct into one beam, so that we need to make use of phase-locked operation.

The experiment to date have been very successful. With phase-locked semiconductor laser arrays, 140 elements give us 5 W cw at room temperature; with 200 elements, 11 W pulsed, pulse width 150μs, which is very reasonable for use as a pumping source for YAG lasers. No doubt in this case the YAG laser will become very compact, and very useful.

The problem people always face concerns the far field pattern; it is very often double-lobed. Of course, we prefer to have single radiation lobe along the normal direction.

In recording and printing applications, we need laser arrays too, but in a different way. For example, the optical disk has high capacity for data storage, but writing and reading rates are not fast enough. For writing, it takes time for the interactions of laser beam with the material usually through some kind of thermal effect. For reading, random access is required. It is easy to imagine using a laser array so that we can have multiple channels for writing and reading simultaneously. We definitely do not need phase-locked operation here; on the contrary, we want to eliminate the coupling effect, so-called cross-talk, (the original error rate for optical disk is 10^{-4} 10^{-6}).

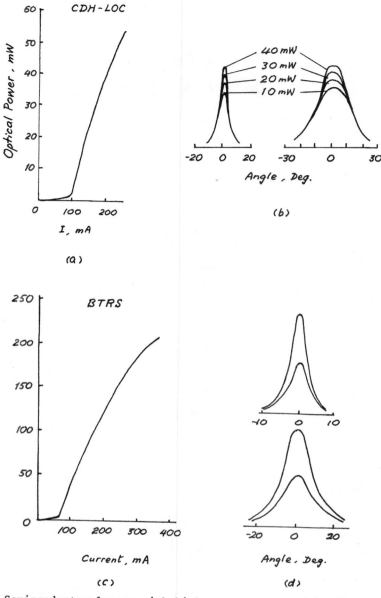

Fig. 11. Semiconductor lasers with high power output: (a) and (b) show the
L-I characteristic and far-field profiles, for CDH-LOC laser,
respectively; (c) and (d) are for the BTRS laser.

On the other hand, in recording, we need to correct errors in real
time. The best design will use a laser array, with one laser for writing
and another for reading and checking. These two lasers must be close to
one another in order to reduce the delay time between writing and reading.
In this case, we see again the need for eliminating coupling between laser
cavities. Figure 13 gives an example. In addition, phase-locked arrays also
show frequency locking and may give us single frequency operation in semiconductor lasers.

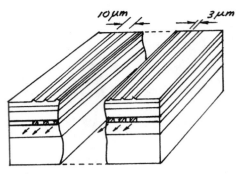

Fig. 12. Semiconductor laser phase-locked array.

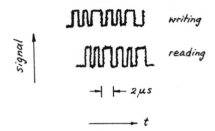

Fig. 13. Incoherent laser array can be used for real-time error correction
in optical recording.

SUPERMODE THEORY OF PHASE-LOCKED ARRAY OF SEMICONDUCTOR LASERS

We take the phase-locked array as an example. As we mentioned above,
it provides a useful means for obtaining high-power semiconductor lasers
with low beam divergences.

A laser phased array is shown schematically in Fig. 14a. When the
array is powered through the common electrical contact, each cavity's
optical modes spread out through the active and cladding layers, overlapping with the optical mode from its neighbors and interacting with them.
Each cavity couples with its nearest neighbors, and all the laser act
together as one powerful source. It is the coupling effect between cavities
which make all the array elements connect with each others and operate as
a whole. In other words, the coupled-cavity effect results in the phase-
locked operation of the semiconductor laser array.

Coupled-mode equations and eigenfunctions

Consider an uniform array consisting of N identical stripe waveguide
cavities, as shown in Fig. 14b. Each stripe guide cavity is presumed to
support a single TE-like spatial mode, and can be described by its electric
field.

$$E_y(x_1 y_1 z) = \mathcal{E}(x_1 y)\exp(i\beta\zeta), \qquad 1=1,2, \ldots N$$

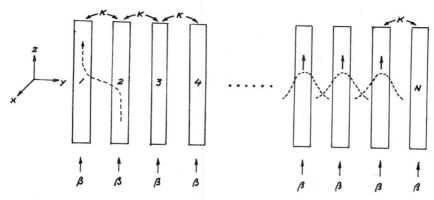

Fig. 14. Model of semiconductor laser phase-locked array.

where $\mathcal{E}(x_1 y)$ is the transverse profile of the mode field, is the complex propagation constant. $\mathcal{E}(x_1 y)$ and β are the same for all the cavities because we have assumed N stripe guide cavities are identical.

Assuming only nearest-neighbor coupling, and a coupling coefficient denoted as K, the coupled-mode equations for the N-element array can be written as

$$
\frac{d}{d\zeta}
\begin{pmatrix}
E_1 \\
E_2 \\
\cdot \\
\cdot \\
\cdot \\
E_N
\end{pmatrix}
= i
\begin{pmatrix}
\beta & K & & & \\
K & \beta & K & & \\
& K & \beta & K & \\
& & & \ddots & \\
& & K & \beta & K \\
& & & K & \beta
\end{pmatrix}
\begin{pmatrix}
E_1 \\
E_2 \\
\cdot \\
\cdot \\
\cdot \\
E_N
\end{pmatrix}
\begin{pmatrix}
E_1 \\
E_2 \\
\cdot \\
\cdot \\
\cdot \\
E_N
\end{pmatrix}
\tag{13}
$$

or $\quad \dfrac{d}{d} \vec{E} = i \, \tilde{M} \, \vec{E}^{\nu}$

in which E_0 is the optical field in 1th cavity.

Consider the eigen-solution of Eq. (13) in the following form:

$$
\vec{E}^{\nu}(\zeta) = \vec{E}^{\nu}(o) \, \exp(i\sigma_{\nu}\zeta).
\tag{14}
$$

We only take the ζ component into account, because the coupling will result in the ζ dependence of E. Also, generally speaking, the propagation constant will be changed due to interaction among the cavities. We call $\vec{E}^{\nu}(\zeta)$ array supermodes, and σ_{ν} the propagation constant of the supermode $\vec{E}^{\nu}$. Substitution of Eq. (14) into Eq. (15) gives:

$$\begin{cases} (\widetilde{M} - \sigma_\nu \widetilde{I}) \, \vec{E}^\nu = 0 \\[2mm] \begin{pmatrix} \beta - \sigma_\nu & K & & & \\ K & \beta - \sigma_\nu & K & & \bigcirc \\ & \ddots & \ddots & \ddots & \\ \bigcirc & & k & \beta - \sigma_\nu & \end{pmatrix} \begin{pmatrix} E_{10} \\ \vdots \\ \\ NO \end{pmatrix} = 0 \end{cases} \tag{15}$$

where $\widetilde{I}$ is the unit matrix.

A solution of Eq. (15) yields the N supermodes that are supported by an array of N single-mode lasers.

$$E_{1o}^\nu = \sin(1\frac{\pi\nu}{N+1}), \quad 1=1,2, \ \ldots \ N \tag{16}$$

and eigenvalues

$$\sigma_\nu = \beta + 2k \, \cos(\frac{\pi\nu}{N+1}) \tag{17}$$

where $=1,2, \ \ldots \ N$.

We can see that $E_1^\nu(\zeta)$ results from the coupling among laser cavities. For describing the field profiles in individual laser cavity of an array, we use

$$E_1^\nu \, (x_1, y_1, z) = E_{1o}^\nu \, \mathcal{E}(x_1, y) \, \exp \, (i\sigma_\nu z). \tag{18}$$

Near-field profiles of phase-locked array

The total electric field of the array is now

$$E_y \, (x_1, y_1, z) = \sum_{1=1}^{N} \, E_{1o}^\nu \, \mathcal{E}(x_1, y) \, \exp(i\sigma_\nu \zeta). \tag{19}$$

Eq. (19) can be used to evaluate the near field of each supermode.

For any convenience, $\mathcal{E}(x_1, y)$ is assumed to be a Gaussian distribution. In Fig. 15a, E_{1o}^ν are shown for N value not too small. Fig. 15b shows the near field profiles for N=5. This is something like an one-dimensional lattice, with supermodes corresponding to the lattice wave.

We see that the supermode describes the relative amplitude between array elements, or describes a phase-locked combination of the individual laser modes with the amplitudes E_{1o}. The near-field intensity profiles thus appear as a series of spots: . . . .

Far-field profiles

We can make use of the Fresnel diffraction integral formula to evaluate the far-field radiation pattern of each supermode. In the junction plane,

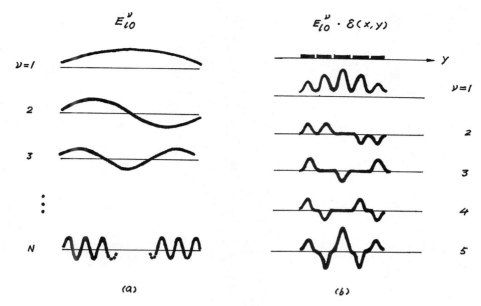

$$E_{lo}^{\nu} \qquad\qquad E_{lo}^{\nu} \cdot \mathcal{E}(x,y)$$

$\nu = 1$

2

3

$\vdots$

N

$\nu = 1$

2

3

4

5

(a)   (b)

Fig. 15

we have:

$$E^{\nu}(\theta) = C \int E_y(x_1 y_1 \zeta)\, e^{ikr}\, dy$$

$$\Gamma = r_0 - (d+y)\sin\theta, \quad y = 1d$$

$$E^{\nu}(\theta) = C\, e^{ikr_0} \int \mathcal{E}(x_1 y)^{-iky\,\sin\theta} \sum_{1=1}^{N} E_{1o}\, e^{-ikd1\,\sin\theta}$$

$$= C\, e^{ikr_0} \int \mathcal{E}(x_1 y)\, e^{-ikd\,\sin\theta}\, dy \sum_{1=1}^{N} E_{1o}\, e^{-ikd1\,\sin\theta}$$

$$= C^1\, E(\theta)\, G(\theta). \tag{20}$$

The problem looks like a grating, now. E (θ) represents the contribution coming from individual cavity; and E (θ) results from the interference between laser light coming from individual laser cavities of the phase-locked

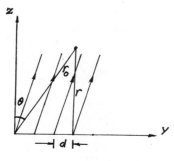

Fig. 16

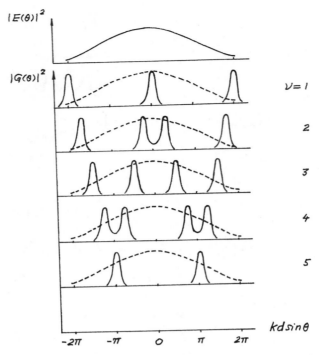

Fig. 17 gives the far-field intensity profiles, with v=kd sinθ as the

array. E (θ) still has a Gaussian distribution and G (θ) can be expressed

$$G (\theta)^2 = \frac{Sin^2(N(u+v)/2)}{Sin^2((u+v)/2)} + \frac{Sin^2(N(u+v)/2)}{Sin^2((u-v)/2)} , \qquad (21)$$

in which u=πv/N+1 and v=kd sinθ .

Fig. 17 gives the far-field intensity profiles, whith v=kd sin θ as the variable, and N=5.

Usually, the v=N mode tends to oscillate. For achieving the fundamental supermode operation, recently several research workers proposed different regimes, for example, a linear modulation structure, an inverted V chirped structure, and so on.

Coupling coefficient and coupled-cavity length

Taking the gain and loss of each cavity into account, we can get a coupled-mode equation similar to Eq. (13). From the threshold condition we can derive a relation between cavity length and coupling coefficient K:

$$L = \pi / | \S_m | \ K \qquad (22)$$

$$\S_m = \sqrt{\frac{N+1}{2}} \quad , \qquad N = \text{odd number}$$

N is the number of array elements. When N=2,

$$L = \pi/K, \tag{23}$$

so that once we know K, we know the cavity length L. In such a cavity, the field amplitude maintains its original value at L. For calculating K, we need only consider the case of N=2.

For the case of index-guiding, we obtain

$$K = 2 \frac{k^2 xo}{k\zeta o} \frac{\xi_5}{a} \frac{\exp(-(C-a)/\xi_5)}{1+k^2 xo \, \xi_5^2} \tag{24}$$

Here we follow Marcatili's notation.

For gain-guiding,

$$K = \frac{2\pi^2}{\lambda^2 T^2 (\beta_{\rho m})} \int_{-\infty}^{\infty} 2n_{10} n_{1v} (\chi+C)(G^*_\rho (x) \, G \, (x+c) \, dx. \tag{25}$$

PROBLEMS OF THE CHARGE COLLECTION

IN LASER MULTIPHOTON IONIZATION EXPERIMENTS

F. Giammanco

Dipartimento di Fisica, Università di Pisa

Piazza Torricelli, 2,   56100 Pisa, Italy

INTRODUCTION

The purpose of this work is to analyze the problems that occur in the charge collection process, in a charge density range typical of the most of the multiphoton ionization (MPI) experiments ($10^8 \div 10^{12}$ el/cm^3) performed with pulsed lasers.

From the point of view of the charge production, the MPI experiments can be divided, for the sake of simplicity, in two classes, the first one including all the experiments in which the final purpose is the collection of a particular ionized species (as for instance isotope separation, trace analysis, photodeposition of special chemical compounds and so on[1]) and the second in which the charge collection constitutes a test of a particular laser-matter interaction (above threshold ionization (ATI), laser and atomic continua interaction, collisions in laser field and so on[2]). In both classes of MPI experiments the collection technique is quite similar. In fact the ionized yields are collected by externally applied electric and/or magnetic fields and the resultant current is either totally detected by the so-called opto-galvanic technique or mass energy analyzed by appropriate spectro-meters.

In general, the response time is much longer than the laser inter-action time (0.1-10 ns), except for particular experiments, performed in the cathode region of the hollow cathode discharge lamps, where the very high electric field accelerates the electrons during the laser pulse[3].

During the collection time, the initially produced yields can be modified by many effects depending on the long range coupled behavior of both ions and electrons (plasma effects) and the binary collisions (re-combination, energy pooling, charge-exchange and so on). Whereas the binary collisions are widely investigated[4] and anyway can be strongly reduced decreasing density, the influence of the plasma effects has been less analyzed, except in the field of the isotope separation, where, owing to the high density required (10^{12} ion/cm^3), the plasma screening[5] of the external fields makes difficult to obtain a fast ion extraction.

Recently, a preliminary study of the plasma effects has been introduced also in a lower range of charge density ($10^8 \div 10^{10}$ el/cm^3) referred to an anomalus behavior of the ATI electronic spectrum as a function of the laser power. In that case, the low energy peaks suppression, that up to day is not completely described by high order Quantum Mechanic calculations, could be assigned to the "electron trapping" phenomenum induced by the mutual electron-ion electric field[6,7,8]. The increasing of the electron confinement time could enhance the rate of recombination collisions or, in what it seems the most effective process, the monochromatic character of the peak could be lost by electron/ion electron/electron thermalizing collisions. Actually, the disappearence of the first peak is related to a growth of a continuum background in the electronic spectrum[6].

An experimental evidence of the collective behavior is given in ref. (9), where potassium vapours were ionized by a non resonant ruby laser at a charge density of 10^{10} el/cm^3. The authors observed an induced emf on an external coil, modulated by fast oscillations at the plasma frequency. An analogous behavior of the current, collected by two plates following the usual optogalvanic technique, will be shown in the last section.

From the previous discussion, it seems important to develop a suitable model for the time-space evolution of the ionized yields, taking into account all the processes that can modify the products of the laser-matter interaction. Owing to the particular kind of the MPI produced plasmas, the approximations, usual in the plasma physics approach, of linearization, cold plasma and so on, do not correspond to the physical reality of the problem. In fact, due to the initial density gradient (in the first approximation gaussian shaped as the laser beam) and the fast evolution, all the non linear terms and all the time/space derivatives must be retained in the exact description. In this work, the plasma evolution is described following the outline of the fluid picture and it is shown that, under some hypotheses well corresponding to the physical reality of the MPI experiments, a solution can be carried out on the bases of analytic methods. The method of solution is firstly tested on the simplest case, i.e. the one-dimensional motion in an external constant and uniform electric field, assuming an initial gaussian density profile and neglecting all the processes changing the number density. The solution is easily extended to the three-dimensional case and it is shown that momentum-exchange collisions, time varying electric field and time varying temperature may be included in the solution. The solution is applied to solve the problem of the coupled two fluid motion under the MagnetoHydrodynamic hypothesis of local neutrality.

It will be shown that the MHD approximation, in the typical density range of MPI experiments, is verified only in the first instants of the plasma motion. Therefore, by using an approximate description of the internal field, the method of solution is extended to the complete system of the two fluid coupled equations.

OUTLINE OF SOLUTION

a) Single fluid

Following the standard fluid picture[10], the one-dimensional motion of a charged species is described by

$$\frac{\partial v}{\partial t} + v \frac{\partial v}{\partial x} = A_o - (KT/m) \frac{\partial}{\partial x} \log (n) \qquad \text{motion} \qquad (1)$$

$$\frac{\partial n}{\partial t} + \frac{\partial}{\partial x} (nv) = 0 \qquad \text{continuity} \qquad (2)$$

where $A_o = (q/m)E_o$. T, m, q are respectively the temperature, mass and charge. The initial conditions are $v(x,0)=0$ and $n=n_o \exp(-(x/\delta)^2)$. Dividing the Eq.(2) by n, the new variables are v and log(n). Both the functions admit a Taylor development in a time interval close to zero, that is

$$v(x,t) = \Sigma b_n t^n \qquad \log(n) = \Sigma a_n t^n \qquad (3)$$

Introducing Eq.(3) in the system (1), (2) and equating the terms with the same power of t, the following recurrence relations are obtained

$$(n+1)b_{n+1} + \sum_{j=1}^{n} b_j \frac{db_{n-j}}{dx} = A_o (\delta(n=0)) - (KT/m) \frac{da_n}{dx} \qquad (4)$$

$$(n+1)a_{n+1} + \frac{db_n}{dx} + \sum_{j=1}^{n} b_j \frac{da_{n-j}}{dx} = 0 \qquad (5)$$

The initial conditions lead to $b_o=0$ and $a_o=\log(n_o)-(x/\delta)^2$ from which it follows that $b_{2n}=0$ and $a_{2n+1}=0$. The x dependence of the a and b coefficients is derived from the system (4), (5).

$$a_{2n}(x) = A_{1,2n} + a_{2,2n}x + a_{3,2n}x^2 \qquad (6)$$

$$b_{2n+1}(x) = b_{1,2n} + b_{2,2n+1}x \qquad (7)$$

Introducing Eq. (6), (7) in the system (4), (5) and equating the terms with the same power of x the $a_{j,2n}$ and $b_{j,2n+1}$ dependence on the physical parameters (A_o, T, δ) is carried out. Thus, the five recurrence relations for the $a_{j,2n}$ and $b_{j,2n+1}$ coefficients is reduced to a four equations system for numerical coefficients (for details of calculation refer to ref. 11).

By using a remarkable relation among the numerical coefficients and rearrangement of the indexes, the solution for v and log(n) can be expressed as a function of three series depending on the variable $\Theta=Bt^2$, where $B=2KT/m\delta^2$.

The decrease of the numerical coefficients, with n increasing, does not assure the convergence of the series for $\Theta>1$. Therefore it needs to find an analytic extension of the series[12]. The series terms are written as unknown functions of the adimensional variable Θ. By using the mass

117

conservation and the other relations obtained from the system (1), (2) equating the terms with the same power of x, the complete solution is given by

$$n/n_o = \phi_2^{1/2} \exp\left(-\phi_2\left(\left(\tfrac{1}{2} A_o t^2 - x\right)/\delta\right)^2\right) \tag{8}$$

$$V = A_o t + \tfrac{1}{2}\left(\tfrac{1}{2} A_o t^2 - x\right)\frac{d}{dt}\log(\phi_2) \tag{9}$$

where ϕ_2 is described by the equation

$$\frac{d^2 \phi_2}{d t^2} = \frac{3}{2\phi_2}\left(\frac{d\phi_2}{dt}\right)^2 - 2 B\phi_2^2 \tag{10}$$

with the initial conditions $\phi(0)=1$ and $\left.\dfrac{d\phi_2}{dt}\right|_{t=0}=0$. The ϕ_2 function is an universal function describing the spatial distribution of the ionized species. Fig. 1 shows the bilogarithmic plot of ϕ_2 versus $(1+\theta)$, obtained by standard Runge-Kutta fourth order numerical integration. At large θ values the ϕ function is well approximated by $(1+\theta)^\alpha$, where $\alpha=1.25$. To avoid the collective pressure induced motion, the external electric field must satisfy the condition $E_o \gg 2KTd/e\delta^2$, where d is the position of the collection point. The above condition means that in the time arrival, depending on the electric field, θ should be small as compared to 1. For instance, assuming KT=1 eV, d=10 cm and δ=.1 cm (typical values for MPI experiments), the above condition gives $E_o \gg 1800$ V/cm.

b) Three-dimensional extension

Since the previous solution depends only on the co-ordinate along the

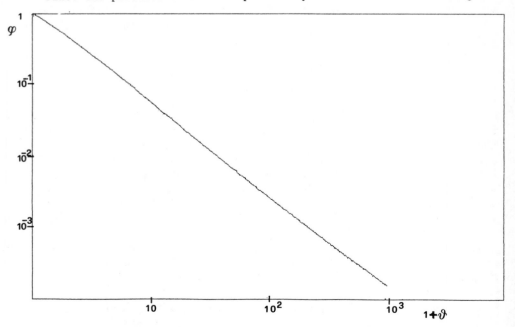

Fig. 1. Bilogarithmic plot of ϕ as a function of $(1+\theta)$.

direction of the motion, a three-dimensional extension is straightforward. Moreover the inclusion of momentum-exchange collisions with neutrals, time varying electric field and temperature, is very easy. In effect, the shapes of n and v are unchanged and the only modification results in the ϕ_2 time behavior and in the time dependent term related to E_0.

Thus the solution is written as (the electric field is supposed directed along the x axis)

$$n/_{n_0} = (\phi_x \phi_y \phi_z)^{\frac{1}{2}} \exp -\{\phi_x((\int v_{o_x} dt - x)/\delta_x)^2 + \phi_y(y/\delta_y)^2 + \phi_z(z/\delta_z)^2\} \quad (11)$$

$$V_x = v_{o_x} + \frac{1}{2} (\int v_{o_x} dt - x) \frac{d}{dt} \log(\phi_x) \quad (12)$$

$$V_y = - \frac{1}{2} y \frac{d}{dt} \log (\phi_y) \quad (13)$$

$$V_z = - \frac{1}{2} z \frac{d}{dt} \log (\phi_z) \quad (14)$$

ϕ_j and v_{o_x} are given by

$$\frac{d^2 \phi_j}{dt^2} = \frac{3}{2\phi_j} (\frac{d \phi_j}{dt})^2 - 2 B_j (t)\phi_j^2 - \beta \frac{d \phi_j}{dt} \quad (15)$$

$$\frac{d v_{o_x}}{dt} = A(t) - \beta v_{o_x}$$

where the index j refers to the directions of the motion.

The different ϕ equations take into account that the initial gaussian widths could be different along the three axes.

c) MagnetoHydrodynamic solution

The first approach to the overall solution of the coupled motion of ions and electrons, which includes the self generated electric field due to the different mobility of the charges, includes the local neutrality hypothesis (MHD approximation) $n_i \approx n_e$.
In that approach[10], owing to the neutrality condition the complete two-fluid system is transformed in a set of equations in the new macroscopic variables

$$\rho = Mn_j + mn_e, \; \sigma = e(n_i - n_e), \; \vec{V} = \frac{M \vec{v}_i + m \vec{v}_e}{M + m} \quad (16)$$

$$\vec{J} = e (n_i \vec{v}_i - n_e \vec{v}_e)$$

by summing and subtracting the motion and continuity equations of the single ionized species. Retaining the non linear term into the motion equation and neglecting all the terms depending on the magnetic field, the complete set of MHD equations is given by

$$\frac{\partial \vec{v}}{\partial t} + (\vec{v}\vec{\nabla})\vec{v} = - \frac{K T_i + KT_e}{M} \vec{\nabla} \log(\rho) - \beta\vec{v} \qquad \text{motion} \quad (17)$$

119

$$\frac{\partial \rho}{\partial t} + \vec{\nabla}(\rho \vec{v}) = 0 \qquad\qquad \text{Mass} \qquad\qquad (18)$$

$$\frac{Mm}{e^2 \rho} \left(\frac{\partial \vec{J}}{\partial t} + \beta \vec{J} \right) = \vec{E} + \frac{1}{e\rho} (M\vec{\nabla}P_e - m\vec{\nabla}P_i) \qquad \text{Ohm} \qquad (19)$$

$$\frac{\partial \sigma}{\partial t} + \vec{\nabla}\vec{J} = 0 \qquad\qquad \text{Charge} \qquad\qquad (20)$$

where P_i and P_e represent respectively the ion and electron pressure terms, that is $P_j = n_j KT_j$ $(j=i,e)$.

The electric field is given by $E = E_o + E_\rho$, where E_ρ is the self-generated electric field. The internal field contributes only in the Poisson equation

$$\vec{\nabla}\vec{E}_\rho = 4\pi\sigma \qquad\qquad (21)$$

The equation (17), (18) admit the general solution of Eq. (11-15). Putting the Poisson equation in the Eq. (19), (20), the system for the total current and the E_ρ field becomes

$$\frac{\partial \vec{J}}{\partial t} + \beta \vec{J} = \frac{e^2}{Mm} (\rho \vec{E} + \frac{KT_e}{e} \vec{\nabla}\rho) \qquad\qquad (22)$$

$$\frac{1}{4\pi} \frac{\partial}{\partial t} \vec{\nabla}\vec{E}_\rho + \vec{\nabla}\vec{J} = 0 \qquad\qquad (23)$$

The one dimensional case admits a straightforward solution. In fact, integrating Eq. (23) between $-\infty$, where both E_ρ and J are zero, and x, and introducing the general solution for ρ , the system (22), (23), is reduced to the time-dependent differential equation for E_ρ :

$$\frac{\partial^2 E_\rho}{\partial t^2} + \beta \frac{\partial E_\rho}{\partial t} + \Omega_{p\epsilon}^2 \frac{\rho(x,t)}{\rho_o} E_\rho = -\Omega_{p\epsilon}^2 \frac{\rho(x,t)}{\rho_o} E_o + 4\pi e B_e \phi_2 \rho x \quad (24)$$

where $\Omega_{p\epsilon} = (4\pi e^2 \rho_o/Mm)^{1/2}$ is the plasma frequency and $B_e = (2KT_e/m\delta^2)$.

The initial conditions are of course $E_\rho (x,0)=0$ and $\left. \frac{\delta E}{\delta t} \right|_{t=0} =0$.

Eq. (24) represents an oscillating rotor-free electric field, whose frequency is time-space varying, as a consequence of the ρ variation. The oscillations are forced by the external electric field as well as by the internal field, due to the different mobility of charged particles. In the three-dimensional case, the rotor-free solution is still valid only if the condition $\delta_x = \delta_y = \delta_z$ is verified. Nevertheless, a typical MPI plasma is cylindrically symmetric and thus also a rotational solution will be expected as experimentally observed in ref. 9. Anyway, the one-dimensional picture allows to point out some features of the MHD model and to test his range of validity. Particularly, since the time scale of variation of ρ is larger than the corresponding E_ρ scale by a factor $(M/m)^{1/2}$, the equation (24) can be solved, assuming $\rho(x,t)=\rho_o$. Therefore

$$E_\rho = (\frac{2KT_e}{e\delta^2} x - E_o) (1 - \cos (\Omega_{p\epsilon}f)t) \qquad\qquad (25)$$

where $f = \exp(-1/2\,(x/\delta)^2)$, and the other relations are given by

$$J = -\frac{1}{4\pi}\left(\frac{2KT_e}{e\delta^2}\,x - E_o\right)\Omega_{p\varepsilon}\,f\,\text{sen}\,(\Omega_{p\varepsilon}f)\,t \tag{26}$$

$$\sigma = \frac{1}{4}\left\{\frac{2KT_e}{e\,\delta^2}(1 - \cos(\Omega_{p\varepsilon}f)t) - \frac{x}{\delta^2}\left(\frac{2KT_e}{e\delta^2}\,x - E_o\right)\right\}\Omega_{p\varepsilon}\,ft\,\text{sen}(\Omega_{p\varepsilon}f)t \tag{27}$$

Fig. 2 shows the spatial profile of $n = \rho/M + \sigma/e$, as a function of x/δ, ad $T = 20\pi/\Omega_{p\varepsilon}$, assuming $KT_e = 1$ eV, $N_o = 10^{10}$ el/cm^3 and $\delta = .1$ cm; i.e. common values to the most of MPI experiments.

Notwithstanding the neutrality condition is initially well verified, what it means that the Debye length $\lambda \approx 7(T/n_e)^{1/2}$ is lesser than the plasma characteristic length δ , the rapid growth of the charge separation suggests that the above mentioned condition is not a suitable parameter. In fact, that condition concerns only with the first term of the Eq. (27). On the contrary, the most important contribute arises from the linear growing second term. Then assuming as $t \approx 3/B_i^{1/2}$ the characteristic time of ρ evolution, the true neutrality condition becomes

$$N_o \gg (KT_e/4\pi\delta^2 e^2)\,(M/m) \tag{28}$$

that it means $\lambda \gg \delta\,(m/M)$. Assuming the previous values and the proton mass, for M; eq. (28) gives $N_o \gg 3.10^{12}$ ion/cm^3. Therefore, it can be concluded that the MHD model does not correctly describe the macroscopic evolution of the charges produced in the most of MPI experiments.

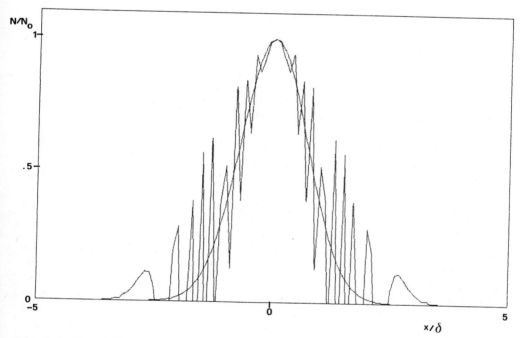

Fig. 2. Spatial profile of n_i and n_e from the MHD picture for $KT_e = 1$ ev $N_o = 10^{10}$ el/cm^3 at $t = 20\pi/\Omega_{p\varepsilon}$.

d) Two-fluid solution

The complete five equation system, coupled by the internal field, cannot be resolved exactly as in Sec.2a). In fact, the previous substitution n---> log(n) is not allowed. Anyway, in the one-dimensional case an useful solution can be carried out by using a linearization of the Poisson equation. The outline of solution is the following one: the previous single fluid shapes for n_i and n_e are assumed, the one-dimensional $E_\rho = 4\pi e \int (n_i - n_e) dx$ integral between $-\infty$ and x is series developed and only the first term is retained. Thus the E_ρ field is given by ($x_j = \int v_{oj} dt$)

$$E_\rho(x,t) = 2\pi e n_o (\phi_i^{\frac{1}{2}} - \phi_e^{\frac{1}{2}})x - \phi_i^{\frac{1}{2}} x_i + \phi_e^{\frac{1}{2}} x_e) \tag{29}$$

Referring to the single fluid solution, the linearly x-dependent term will modify the ϕ_i and ϕ_e time evolution and the time dependent term will modify the like single particle motion. Putting Eq. (29) in the complete system and following the same procedure as in Sec.2a), the following differential system is obtained

$$\frac{d^2\phi_i}{dt^2} = \frac{3}{2\phi_i}(\frac{d\phi_i}{dt})^2 - \Omega_{pi}^2 \phi_i (\phi_i^{\frac{1}{2}} - \phi_e^{\frac{1}{2}}) - 2 B_i \phi_i^2 \tag{30}$$

$$\frac{d^2\phi_e}{dt^2} = \frac{3}{2\phi_e}(\frac{d\phi_i}{dt})^2 + \Omega_{pe}^2 \phi_e (\phi_i^{\frac{1}{2}} - \phi_e^{\frac{1}{2}}) - 2 B_e \phi_e^2$$

$$\frac{d^2 x_i}{dt^2} = \frac{e}{M} E_o + \frac{1}{2}\Omega_{pi}^2 \phi_e^{\frac{1}{2}} (x_e - x_i) \tag{31}$$

$$\frac{d^2 x_e}{dt^2} = \frac{e}{m} E_o - \frac{1}{2}\Omega_{pe}^2 \phi_e^{\frac{1}{2}} (x_e - x_i)$$

The above solution well approximates the E_ρ field in a zone whose length is roughly δ. For $x>\delta$ the E_ρ field fast decreases to zero, in a very narrow transition zone. Thus the time-space behavior can be divided in an inner zone, where the overlapping of the charge profiles makes the coupling term important, and an outer zone where the motion is like single fluid. In fact, when $\Omega_{pi} = \Omega_{pe} = 0$, the coupling term disappears and the system of Eq. (30) is reduced to the Eq. (10) for both ions and electrons.

The influence of the coupling term is shown in Fig. 3.(a). The initial charge density is 10^{10} ion/cm^3. Fig. 3.(b), shows, as a reference, the unperturbed behavior. The oscillation frequency of ϕ_e is $\Omega \approx \Omega_{pe} \phi_i^{1/4}$. The decreasing of ϕ_e stops when the coupling term equates the term $2B_e \phi_e^2$. The coupling term also affects the ϕ_i behavior as Fig. 4 shows, where is plotted the ϕ_i value at a fixed time as a function of the initial density and of the initial electron energy. The faster decreasing of ϕ_i indicates an unstable evolution of the plasma, in which a fraction of electron energy is transferred to the ions. The fig. 5 shows the time evolution of the

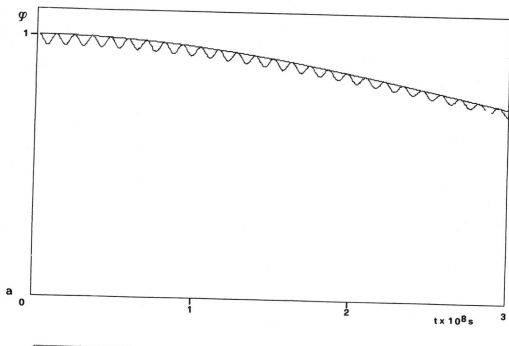

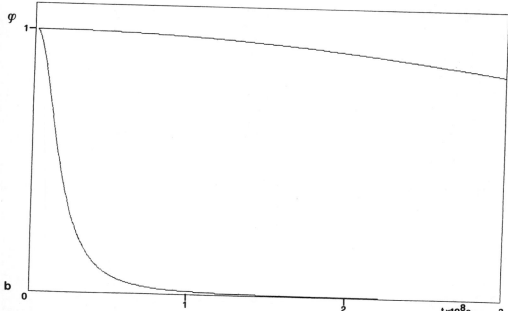

Fig. 3. ϕ_i and ϕ_e behavior from the two-fluid picture for KT =1 eV.
(a) coupled behavior for $N_o = 10^{10}$ el/cm^3; (b) uncoupled behavior (Ω_{pe} =0).

ratio n_e/n_i in the internal (coupled) zone, compared with the uncoupled single fluid behavior. It results evident the increasing of the electron collection time due to the internal field. The influence of the "electron trapping" on the electronic energy spectrum is shown, as an example, in Fig. 6. The peaks are supposed to be produced initially at the same rate

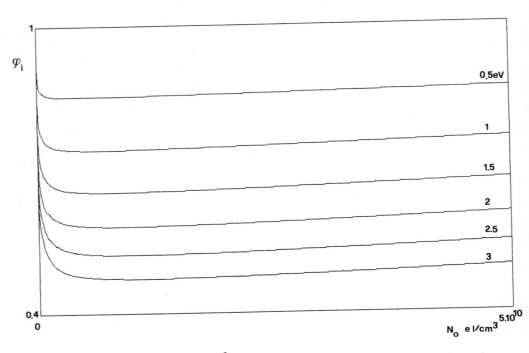

Fig. 4. ϕ_i value at $t = 3.10^{-8}$ sec as a function of the initial density, at different values of the initial electron energy.

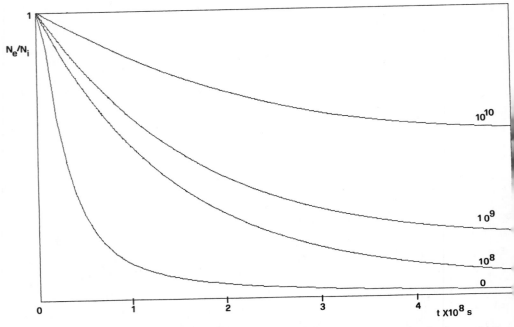

Fig. 5. Time evolution of the "electron trapping" factor (n_i/n_e) at different initial densities for $KT_e = 1$ eV. The 0-line corresponds to the uncoupled behavior.

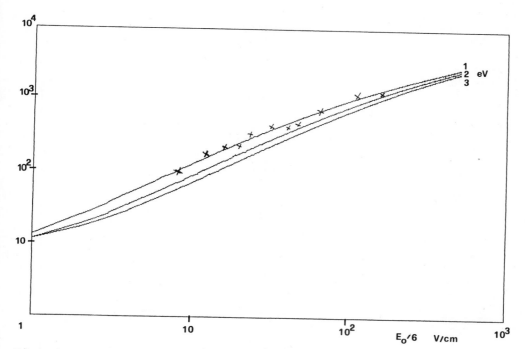

Fig. 6. Electron energy peaks (1 eV, 2 eV, 3 eV) at a fixed position and initial density $N_o=10^{10}$ el/cm³. The y-axis is normalized to 1 and the x-axis indicates the collection time.

and propagate as a single particle gaussian profile only in a fixed direction. No external electric field is introduced and thus the collection scheme is quite corresponding to that of ref. 6. Although Fig. 6 derives from a simplified picture, both the lowering of the low energy peaks and the growth of the background appear clearly.

AN EXPERIMENTAL TEST OF THE MODEL

Fig. 7 and Fig. 8 show two preliminary tests about some features of the above-explained model. The results are obtained by a standard experimental set-up for multiphoton ionization experiments. A Sodium beam was ionized by a pulsed dye laser (10 ns), pumped by an excimer laser, tuned at the D_1 or D_2 resonant lines. The total current was detected by the opto-galvanic technique. An external constant electric field was applied between the plates, one of which was grounded. The signal was collected at the ends of a resistor (500 kΩ) on the polarized (+ or -) plate. The signal was time/amplitude analyzed by a boxcar, triggered by the laser pulse, and finally recorded by an x-y recorder. Fig. 7 shows the behavior of the maximum of the signal as a function of the external electric field. To compare the experimental behavior with the theoretical predictions of the single fluid model, the charge density was about 10^7 el/cm³. In fact, in such conditions, the electron-ion interaction can be neglected compared with the externally applied electric field.

Fig. 8 shows the time evolution of the signal under opposite conditions,

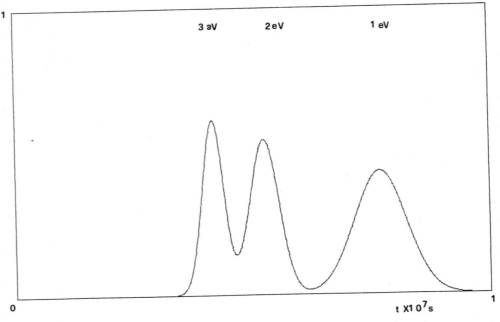

Fig. 7. Bilogarithmic plot of the peak of the current versus the external electric field from the single fluid picture. The dots represent the experimentally observed behavior.

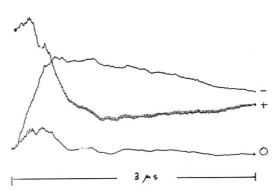

Fig. 8. Time behavior of the total current detected by the optogalvanic technique. The sign +/- indicates the polarization of the collection plate, 0 corresponds to the Laser off. The noise depends on the integrated R.F. noise of the excimer pump.

i.e. weak electric field and high charge density. The signal corresponding to the positive polarization shows a modulation at a frequency $\Omega \approx 6.3 \ 10^8$ rd/sec. Owing to the time constant of the electric circuit (1 μs) the signal is partially integrated. The measured frequency corresponds to a charge density of $1.3 \ 10^8$ el/cm^3 at the collection point. From the above-mentioned model, the initial density results $6.5 \ 10^{10}$ el/cm^3. Considering that the initial beam was $\approx 10^{11}$ at/cm^3 and that the laser was not so powerful to completely saturate the ionization, the results are in a good agreement with the theory. On the contrary, the signal obtained reversing the plate polarization does not exhibit any modulation. In fact, the plasma oscillations are enhanced when the electric field trails the electron (see two fluid solution).

CONCLUSIONS

Following the previous discussion, the plasma effects can seriously affect the charge propagation also in the low density region. Hence, it seems important that the plasma parameters are measured in an independent way, to separate the effects of the laser interaction from the collective ones.

On the contrary, from the plasma physics point of view, the MPI produced plasma constitute a very suitable sample to study non-linear phenomena, as for instance internal instabilities, due to the different charge mobility, or induced by the external electric field.

Finally, it will be noted that many other phenomena must be taken into account in a theoretical approach, as for instance motion in a magnetic field, in a spatially non uniform electric field and motion in presence of collisions changing the number density in the continuity equation. The previous solution could be considered as a "homogeneous" solution, from which the influence of the added terms could be deduced by perturbative methods.

REFERENCES

1. "Proceedings of the Colloquium on Optogalvanic Spectroscopy", Aussois, France, June 1983, Published in J. Phys. (Paris) C7 -44 (1983) "Third International Symposium on Resonance Ionization Spectroscopy & Application", Swansea, UK, Sept. 1986 (Adam Hilger, 1987)
2. "Photons and Continuum States of Atoms and Molecules" Proceedings of Workshop - Cortona, Italy , June 1986, Published by Springer Verlag (1987)
3. M. Broglia, F. Catoni, A. Montone, P. Zampetti and F. Giammanco, "IV Congresso Elettronica Quantistica e Plasmi", Capri (1984) "Proceedings of XVI EGAS Conference", London (1984) see also ref. 1
4. "Atomic and Molecular Collisions in a Laser Field", Les Editions de Physique, Les Ulis (1985)
5. G. Sargent Janes, H. K. Forsen and R. H. Levy, AIChE Symposium Series, 73, n. 169 (1980)
6. F. Yergeau, G. Petite and P. Agostini, J. Phys. B:At. Mol. Phys., 19, L663 (1986)
7. B. Carre, F. Roussel, G. Spiess, J. M. Bizeau, P. Gerard and F. Wuilleumier, Z. Phys. D., 1, 79-90 (1986)
8. M. Crance, J. Phys. B: At. Mol. Phys., 19, L267 (1986)
9. A. M. Davtyan, R. Kh. Drampyan and M. E. Movsesyan, Sov. J. Quantum Electron., 14 (5), May (1984)

10. L. Spitzer, "Physics of Fully Ionized Gases", Interscience, NY (1956)
11. F. Giammanco, Collective Effects in Laser Multiphoton Ionization,
    Phys. Rev., A 36, 5658 (1987)
12. A. Erderlyi, "Asymptotic expansions", Dover Pubbl. Inc., NY (1956)

ITALIAN PROGRAMS FOR LASER SOURCES DEVELOPMENT AND APPLICATIONS

A. Gilardini

Selenia S.p.A.

Rome, Italy

A few years ago it was common to hear people saying that lasers were a solution in search of a problem. The fact that I am here today to talk about lasers and their applications, and that applications will be the driving motif throughout my entire presentation, indicates that the situation is changed, so that we can state that now lasers provide new and efficient solutions for many technical and industrial problems.

As we all know there are today many kinds of lasers, and a large variety of problems are solved by systems incorporating these lasers. As a background information for a better understanding of why and how the Italian national programs in this area have been conceived and structured, I have represented in table I a matrix of each laser type versus the industrial application sectors, where a dot indicates appropriate and significant (at least potentially) use of the laser in that sector. Both lasers and industrial sectors have been listed in decreasing laser market importance; the corresponding market percentages, as available for the year 1984, have also been given, except when the individual figure was of the order or below 1%.

It has also to be remarked that approximately 22% of the total laser market is not for industrial applications, but is for research and development activities. On this regard please note that, also in order to maintain this presentation within reasonable limits, uses of lasers for research and in fields of far future applications, which require specific and peculiar developments, as nuclear fusion, energy production and strategic weapons, are not included in this presentation. In order not to overcrowd the table, some of the less used lasers have not been included; these lasers, as the pulsed nitrogen laser, the mercurous bromide laser, the HF, DF and CO TEA lasers, can find application particularly in remote sensing and in photochemistry, since in these cases specific wavelengths, appropriate for the interaction with given chemical species, are sometimes required.

Coming now to the italian situation, we must first of all remind that in the electro-optic area our industry is essentially a system industry,

Table 1.

Lasers / Industrial sector	Market %	Solid state 28	CO$_2$ 19	Ion 17	Semicond. diodes 16	HeNe 10	Dye 6	Excimer 3	HeCd	Metal vapors	FEL
Metalworking and material processing	20	○	○								
Biomedical	14	○	○	○	○	○	○			○	
Tactical military	14	○	○		○						○
Information printing and reading	12			○	○	○			○		
Communication	11				○					○	
Industrial metrology	5			○	○	○			○		
Remote sensing		○	○				○	○			○
Photomedical processing and diagnosis		○	○				○	○		○	○

and not a component one. Then it is worth establishing first the relative importance, for our country and from the system strategy point of view, of the industrial sectors of table I. The sectors can then be grouped in decreasing order of strategic importance, taking into consideration both the strength of our industry in each sector and the national community interests. In my opinion the groups could be as follows: tactical military, metal-working and material processing, communication are the sectors of top priority; industrial metrology, information printing and reading are the less strategic ones. Looking now to lasers, we can split them into two categories: those lasers which are easily available commercially with the system required performances, and those which are not. In the first category I should include the ion, HeNe and HeCd lasers, and the semiconductor diode laser of low characteristics and price; all other lasers fall in the second category.

These priorities and distinctions are important, since I believe that the Italian national programs must promote research and development activities for the benefit of the industrial sectors, according to their strategic importance, as listed above, and to their necessity of specific, not available lasers. At the end of the Seminar we shall verify whether these criteria have been actually met.

If we see the national programs as a form of public support, which in many cases is devoted mostly to industrial and industry-related research activities, two natural questions arise: what reasons justify the public support by a government to the industrial research and development activity, and why this support has been structured into national programs and projects. The answer to the first question is very simple: our industry has to compete in the international market and in each country public support is provided to all our competitors. According to data collected by the Battelle Institute for AIRI, the Italian Industrial Research Association, this support amounted in 1983 to 0.71% of the Gross Domestic Product in USA and in United Kingdom, to 0.43% in France, 0.41% in West Germany and only 0.17% in Italy.

The government support to industrial research may take various forms: direct financial contributions, reduced interest rates on financial loans, contracts for new public services, and so on. A national program is one of these forms, which is particularly effective since it collects all the best competences of the country, in the universities, in the research centers and in the industries, for the purpose to achieve important scientific and technical common goals, which are useful to the country economy and rewarding for all the participants. These national programs can be launched in Italy by the Ministry for the Scientific and Technological Research, by the National Research Council (CNR) as Finalized or Strategic Projects, by the ENEA Agency in the frame of its Five-years Programs. Contracts, which cover totally or partially the cost of the study, are given to each participant: each program is supervised by the supporting Institution through its staff and consulting committees.

The first program, launched on a national basis, for the development of lasers and of their applications has been the CNR Finalized Research

Project "High power lasers". It is worth recalling that ten years ago, when the project started, the industrial activities in the high-power laser field in Italy where essentially limited to two Companies (Selenia and Fiar) working for military applications and to one Company (Valfivre) working for mechanical and medical applications. Scientific experience and qualified competence, however, had been significantly developed during the previous years in CISE, ENEA and various CNR Institutes.

The following three objectives were then stated for the purpose to assist the Italian industry in exploiting laser processing technologies and in developing laser manufacturing systems:

(1) to investigate and promote the use of lasers in material processing for automated manufacturing systems;
(2) to study medical applications of lasers mainly in the fields of general and endoscopic surgery and of phototherapy;
(3) to study high-power lasers technologies and to develop prototypes for mechanical and medical applications.

The first two objectives could be attained using already commercially available lasers, a particularly favorable situation that explains the achievement of significant project goals, as the collection of consistent and reproducible data for metalworking and material processing (welding, cutting, drilling, scribing, surface treatments) and for the automated manufacturing systems design, as well as the assessment of the appropriate procedures for laser treatments in various clinical fields (surgery and microsurgery , endoscopic photocoagulation, phototherapy and photochemiotherapy). On this regard a particularly significant result is represented by the fact that more than 200 industries, mainly small Companies, have conducted feasibility studies for the use of high-power lasers in five, properly equipped, application Centers of the project.

Carbon dioxide, neodymium in YAG, excimer and dye laser sources have been developed with various characteristics and power levels; know-how transfer to industries interested in producing and selling these sources has in general taken place, but the related business has been up to now rather disappointing. It is, however, worth noting that because of the project the number of laser manufacturers in Italy has grown up to about ten and that a much better knowledge of the industrialization problems of laser sources has been spread throughout the country.

Finally, I would like to recall that the project was finished in 1983, that more than 300 researchers from Universities, CNR Institutes and industrial laboratories participated to its activities and that the total cost was of 23.8 billions Lire.

A general feeling of all the participants to the "High power lasers" Project was that a new CNR Finalized Research Project, more system oriented, had to be launched as a continuation of the research effort on high-power lasers developments. However, the new project, which was called "Electro-optic technologies" and which will be later described, is still on the starting line. In the mean time lasers of various kinds, related optical devices and their uses have been included as a research subjects in the National Research Program "Microelectronics", in the CNR Finalized Research Projects "Materials and devices for solid state electronics", "Biomedical technologies" and "Oncology", in the CNR Strategic Project "Optoelectronics", and in the ENEA Five-years Project "Optic and electro-optic technologies".

The "Microelectronics" National Research Program has been conceived and launched, within the frame of a law for the promotion of applied research and technological innovation, by the Italian Ministry for the Scientific and Technological Research in the years 1983-5. It covers two main areas: technologies for VLSI devices and technologies for compound semiconductor devices. A contract for performing the activities listed in the second area, which regard the development of both active optical components for fiber optic communication and of microwave analog and digital components, has been given at the beginning of this year to the "Consorzio per la ricerca sui semiconduttori composti", a consortium formed for this purpose by some of the most important italian electronics industries (Elettronica, Italtel, Selenia, Telettra) and by the CNR, as a coordinator of participating Universities and public research institutions.

More precisely the development of active optical components, namely sources and detectors, is restricted to devices for the wavelength region of 1.3 μm, where second generation fiber communication systems operate, and its main object is to transfer the results of the feasibility studies, already performed in our country, into reliable products manufactured according to well defined and controlled processes. In fact, after the design and development of the basic technologies, active optical components have to be developed and characterized, and transmitting-receiving modules, suitable for field testing, must be assembled.

Components and modules for use in multimode and monomode fiber optic systems have to be designed and fabricated. In both cases the lasers are using indium gallium arsenide phosphide/indium phosphide heterojunctions, fabricated by multiple epitaxy techniques, and the final assembled modules will include optical feed-back and a pig-tail interface to optical fibers. However, the laser for the multimode system is of the gain-guided type with a mesa stripe structure, whereas the one for the monomode system is of the index-guided type with a ridge waveguide structure.

In the consortium the research on the active optical components is being conducted by Italtel and Telettra, with CSELT as the main subcontractor, and with the participation of Milano University and Torino Polytechnic from the academic side. This part of the program is expected to be completed by the end of 1989 and its contractual cost is 12.2 billions Lire.

Active optical components for fiber optic communication is also the theme of a subproject of the CNR Finalized Research Project "Materials and devices for solid state electronics". This project, which started in 1986, will last five years and it is then expected to finish by the end of 1990. On the specific subject of active optical components for fiber optic communication it will spend much less than the National Program, something of the order of 4 - 4,5 billions Lire. Research objectives, however, are quite demanding, also if we are aiming here at feasibility results more than at fabrication procedures, as it is general characteristics of the Finalized Projects, which must provide results applicable after about 5 years, compared to the National Programs, whose application time scale has to be much shorter.

CSELT is by far the leading center of this activity in Italy and its research goals for the first year are:

(1) The fabrication of a distributed feed-back (DFB) laser with a ridge waveguide structure, that in particular implies to have attained a capability of growing high quality epitaxial layers on corrugated surfaces;

(2) The construction and characterization of monomodal planar strip guides, to be considered the first essential step for the design of a wavelength division multiplexer in monolithic form;

(3) Laboratory experimentation of new very advanced structures, like the strained layer superlattice and multi quantum well, which are very important for the new generation of low threshold lasers and of low noise photodetectors.

On the academic side, Ancona University is developing autoconsistent models of the DFB lasers, Modena University is studying the metallization processes on binary and ternary compound semiconductors, and in particular the stability problems of platinum and silicide contacts, and the CNR Institute MASPEC is working on the techniques of growing, cutting and lapping indium phosphide crystals and on their characterization.

It is also worth mentioning that two other optoelectronic themes are included in the same subproject of the CNR project. One considers passive optical components for fiber optic communication, and in particular two subjects strictly related to the previous ones: the development of micro-optic components for a wavelength division multiplexer and the study of planar optical waveguides. The second one concerns the development of devices and technologies for optical signal processing, and its first objective is the construction of the hybrid integrated version of an acusto-optic spectrum analyzer.

Biomedical laser applications are being pursued in the two other CNR Finalized Research Projects "Biomedical technologies" and "Oncology". In the first project the use of laser light, carried through an optical fiber catheter to the heart for angioplastic treatments in cases of cardiovascular diseases, is investigated. The financial effort of this research is of the order of 250 millions Lire per year. In the second project the laser phototherapy of tumors is studied along a line previously established in the "High power lasers" Project, namely by the use of the hematoporphyrin derivative, a photosensitive substance which properly injected concentrates in tumor tissues.

The development of specific demanding lasers of near future use is

included in the two-years, 800 millions Lire Strategic Project "Optoelectronics": a quasi continuum, multimodal dye laser at the CNR IFAM Institute
in Pisa, vapor metal lasers at the CNR IEQ Institute in Florence, femtosecond
lasers and a solid state laser with a properly designed resonator for large
mode volume and low divergency at the CNR CEQSE Center in Milan.

Before presenting now the ENEA Project "Optic and electro-optic technologies" and the future national projects, which include studies on lasers
and on their applications, well defined on the basis of already performed
preliminary analyses, I would like to highlight briefly two major areas of
growing research interest for laser applications, which have determined many
of the specific choices in the above projects.

Geometrical and temporal characteristics of the laser radiation have
been widely utilized in most applications up to now. Examples of exploitation
of the geometrical characteristics are the propagation in comparatively (relative to microwaves) very narrow beams and the capability of this radiation
to be focused on very small spots. Among the temporal characteristics of
concrete interest, I like to mention the possibility of generating, with the
Q-switching technique, very short but intense light pulses and the sensing
significance of the light pulse amplitude shape after reflection, absorption
or scattering by obstacles or atmospheric constituents. Exploitation of the
laser radiation spectral characteristics, made practical by the possibility
of using nearly monochromatic light sources, has been instead rather limited,
but it is now gaining increasing attention in view of many possible applications. Frequency and phase modulations of optical signals, frequency
shifts due to the doppler effect or to inelastic scattering by molecules,
laser induced fluorescence spectra are significant examples of spectral
features of interest for communication and remote sensing uses.

Spectrum analyzers suitable for measuring these features can be based
on passive interferometric or on active homodyne or heterodyne detection
techniques. In fiber optic communication systems the use of heterodyne detection is most promising and improvements of more than 10 times in sensitivity
and up to 1000 times in selectivity are expected; the problem here is on the
laser side, since semiconductor lasers are essential for size and low power
consumption, while their emission is at present non particularly coherent.
Wide tunability and improved coherence are then the next major goals of
research activities in semiconductor lasers.

In remote sensing applications instead, the received wavefronts are
rendered imperfect by atmospheric inhomogenities and speckle effects, and
thus it is difficult to match signal and local oscillator wavefronts, as
required for a good heterodyne detection. This situation is less critical
at longer wavelengths, that is one of the reasons why heterodyne detection
has been practically attempted only for receiving 10 μm CO_2 laser radiation.
Anyhow, when the laser radiation is received by large aperture optics after
atmospheric propagation, heterodyne detection does not appear to be a satisfactory solution, and passive interferometric techniques are more effective. Optimization of spectrum analyzers not requiring a local oscillator
is thus an important research theme for laser remote sensing applications.
It is also worth mentioning that in this case the laser frequency stability

and the optic quality are less demanding than in the heterodyne case.

Another interesting, quite recent approach, which also does not require a local oscillator, is the autodyne technique, which is based on the self-beating of the various frequency components of the received signal, and can be used, for instance, for measuring the relative doppler velocities between the components of a multi-element target.

Another research area of growing importance in a large number of applications, particularly in high-power cases, is that of the techniques of laser beam handling. Typical problems in this area are: splitting a beam into many others of specified characteristics, adding up many beams coherently, transmitting or focusing a beam through a changing path or medium, also when it depends on the power of the beam itself, correcting by means of an adaptive optics a distorted beam wavefront. In most of these cases an high quality beam is required, and it must be maintained through all the handling.

The already running ENEA Project "Optic and electro-optic technologies" and the ready to start CNR Finalized Research Project "Electro-optic technologies" have many common and complementary features. Let us examine first the common aspects of the two projects. Both give priority attention to the system choices: lasers and optical components are developed only insofar as they are necessary for the system activity. At the system level both projects include metrology and remote sensing applications, as the detection of pollutants in the atmosphere and in the sea and the measurement of wind patterns. In these cases, however, the approaches to the problem are different and, in a certain sense, complementary. The ENEA Project is aiming to the realization of specific instruments, which can also be used for the control and supervision functions of the Agency, whereas the CNR Project has chosen a more general approach, namely the development of a set of functional modules, which can be assembled in different ways, so to test and optimize the performance of possible various measurement techniques.

In the area of industrial processing the two projects are complementary: while CNR is mainly considering metalworking and material processing, ENEA has a research line on photochemical processing. Also, as far as the laser sources are concerned, the two projects are complementary; ENEA is conducting the development of excimer, color-center and free electron lasers, CNR is planning the development of high power solid state (Nd) lasers and of metal vapor lasers. Both projects consider the development of CO_2 lasers, but ENEA is emphasising the tunability features, while CNR is considering high power, modularly structured sources. As for the optical components both projects consider the development of the technologies for their fabrication; furthermore the CNR project includes studies on fiber optics for high power transmission.

Due to the scientific and technical importance of these two projects, a few more details on specific research themes seem worthwhile. Since it is here impossible to give an exhaustive list of all these themes for the ENEA project, I shall indicate only some of the most significant ones.

For the remote sensing of pollutants three major kinds of systems will
be fully developed in the ENEA project:

(1) an infrared differential absorption lidar (DIAL) using a pulsed
    CO_2 laser source;
(2) UV-VIS lidar systems using excimer high-power laser sources with
    Raman cells to produce convenient output frequency shifts, suitable
    for fixed or mobile stations;
(3) an air-borne UV laser fluorosensor using a Raman-shifted excimer
    laser and pulse compression techniques.

Advanced spectroscopic systems, first of all those relying on the coherent
anti-Stokes Raman scattering (CARS) method with the use of an excimer + dye
laser and of a Raman shifted excimer laser, will be finalized to monitor
"in situ" combustion gases compositions and temperatures, pollutant concentrations, etc. over limited volumes but with good spatial resolutions. Among
the photochemical processes considered in the ENEA project we find the production of Si and Al powders to be sinterized and the deposition of high-
quality thin film coatings.

Electro-optic sensors and interferometric holographic systems are considered for the inspection and the precise localization of objects or of
profiles. This metrological capacity, as well as that of communicating over
limited distances and of making micromovements also in an hostile environment, two other topics included in the ENEA project, form the bases for the
development of new, more autonomous robot systems, a very important task
for the ENEA Agency.

To accomplish all the objectives of the program in a five year period,
1983-89, a total effort of 415 men-years of ENEA personnel plus and additional cost of 45 billions Lire for equipment, materials an   external
research and manufacturing activities has been forecasted.

Coming now again to the CNR Finalized Research Project "Electro-optic
Technologies", the best way to describe the planned activities is to list
the research themes and goals proposed by the feasibility study committee
for each of the five subprojects in which the entire project has been split-
ted.

Subproject I. Manufacturing and industrial diagnostic electro-optic systems

I. 1. Manufacturing systems for the electronics industry (laser
      drilling of non-homogeneous printed cards; automatic control
      systems with multiple working heads, and focus and position
      sensing; automatic systems for direct laser writing).
I. 2. Laser-robot systems for three-dimensional cutting and welding
      and for thermal treatments, using optical powers up to 5 KW
      (design and construction of an assembly cell for testing
      systems and components; standard and modular design of com-
      ponents and subsystems; automatic control systems with focus
      and position sensing).
I. 3. Flexible multi-process and multi-piece systems for laser

welding, thermal treatments and surface coatings, using optical powers larger than 5 KW (laser beams summing; other research subjects as for theme I.2.).

I. 4. Flexible and programmable equipment for laser beam handling and positioning (space and time beam transport and shaping; curve and surface profile following techniques).

I. 5. Beam and process diagnostic equipment (space and time control of laser beam characteristics; control of working parameters).

I. 6. Electro-optic systems for measurements of surface quality, dimensions, positions and dynamic parameters (on line controls; piece selection; vibration detection and measurements).

Subproject II. Electro-optic systems for informatics, environment control and national defense

II. 1. Coherent systems for the analysis of the light reflected by an obstacle or by a scattering medium (modular structure: highly coherent laser sources, passive and active spectrum analyzers, scanning and projection optics, high sensitivity receiver; laser imaging; doppler-detection of vibrational signatures; wind pattern measurements).

II. 2. DIAL and CARS systems for gas detection and identification and temperature measurements (modular structure as in II.1., but engineered for optimization of weight, power consumption and cost).

II. 3. Heterodyne radiometer for spectrum analysis and identification of optical signals (modular structure as in II.1., but with fast frequency scanning of the laser source).

II. 4. Optical fiber sensor systems for physical and physical-chemical measurements (phase detection for vibration measurements; amplitude detection for physical quantities measurements and for fluids monitoring; multiple and intelligent sensors).

Subproject III. Active electro-optical components

III. 1. CO_2 lasers (modular structure with 1 KW basic module of low weight and size; modular structure with 2.5. KW basic module of high efficiency and reliability; monomodal, compact and reliable 100-200 W laser).

III. 2. Neodymium lasers (modular structure with 50 W basic compact module with high efficiency and beam quality; modular structure with 500 W basic module, pulsed and with high beam quality).

III. 3. Vapor metal lasers (1-10 W copper and gold lasers).

Subproject IV. Passive electro-optical components

IV. 1. General facilities for integrated optic technologies.

IV. 2. Lenses and mirrors, beam mixers, polarizers (thin film coatings with high damage threshold and excellent chemical and mechanical hardness; metal mirror machining; component diagnostics).

IV. 3. Optical fibers for high-power uses in the infrared (up to 10

/um; fabrication and characterization of fibers and components).
IV. 4. Adaptive optics and phase conjugation mirrors (correction of
wave-front distorsions; improvement of high-power beam quality).

Subproject V. Radiation-biological structures interaction and systems for
biomedical applications (specific research themes will be identified later,
after a careful review and analysis of the area scientific and application
interests).

The finalized project will last five years and will require an effort
estimated in 825 men-years and 66 billions Lire (training included).

It is worth remarking at this point that the laser project in the
Eureka international program, the so-called Eurolaser project, to which
Italian industries and scientific institutions participate in a coordinated
way, is also pursuing the goal of developing high-power CO_2, solid state
and excimer lasers, mostly for material processing applications. A modular
approach, like the one of the CNR project, will be probably chosen also for
Eurolaser, but with a power scaling-up. In fact, Eurolaser plans to develop
a 10-100 KW CO_2 laser system, a 1-5 KW solid state laser system, and an
excimer laser with an output power up to 10 KW.

In order to complete the picture, I must also mention two other future
CNR Finalized Research Projects, which plan laser-related activities. While
project "Robotics" only mentions vaguely the use of lasers for sensing purposes, the other project "Telecommunications" stresses clearly the importance
of coherent optical communications, and then proposes specific developments
of very stable monomode laser diodes (probably DFB lasers), amplitude, frequency and phase modulators, polarization compensators, optical amplifiers.
Here too it is worth remarking that similar developments are also among the
research objectives of the new EEC international program RACE for use in
high bit-rate optical transmission, as well as in the future customer optical
loops.

At this point, as a conclusion, let me try to synthesize national
program activities in a schematic plot. In table 2) I have considered the
groups of lasers and of industrial sectors for which national developments,
as previously discussed, are most important. The first row indicates for
each type of laser the projects which include developments of that laser,
but not of related systems. Projects considering the development of laser
systems, including the laser itself whenever necessary, are shown in the
following rows, one for each industrial sector. The dots of table 1) are
shown too.

If we keep in mind also the actual state of art of each case, the
resulting picture seems rather satisfactory. In fact, if present and
future national programs and projects will be timely and properly implemented as planned, all the lasers of strategic value and most of the important laser-based systems will be adequately studied, realized and opera-
tionally tested in Italy. The achievement of these goals will strongly
depend, however, on the success of the coordination efforts between and
within the programs, a not-easy task considering the large number of re-

Table 2

	Solid state	CO_2	Diodes (high value)	Dye	Excimer	Metal vapors	FEL
Laser development	OE		/uE/MD	OE		OE/EO	EN
Laser-based system development :							
Tactical military	°EO	°EO	°				°
Metal working and material processing	°EO	°EO					
Communication			°TLC			°	
Biomedical	°	°		°ON		°	
Remote sensing	°EO	°EN/EO		°EN/EO	°EN		°
Photochemical processing and diagnosis	°EO	°EN		°EN	°EN	°	°

Legend:

MD – Finalized Project "Materials and devices for solid state electronics"

OE – Strategic Project "Optoelectronics"

EN – ENEA Project "Optical and electrooptical technologies"

EO – Finalized Project "Electrooptical technologies"

TLC – Finalized Project "Telecommunications"

/uE – National Research Program "Microelectronics"

ON – Finalized Project "Oncology"

search units, of different structure and mentality which participate to the projects. I like then to conclude this Seminar sending my best wishes to all the scientists and engineers involved in the national projects and in particular to those of them who will have the task of their management and coordination.

LASERS IN MODERN INDUSTRIES

A. J. De Maria

United Technologies Research Center

East Hartford, CT 06108

INTRODUCTION

Selected applications of lasers in semiconductor integrated circuits manufacturing, radar systems, material cutting and drilling, and inspection of electric power cables will be briefly discussed to serve as illustrative examples of laser applications in such modern industries as microelectronics, avionics, machining, and electric power. Reference 1 provides a discussion of worldwide laser markets and their growth for those readers who desire a broader overview of this subject.

SEMICONDUCTOR IC MANUFACTURING

The semiconductor integrated circuits technology has contributed greatly to the "electronic revolution". This technology has been credited as being primarily responsible for the electronic or information revolution which has been labeled as having a more enduring consequence on mankind than the industrial revolution[2]. Among the large number of different integrated circuits, the dynamic random access memory (DRAM) chips have the largest unit sales and volume, as well as the most dependency on the manufacturing learning curve for reducing cost and increasing yields in order to meet the aggressive pricing strategy of competitors. It is also generally agreed that DRAM chips utilize the most advance processing technologies to achieve the highest density of semiconductor devices per chip. They also have one of the shortest product life cycles within the semiconductor industry. The rapid increase in complexity of these chips from the first 4K-bit product, up through the 16K, 64K, 256K and present 1M-bit DRAM chip has been breathtaking over the last 25 years from the standpoint of increased number of mask set per wafer, rapidly decreasing layout rules down to the present micron to submicron dimensions, increase in die sizes, decreasing number of dies per wafer and increased capital cost for a wafer fabrication factory. Consequently, a manufacturer's ability to rapidly ramp up DRAM chip yields early in the production cycle usually determine success or failure of the product in the marketplace. Because of the insatiable appetite of computer manufacturers for ever increasing money storage capacity per chip, competition is presently fierce in being first to market with 4M-bit and 16M-bit DRAM products.

Laser technology has made it possible for DRAM chip manufacturers to sharply increase their yields and therefore volume in the early phase of their production ramp cycle by use of a technology often referred to as laser redundancy[3,4]. Laser redundancy enables manufacturers to design spare, normally inactive address encoders into their memory chip. When a portion of a memory chip is found to not meet specifications during the die functional probe testing, a pulsed 1.06 micron YAG:Nd^{3+} laser is used to "explode" away the polysilicon conductor connecting the encoder addressing that portion of the circuitry; thereby, open circuiting the defective portion of the chip. The laser system is then used to open the polysilicon conductor shorting out the spare inactive address encoders; thereby, connecting the spare encoder to the other circuitry of the memory chip.

Figure 1(a) is a photograph of a small portion of an IC. A functioning polysilicon film conductor interconnecting line is highlighted in the photograph along with the crystal silicon substrate, the silicon dioxide insulator film, and the aluminum film interconnecting line. Figure 1(b) illustrates a polysilicon film interconnect line cut with a laser. The polysilicon has a higher absorption coefficient to the 1.06 micron radiation than the crystal silicon, thereby causing it to selectively melt and be evaporated away under high intensity, pulsed 1.06 micron laser irradiation. The chip is then treated with a chemical etch to remove the scattered polysilicon residue deposited by the laser treatment which is in turn followed by a reoxidation process.

Using laser redundancy in the manufacturing of 64K-bit DRAM, manufacturers have obtained 2 to 3 times yield improvements during the early DRAM manufacturing (i.e., start-up) phase, obtained on the average 40 to 50% good dyes per wafer during the start-up manufacturing phase, and obtained an equivalent number of chips from one fabrication facility as previously obtained from two facilities.

Mostek was one of the early users of laser redundancy. During the production start-up of 4K-bit DRAM, it took Mostek 1 1/2 years to ramp production up to 2 million chips. Because of the added complexity of the 16K-bit DRAM chip; it took Mostek 2 years to reach the 2 million chip production. In spite of the increased complexity of the 64K-bit chip over the previously mentioned chips, the use of laser redundancy techniques resulted in Mostek ramping up to a production of 2 million 64K-bit chips in only 3/4 of a year.

Based on such outstanding results, laser redundancy techniques are now used extensively in the manufacturing of complex semiconductor chips.

LASER MATERIAL WORKING

The ability of a laser to deliver a high intensity beam of radiation

Fig. 1(a) Photograph of small portion of an integrated circuit showing the single crystal silicon substrate, the silicon dioxide insulator film, aluminum film interconnecting conductor lines and polysilicon film interconnecting conductor lines.

through the atmosphere and heat an absorbing material has focused attention on its use in the material working industry for cutting, drilling, welding, heat treating, melting, etc., type applications. It is recalled that the Stefan-Boltzmann relationship for the total energy radiated per unit area by a perfect thermal source is equal to the fourth power of the temperature times the Stefan-Boltzmann radiation constant. For reference, it is noted that a power density of one million watts/cm^2 corresponds to a thermal source operating at 20,500K. By means of optical focusing, the laser can achieve such temperatures and is therefore capable of providing such high energy concentrations that its focused radiation can melt or vaporize any known material.

Figure 2 illustrates important material processing tasks along with the laser radiation power density, and the interaction time of the radiation with a material required to accomplish the indicated tasks[5]. In the case of a laser beam moving continuously across a material, the interaction time can be defined as the time that it takes the incident laser spot to move one diameter relative to the workpiece surface. For a material process requiring short pulses of laser radiation, the interaction time can be considered to be simply the pulse duration because the material can be assumed to be stationary during the short irradiation process.

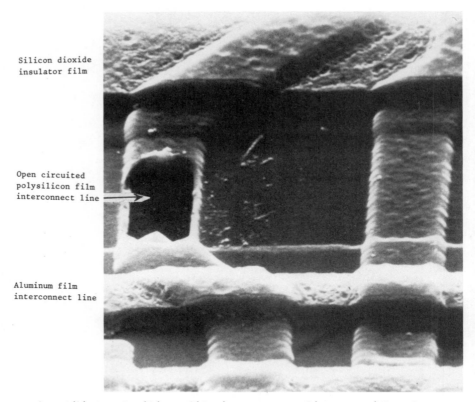

Silicon dioxide
insulator film

Open circuited
polysilicon film
interconnect line

Aluminum film
interconnect line

Fig. 1(b) A polysilicon film interconnect line cut with a laser.

Figure 3 illustrates photographs of typical laser welds performed with CO_2 laser, as well as relevant data relating to power density, type of material, and process speed[6]. Figure 4 is a photograph of a 3-stage, gas recirculating closed cycle, electric discharge CO_2 laser. Each stage is rated at 3 kW to 4 kW average power. Such a device can yield between 9 to 12 kilowatts of continuous output power. The use of fast gas flowing techniques to

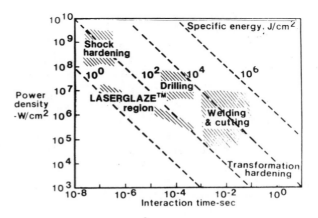

Fig. 2. Laser power density (W/cm^2) and material interaction time (sec.) required by various material working tasks.

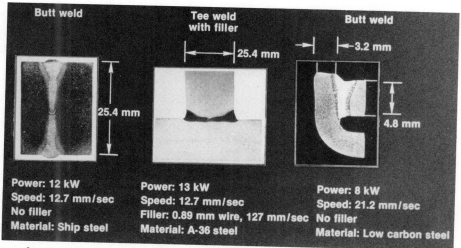

Fig. 3. Photographs of typical welds performed with a CO_2 laser under the conditions indicated.

achieve several tens of kilowatts of continuous power from electrically excited CO_2 lasers[7] has been responsible for placing CO_2 lasers in a dominance position for large material working applications.

Within the next decade, it is anticipated that laser material processing utilization in manufacturing will increase dramatically. Lasers such as $YAG:Nd^{3+}$, ruby, $glass:Nd^{3+}$, and CO_2 are expected to address most applications.

Fig. 4. Photograph of a 3-stage, gas recirculating, closed cycle, electric discharge CO_2 laser. Each stage contributes 3 to 4 kW of laser output power.

Laser radar technology is an obvious progression of radar technology from the RF, microwave, and millimeter wave region of the spectrum into the optical region (i.e., IR, visible, near UV). Laser radar technology has its own advantages and disadvantages when compared with conventional radar technology. When one takes these advantages and disadvantages into consideration, one concludes that laser radar systems will complement and not compete with conventional lower frequency radar systems. Laser radar systems will find usage predominantly for those applications that cannot be addressed by conventional radar systems.

Rangefinders are the most basic radar systems of either the microwave or laser variety. They basically measure range to a target by measuring the time of flight of a transmitted and an echo pulse of electromagnetic radiation. Rangefinders can also provide azimuth information to the target. Information on the speed of the target can also be obtained by measuring the change in range as a function of time (i.e., range rate). These radar systems are known as incoherent radar systems.

The more sophisticated radar systems are of the coherent variety which have the ability to also measure the velocity of the targets by means of the Doppler effect. This type of radar was originally extensively used in the early development of radar technology to detect moving targets against stationary background clutter. A block diagram of a coherent radar system is shown in Fig. 5. If one were to replace the term optics for the boxes shown in the figure with a general "transmission line" term, one could not distinguish whether the block diagram of Fig. 5 represented a coherent microwave or a coherent laser radar system.

Coherent radar systems measure the Doppler shift of the echo radiation by comparing the frequency of the received echo signal with the frequency of the transmitted radiation. This comparison is accomplished by heterodyning (i.e., mixing) the returned signal with the signal from the frequency reference of the system (i.e., called the local oscillator) on a detector. If one maintains the frequency of the transmitter signal at a fixed value either above or below the local oscillator signal by electronic control circuits and then superimposing on the detector the local oscillator signal

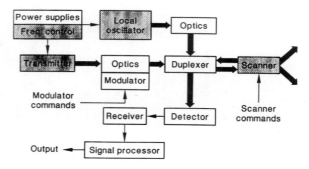

Fig. 5. Block diagram of a coherent laser radar.

with the return signal from a stationary target, an interference pattern is generated on the detector which amplitude modulates the laser radiation falling on the detector. This "beat" signal is equal to the difference between the frequencies of the transmitted and local oscillator signals. Since this beat signal is arranged to occur within the radio frequency range (i.e., tens of MHz to 100's of MHz), electronic amplifiers tuned to this frequency can be utilized to obtain the same signal-to-noise benefits well known in conventional heterodyne radio receivers. Measurements of the deviation of this known "beat" signal provides the Doppler shift and thus the speed of the target.

An additional advantage of the coherent radar system is that one can increase the local oscillator signal power on the detector to achieve the theoretical detector performance which is the quantum-noise-limited sensitivity.

A laser radar, utilizing CO_2 lasers, typically operates in the 10.6 micron wavelength region[8]. At this wavelength, HgCdTe detectors are presently optimum from a sensitivity viewpoint. A figure of merit for detectors is usally given in terms of noise-equivalent power or NEP. Heterodyne NEP's of $2x10^{-19}$, $5x10^{-18}$, and $2x10^{-17}$ W Hz have been measured with HgCdTe detectors operating at 1 GHz at 77°K, 195°K, and 300°K, respectively.

Table I makes a comparison of some of the relevant parameters between an x-band and a CO_2 laser radar. One notices a 3000 times higher operating frequency or a 3000 times shorter wavelength for the CO_2 laser radar when compared with an x-band radar. The large difference in wavelengths between the CO_2 laser radar and the x-band radar results in large differences in reflection characteristics of targets for the two technologies. Variation in target surface dimensions (i.e., surface roughness) typically are greater than the wavelength for CO_2 laser radar, but less than the wavelength for x-band radars. Since man-made targets usually have smoother surfaces than natural targets, even small man-made targets such as wires have a larger cross-section than natural targets for laser radars. Figure 6 presents detector signal current to noise current ratio (i_s/i_n) as a function of range for various natural and man-made targets irradiated with a pulsed, coherent CO_2 laser radar system having 400mW average power and a HgCdTe detector.

Since the beam divergence varies directly with wavelength and indirectly with transmitting aperture, CO_2 laser radars have 3000 times smaller beam divergence for the same aperture than x-band radars. Since the Doppler shift varies inversely with wavelength, the CO_2 laser radar has three order of magnitude higher Doppler sensitivity than an x-band radar. For a radar frequency of 30 THz (i.e., the CO_2 laser frequency), a target moving at 0.5 km per hour (about 1/10 the speed of a person walking) yields a Doppler signal of 100 kHz, whereas a 30 GHz microwave radar would yield a Doppler signal of 10^2 Hz.

Since the photon energy of the CO_2 laser radar is 3000 times higher than for the x-band radar, the laser radar beam has 3000 times fewer photons per unit of energy than the x-band radar. If one photon in unit time is the

Table I. Comparison of basic radar parameters

	CO_2 laser radar	X-band radar
Frequency, Hz	3×10^{13}	10^{10}
Wavelength, cm	10^{-3}	3
Beamwidth, λ/D, radians	10^{-3}/dia.	3/dia.
Doppler sensitivity, $2V/\lambda$, Hz	2,000 x velocity	2/3 x velocity
Photon energy, Joules	2×10^{-20}	6.6×10^{-24}

minimum detectable signal, then the operation of a CO_2 laser radar is limited to smaller field of view coverage than the x-band radar for the same trans-mitted power. Consequently, the laser radar is not well suited for wide area search applications. Laser radars are well suited for applications requiring ultra-high sensitivity in range, azimuth, Doppler, image resolution, and saml1 field of view.

It is important to point out that laser radar suffers from poorer pro-pagation characteristics through the atmosphere than convention microwave radar because of higher backscatter from rain, snow, haze, fog, and higher absorption from water in the atmosphere. Consequently, in the earth's atmo-sphere, laser radars have shorter range than microwave radars. Fortunately the wavelength of CO_2 lasers falls within one of the best atmosperic windows when compared with other laser wavelengts. Consequently, for applications within the atmosphere, the relative long wavelength of 10.6 micron for CO_2 lasers over other lasers such as Nd^{3+}:YAG, ruby, semiconductor lasers, etc.,

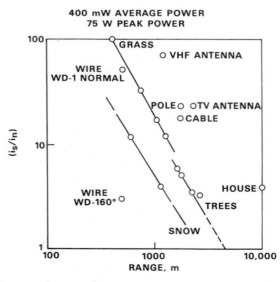

Fig. 6. Signal-to-noise ratio of a CO_2 laser radar for various natural and man-made targets.

makes coherent CO_2 laser radar the optimum coherent system of choice for most applications.

Figure 7 provides general areas of applications of various radar technologies from an angular resolution and field of view prospective. Ladar is a commonly used acronym for laser radar and it was formed from laser detection and ranging following the example of the word radar which was originally an acronym formed from radio detection and ranging.

ELECTRIC CABLE INSPECTION

When it was established that polyethylene (PE) and cross-linked PE (XLPE) material had intrinsically high dielectric strengths, on the order of 800 kV/mm, the electric power industry anticipated 40 years life for underground electric power cables for distribution systems fabricated from such materials. Consequently, in the 1960's, the electric power industry began to make extensive usage of underground cables using PE and XLPE as the dielectric between the inner and outer conductors. This expected life was not achieved even at average stress levels of 2 to 4 kV/mm even though such stress levels provided 200 to 400 less voltage gradients than the intrinsic dielectric strength of the material would allow. The failure rate for cables put into service since the 1960's reached a level that became disturbing to the electric power industry and lead to the Electric Power Research Institute, the Department of Energy, and cable manufacturers to launch an R&D program to determine the causes and find solutions to the premature failures of EP and XLPE cables.

The cables are produced in a continuous operation. A central conductor of stranded copper wire passes through an extruder which coats it with a smooth thin semiconducting shield consisting of carbon-black filled PE. Over this opaque semiconducting surface, the white PE insulation is extruded and then crossed linked with either heat, UV, or e-beam bombardment.

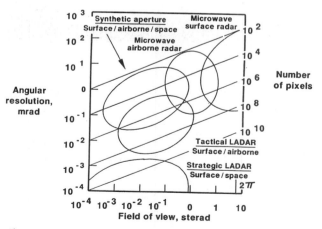

Fig. 7. General areas of applications of various radar technologies from an angular resolution and field of view perspective.

The R&D programs indicated that the aging progress of the insulation generates branched channels caused by dielectric breakdown which in turn causes an electrical short circuit between the outer ground conductor and the inner conductor (see Fig. 8). The dielectric breakdown branched channel structure in the insulation, now called "trees" in the industry, is believed to be caused by surface imperfections at interfaces within the cable and by irregularities within the insulating materials. Irregularities within the dielectric are caused by gas or vapor filled voids, contaminating particles, inhomogeneous density variation within the materials, etc. All of these general forms of defects are caused during the manufacturing process. Unfortunately, visual inspection is not possible during the manufacturing process because polyethylene is normally a milky white opaque material except when immersed in hot oil.

One present form of nondestructive inspection techniques involve corona testing which can detect 50 micron diameter voids in 500 ft. length of cable. The disadvantage of the technique is that it cannot detect contaminants and flaws, voids filled with liquids or vapors, and microvoids of 1 to 10 micron diameter. The technique also does not provides on-line inspection during the manufacturing process nor can it locate the position of the defect.

The present commonly used inspection procedure is to cut out a 2 inch piece of manufactured cable every 10,000 feet, slice it into 1/2 inch portion portions, slicing these 1/2 inch portions into 0.020 inch thick wafers, making these wafers transparent by immersing them in hot oil, and then vis visually inspecting the wafers under a 15 power microscope. The obvious disadvantages of this technique are that it is not an on-line, real-time inspection process nor is it a nondestructive testing procedure.

One of the promising approaches for developing an on-line, real time, nondestructive inspecting technique is based on the fact that PE and XLPE are relatively transparent in the far infrared. Consequently, it appeared fruitful to investigate far infrared lasers[9] for continuously monitoring the quality of electric power cables in real time in a nondestructive manner[10].

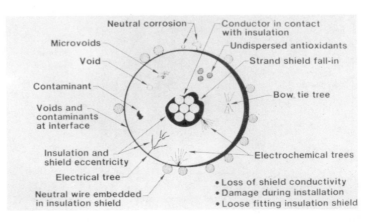

Fig. 8. An illustration of common causes of electric power cable failures.

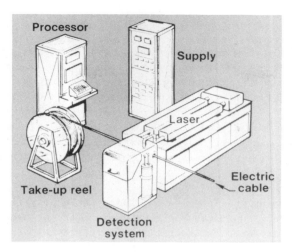

Fig. 9. Schematic of a far infrared laser system for electric power cable insulation inspection.

Figure 9 is a sketch of the operation of such a unit in a cable factory. Far infrared laser radiation is scanned around the cable before the outer ground conductor and its protective coating are extruded onto the cable. The laser radiation scattered by voids, contaminants, etc., within the dielectric is collected, detected, and its magnitude digitally recorded. The speed of the cable through the system is monitored and also recorded so a complete record of signal amplitude caused by the scattered radiation versus cable position is maintained. Figure 10 illustrates typical signals obtained at 118 micron laser wavelength and 0.1 W laser power with a liquid helium cooled Ge doped silicon detector.

It is still too early to determine the practical impact far infrared laser dielectric inspection instrumentation is going to have on the factory floor of the electrical power cable industry, but it appears safe to predict at this point in its developmental cycle that the technology will provide very useful research instrumentation for the industry.

25 KV Moving Cable

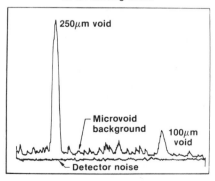

Fig. 10. Signals of a far infrared laser inspection system from voids in cross-linked polyethylene insulation used in electric power distribution cable.

CONCLUSION

Laser technology is young and robust, with a highly promising and exciting future. It is now spawning new products and opening major new segments of basic industries that will ensure its growth well into the next century. The fields of fiber optic telecommunications; optical audio and video disks; optical data storage; opto-electronics; lasers for material working (cutting, welding, heat treating, hole drilling, and scribing); laser applications in medicine; laser instrumentation. and military applications are still in their infancy, so considerable growth is yet to come.

The most serious challenge facing laser technology is the continuing shortage of photonic engineers required to develop the numerous new and rapidly evolving products the technology is generating, and to continually advance the state of the art required to meet new product needs, and to work at the interface between electronics and photonics technologies. An engineer in this field needs a background in optics and electronics and in quantum electronics. Most engineers working in the field today are either physicists who have learned some electronics or electronic engineers who have learned some optics. The offering of a formal undergraduate engineering curriculum in photonic engineering would be a big boost to this important emerging field of technology.

REFERENCES

1. A. J. De Maria, Optics News 10:15, 1985
2. P. H. Abelson, Editorial; AAAS Science Journal, March 1977
3. J. G. Posa, Electronics, 28, July 1981, pp. 117-120
4. R. T. Smith, Electronics, 28, July 1981, pp. 131-134
5. M. Banas Conrad and R. Webb, Proceeding of the IEEE, Vol. 70, June 1982, p. 556
6. P. F. Dumal and C. M. Banas, 1983 ASM Conference on Applications of Lasers in Material Processing, Los Angeles, CA, 24-26 January 1983 Reprint 8301-020 from American Society of Metals
7. A. J. De Maria, Proceedings of the IEEE, Vol. 61, June 1983, pp. 731-748
8. B. B. Silverman, Proceedings of the IEEE 1982 National Aerospace and Electronics Conference, Vol. 2, pp. 569-575
9. T. Y. Chang, T. J. Bridges, E. G. Burkhardt, Applied Phisics Letters, Vol. 17, September 15, 1970, pp. 249-251
10. A. J. Cantor, P. K. Cheo, M. C. Foster and L. A. Newman, IEEE Journal of Quantum Electronics, Vol. QE-17, April 1981, pp. 477-489

INTERACTION OF PULSED LASER BEAMS WITH SOLID SURFACES:

LASER HEATING AND MELTING

A. M. Malvezzi

Division of Applied Sciences, Harvard University

Pierce Hall, 29 Oxford Street, Cambridge, MA 02138, USA

INTRODUCTION

Experiments on laser interaction with solid surfaces have been performed since the development of laser sources. It was readily recognized that heat was efficiently generated within an absorption depth from the surface. In this way extremely high temperature could be generated with little energy, leading to melting, evaporation and ultimately ionization of thin layers. These phenomena are exploited in a great variety of ways in many fields of science[1-9] and technology. Laser cutting and drilling and surface heat treatment find growing industrial applications. Laser annealing of semiconductors has opened new possibilities in the fabrication of highly controlled high speed electronic components. Rapid resolidification (thermal quenching) after pulsed laser melting of metals and compounds has provided a mean of obtaining new exotic materials. This field of studies has taken full advantage of progress in ultrashort laser pulse generation, high temporal resolution being mandatory in separating these effects occurring in a wide range of time scales.

The scope of this paper is to provide a general account of the microscopic mechanisms responsible for the transformation of laser radiation impinging onto a solid surface into heat. The discussion is restricted to strong absorbers, thus disregarding non linear electron excitation mechanisms. Intermediate laser intensities are here considered, sufficiently low to avoid ionization and plasma generation on the surface of the sample. The characteristics time scales range from femtoseconds to nanoseconds. Also, the hydrodynamical effects related to strong vaporization and material removal from the irradiated area are not considered here. Some relevant aspects of the energy path from the optically coupled electronic states to the lattice are presented. The discussion deals first with the elementary mechanisms occurring at the fastest rates. The presence of electron hole plasmas in semiconductors and their relevance in the energy transfer process is stressed. Techniques based on the transient optical properties of materials under laser illumination are discussed in some details. Also, some examples of numerical simulation of the interaction in the various temporal regimes are described. For the treatment of the basic notions of solid state physics and of the optical properties of matter in general, the reader is referred to standard textbooks[10-12]

PRIMARY INTERACTIONS IN METALS AND SEMICONDUCTORS: ELECTRON EXCITATION AND THERMALIZATION

Absorption of light by matter is well described and understood both in terms of classical models as well as via quantum mechanical treatment. They provide accurate explanations for the optical properties of solids, both metal and insulators. The classical model of Lorentz for the optical response of bound electrons in solids leads to describe the effects in terms of a complex dielectric functions which fully accounts for the effects of the lattice. In the same framework, oscillations of the ions, described by an atomic polarizability and phonon effects can be treated with fair accuracy. Conduction electrons in solids are similarly described by the classical model of Drude which accounts for many of the effects related to the outermost, less tightly bound electrons in a metal. The model can be applied with success, as it will be apparent below, also to free carriers in semiconductors. It is surprising that the extremely simplified hypotheses underlying these models are still well applicable to situations far out of equilibrium. The main result of these approximate models is given by the frequency dependent complex dielectric constant which adequately describes the optical response of a solid interacting with an electromagnetic field at frequency ω:

$$\hat{\varepsilon}(\omega) = \varepsilon_\infty(\omega) - \frac{4 \, Ne^2}{m^*} \frac{1}{\omega^2 + i\omega\tau} \tag{1}$$

The first term $\varepsilon_\infty(\omega)$ which in the visible part of the spectrum is in general real describes the optical response of the "ions" (bound electrons plus nucleus) in the lattice and their oscillations (displacement polarizability in ionic crystals). The second term describes, in metals, the optical response of their conduction electrons. In semiconductors two such terms exist, one for electrons in the conduction band and one for holes in the valence band when an electron - hole plasma is present. N is the electron (carrier) density, m^* is the optical reduced mass which approximates the free electron mass with corrections related to the curvature of the bands, τ is the Drude relaxation time accounting for the presence of carrier collision. Its value is in the 10^{-14} to 10^{-15} rs range. The optical properties of solids are then determined by the complex index of refraction $\hat{n}$:

$$\hat{n} = \sqrt{\varepsilon} \equiv n + j \, K \tag{2}$$

which is related to the reflectivity R of the surface of the material through the Fresnel formulas. In the case of an isotropic material in vacuum, light impinging at normal incidence,

$$R_n = \frac{(n-1)^2 + K^2}{(n+1)^2 + K^2} \tag{3}$$

and K is related to the absorption coefficient α of the medium by

$$\alpha = \frac{4 \pi k}{\lambda} \qquad (4)$$

being λ the wavelength of the radiation.

The interaction of visible photons with matter proceeds mainly via electronic excitation. A notable exception occurs in the long wavelength region, where direct coupling with optical phonons may be possible due also to the matching of the energies involved.

In metals, electrons are excited by visible radiation via inelastic free-free transitions. Simultaneous energy and momentum conservation in the interaction of a photon with an electron require the presence of a third particle, i.e. phonons. These are also referred as intraband transitions, as illustrated in figure 1a). Above a threshold value, however, direct (i.e. without phonon assistance) interband transitions become energetically possible. These can be originated either from deep lying bands or at the Fermi energy. Since the threshold energy for direct interband transitions is in the visible, they are mainly responsible for the color of metallic surfaces.

In semiconductors and insulators interband transitions occur when the photon energy $h\nu$ exceeds the band gap energy E_g. Light is absorbed within a thickness $d \ll \alpha^{-1}$ being α the absorption coefficient. Momentum conservation in indirect band semiconductors is satisfied by the emission or absorption of a phonon. However, due to the small phonon energy as compared to the one of the photons, very little energy is provided at this stage to the lattice. The main photon absorption mechanisms are sketched in figures 1b) and c) for direct and indirect band semiconductors. A plasma of electrons and holes may also be generated when $h\nu < E_g$ by multiphoton absorption at sufficiently high laser intensities I. The absorption coefficient $\alpha_{nI} = \beta_n I^{n-1}$ becomes dependent on the laser intensity (W/cm^2), n being the number of photons necessary to bridge the gap. When a sufficient density N of carriers is accumulated in the conduction band, free carrier absorption, $\alpha_{fca} = \beta N$, and intraband transitions become operative and the interactions proceeds as in metals. At very high laser intensities, multiphoton absorption can directly ionize the surface. Also, ionization of impurities and, at extremely high intensities, electron tunneling in half a light cyrcle generate free electrons. In this case, a plasma on the solid surface is formed which absorbs light and promotes further ionization in the material.

ENERGY COUPLING TO THE LATTICE

The electron excitation process by laser pulses populates selectively the electronic states located in narrow energy intervals in unoccupied bands. A nonthermal hot electron (electron-hole) distribution is thus first generated in metals (semiconductors).

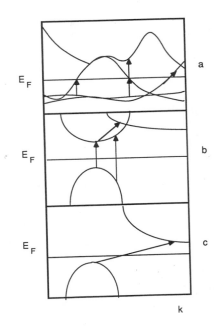

Fig. 1. Schematics of photon absorption processes in metals (a), direct
band semiconductors (b), and indirect band semiconductors.
Oblique arrows indicate phonon assisted transitions.

In metals, the photoexcited hot electrons rapidly thermalize by in-
terparticle collisions and plasmon coupling, giving rise to a hot electron
thermal distribution at a temperature T_e. Due to the small electron heat
capacity, very high temperatures are easily reached with modest laser
fluences in the femtosecond and picosecond domain. Since the energy relaxation time to the lattice is larger than the time required for electrons
to thermalize, these very high electron temperatures T_e, of the order of
several thousands °K may be maintained for short times. The hot electron
gas may give rise to blackbody radiation emission in the visible. Once a
thermalized electron distribution for the electron is achieved at a temperature T_e, energy transfer to the much cooler lattice via electron phonon
coupling occurs. The transfer can be described by the following set of nonlinear coupled equations for the electron and lattice temperatures T_e and
T_1

$$c_e(T_e) \frac{\partial T_e}{\partial t} = -g (T_e - T_1) + S(t) \tag{5}$$

$$c_1(T_1) \frac{\partial T_1}{\partial t} = +g (T_e - T_1) \tag{6}$$

Here c denotes the specific heat, that for electrons is proportional
to T_e, and S the hot electron generation rate by the laser pulse. Thermal electron conductivity has not been considered in this time scale. The
electron phonon coupling constant g is expressed, assuming a thermal popu-

lation distribution for the phonons, by[13]

$$g = \frac{p^2 \, m_e \, N_e \, u_s^2}{6 \, \tau_r \, T_e} \tag{7}$$

where m_e, N_e are the electron mass and density, u_s is the sound speed and τ_r is the intercollision time for the electrons at temperature T_e, which is proportional to $1/T_e$[14]. It is the equivalent of the classical Drude relaxation time. The coefficient g is thus independent from electron temperature T_e in metals. Typical values for g are in the 10^{11} W/cm^3 °K range. The source term S(t) is given by

$$S(t) = (1 - R) \, \alpha \, I(t) \tag{8}$$

if a sample of total thickness $d \ll \alpha^{-1}$ is considered . R is the reflectivity of the surface at the excitation wavelength and I(t) represents the laser intensity (W/cm^2). As an example, a numerical solution of the coupled equations (1-2) is illustrated in Fig. 2 for a 10 ps visible laser of 2 x 10^{-3} W/cm^2 impinging on a silver thin film. The electron temperature T_e is seen to exceed the lattice temperature T_l by several hundred degrees during the laser pulse. This is due to the difference in specific heat for free electrons and for the lattice, $c_e \ll c_l$, which determines time constants $\tau_e \cong c_e/g$ for the electron to cool shorter than $\tau_l \cong c_l/g$ for the lattice to be heated. Obviously, the whole process is critically dependent on the value of the coupling constant g.

To present, non thermal transient electron distributions in solids have not been observed. However thermal electron distributions at a temperature T_e far in excess of the ones of lattice have been observed through photoemission[15] reflectivity thermomodulation[16], and transmission experiments[17]. Direct observations of the electron temperature in gold film irradiated by 65 fs laser pulses were performed by Schoenlein et al.[18] by monitoring the electron population in the vicinity of the Fermi energy E_F. This is done by probing the reflectivity of the surface at a photon energy corresponding to transitions from the d-bands to the room temperature Fermi energy of gold. An increase in the temperature corresponds to a lower occupancy of the states immediately belw E_F. This corresponds in turns to a decrease of the reflectivity. The opposite is true for states immediately above E_F. Using this technique, transient electron temperatures in excess of 1000 °K are measured with a total lattice heating of few tens degrees.

Contrary to metals, excited states are described in semiconductors in terms of electron and hole density distributions in the conduction and valence band, respectively. Population distributions in these "quasi - equilibrium" states are described by separate "quasi" Fermi energies for electrons and holes and by different carrier temperatures T_e and T_h, respectively. Note that, in contrast to the metal case, due to non conservation of the total number of particles, the quasi Fermi levels are mobile, their splitting increasing with carrier concentration. In semiconductors, photoexcited carriers decay from the optically coupled states, giving rise at

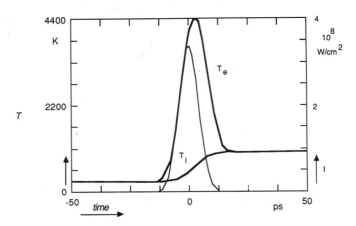

Fig. 2. Calculated time evolution of electron, T_e, and lattice, T_1, temperatures in a thin metal film excited by a 10 ps laser pulse (heavy lines, scales at left) and laser pulse intensity (light line, scale at right)

first to a thermalized hot carrier distribution, characterized by a temperature T_c. Different temperatures are reached for electrons and holes, $T_e \neq T_h$, due to the different values of their reduced masses. The primary scattering channels out of the optically coupled states are carrier carrier collisions, plasmon production and phonon emission. The first two processes do not involve energy transfer to the lattice, thus they redistribute the energy within the carrier gas. Phonon emission, on the other hand, promotes lattice heating.

Intercarrier collision rates are effective in redistributing the energy among carriers in extremely short times. Equal temperatures $T_e = T_c$ for electrons and holes are readily achieved. Their scattering rates increase with carrier density. However, when the screening distance $1/q_0$ exceeds the intercarrier distance $1/N^{1/3}$ at a carrier density N, the binary collision picture breaks down and energy is transferred to plasmons, i.e. longitudinal collective excitations carrying quanta of energy h $(4 \pi N e^2/\varepsilon_\infty m^*)^{1/2}$. Under equilibrium conditions the screening parameter q_0 is given by[19] $q_0^2 = 4 \pi N e^2/\varepsilon \kappa T_c$. It is questionable, however, whether this expression is still meaningful in the presence of quasi monoenergetic electron distributions. Contrary to the case of metals, where an ab initio free electron density exists, the buildup of carrier population during the first phase of the interaction make the plasmon energy sweep through the values for phonons and, at high densities, for the band gap. Each of these resonances provides enhancement in the corresponding coupling. Carrier relaxation via plasmon emission occurs when the kinetic energy of the carriers exceeds the plasmon energy. At high carrier temperatures, plasmon rapidly decay into single particle excitations. These cascade mechanisms are extremely efficient in thermalizing the carrier gas at high densities. The transition to a hot thermal carrier distribution is completed within few tens of femtoseconds.

Among the electron phonon scattering processes, the decay of the carrier energy via longitudinal optical phonons appears to dominate at low carrier densities[20]. Decay constants for polar optical phonons have been measured by monitoring the temporal evolution of the phonon population through time resolved Raman scattering in GaAs crystals by several groups groups[21,22], in the 10^{17} cm^{-3} carrier density regime. Values of $\tau_{LO} \cong 0.15$ ps have been obtained. These same experiments indicate that phonons reach an equilibrium distribution in several picoseconds in this density regime. The excited phonons of the main decay channels distribute their energy among other modes via anharmonic interaction. At high carrier densities the phonon thermalization process is speeded up. For n $\cong 10^{20}$ cm^{-3} relaxation times of $\cong$ 200 fs have been deduced for the optical phonons. At high excitation levels, when the energy of the carriers is sufficient to overcome the energy difference between valleys, intervalley scattering via deformation potential further promotes carrier relaxations. Large momentum phonons are involved in the process. In GaAs this process is an order of magnitude faster than optical phonon scattering. These electron − phonon scattering mechanisms are modified at higher carrier concentrations mainly by screening effects. This effect should delay the energy transfer process, though lack of femtosecond data for III-V semiconductors at high excitation levels preclude at present a more quantitative picture. However, the extensive investigations in the picosecond domain indicate an extremely fast energy relaxation to the lattice. The reduction in carrier phonon scattering rates at high excitation is thus expected to be modest.

Since the rates for carrier − carrier and carrier − plasmon scattering increase with carrier density, whereas carrier − phonon scattering is intrinsically density independent, the former processes will dominate at high carrier concentrations. Intra and inter carrier thermalization will occur before a substantial fraction of the stored energy is delivered to the lattice. The emerging picture from the available ultrafast optical experiments[23] is that carriers rapidly evolve from a monoenergetic distribution at the optically coupled levels to a hot thermal distribution, $T_e = T_h = T_c \gg T_1$ which then relaxes to a cooler distribution in equilibrium with the lattice, $T_c = T_1$. At high excitation levels carriers are essentially at first a classical gas, owing to the extremely high temperatures (up to few eV). In the cooling process more and more carriers accumulate at the bottom of the respective bands. The gas then becomes degenerate and effects related to the occupation statistics become important.

In indirect band semiconductors, light being absorbed mainly by phonon assisted interband processes, carrier excitation occurs over a relatively wide energy region of the order of $\hbar\omega - E_g$. On the contrary, in direct gap materials, direct interband absorption selectively populates well defined states in conduction and valence bands at low excitation regimes. In these materials, when illuminated by femtosecond pulses, saturation of the optically coupled states occurs at relatively low laser energies. The rate of decay from the excited states is less than the rate of resonant photons provided by the laser. In these conditions, a sufficiently thin sample will become transparent to the incoming laser radiation. This bleaching in the femtosecond time regime has been exploited to monitor the scattering rate out of the excited states distribution in GaAs and

AlGaAs by transmission correlation experiments and time resolved transmission experiments.

The first technique[24] uses two equal intensity, orthogonally polarized, collinear pulses with the target at variable relative delay times. The autocorrelation signal observed in transmission may detect temporal relaxations shorter than the duration of the exciting pulses themselves. A 40 fs decay constant in GaAs, independent of carrier density up to the maximum available density of 10^{19} cm^{-3}, has been attributed to intervalley scattering off the optically coupled states[25]. A two decay constant behavior in the time resolved transmission response of thin GaAs and $Al_x Ga_{1-x} As$ thin samples under 35 fs, 2 eV laser pulses has been observed in pump and probe experiments by Lin et al.[26] at carrier densities ranging from 10^{17} to 10^{18} cm^{-3}. The instantaneous transmission of the samples, probed with a variably delayed sampling pulse at the same frequency but with orthogonal polarization to the pump pulse, exhibits a sharp peak in correspondence of the excitation. Its decay becomes progressively faster with increasing pump fluence. Measured values of the first time constant are in 10 – 30 fs range. The peak is followed by a longer ($\tau = 1.5$ ps) decay profile which is insensitive to the fluence level of the pump. Thus, one is led to identify the first transmission peak as due to (density dependent) carrier – carrier or carrier – plasmon scattering. The slower decay of the transmission is consistent with a carrier-phonon scattering (both intervalley and optical phonon scattering). Carrier thermalization and transition to "cold" carrier distribution has been clearly resolved in this excitation regime. A useful variation of the pump and probe technique is to use broadband probing pulses generated by continuum generation in liquids. Moreover, tuning the pump frequency in GaAs just above the band gap energy so that the excess carrier energy is insufficient for phonon emission, intercarrier scattering has been studied[27]. Since the pump and probe pulses originate from mutually incoherent, though temporally synchronized sources, coherence coupling effects in the resulting spectra are avoided.

Another aspect of the interaction here discussed is hot electron transport phenomena. It has been speculated in the past that photoexcited carriers could travel considerable distances before releasing their energy by scattering to the cold environment[28]. Indirect evidence of this effect was obtained in UV picosecond excitation of silicon[29]. An effective absorption coefficient $\alpha/(1 + L\alpha)$, L being a diffusion length for hot electrons, had to be introduced in order to explain the observed values of the critical laser fluence for surface melting. Very recently, femtosecond experiment on metals[30] specifically addressed to this problem indicate extremely fast energy transport by electrons in metals. The notion of local energy deposition in a thickness of the order of the absorption length of the exciting radiation would then need revision in the ultrafast interaction regimes.

EXPERIMENTS ON HEATING AND MELTING

The energy transfer from the electronic system to the lattice may pro-

162

duce sufficient energy to heat, melt and vaporize the sample surface in short times. The quantitative understanding of laser induced melting has been of central importance in the development of this field of investigation and has triggered a wide experimental effort to establish the relevant time scales of the process.

Many semiconducting materials become metallic upon melting. This naturally suggests to monitor the onset of the phase transition by optical means. Transient reflectivity measurements on silicon under nanosecond irradiation indicated that melting of the surface occurs within the laser pulse[31]. This result confirmed the morphological studies on nanosecond irradiated samples indicating a rapid propagation of a melting front towards the bulk, followed by a solidification front back to the surface. The high velocities of the latter, which have been extensively investigated, are responsible for the formation of metastable structures and impurity segregation at the surface[32].

Clearly, a more stringent test to the notion of fast energy release from the electronic system to the lattice comes from picosecond irradiation of solid surfaces. Picosecond time resolved measurements of the reflectivity and/or transmission provide the required information. By using appropriate probe frequencies, the optical response of the irradiated material can be made more sensitive to the lattice temperature or to the carrier density. The situation is illustrated in Fig. 3 where the reflectivity of crystalline silicon surfaces at 0.53, 1.06 and 2.8 μm pulses is plotted as a function of the 0.53, 20 ps pump beam fluence. The data are taken 15 ps after the peak of the heating pulse. Both visible and near infrared probing wavelengths show a sudden jump of the reflectivity at the critical pump fluence of 0.2 J/cm^2, where the formation of amorphous spots is first observed. The high reflectivity values correspond to the reflectivity of molten, metallic silicon. At lower fluences, however, different behaviors are observed. In the visible a slight reflectivity increase versus pump

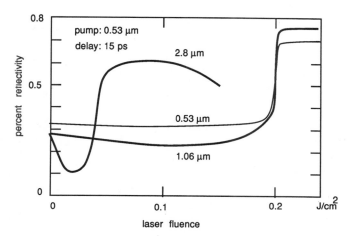

Fig. 3. Reflectivity of a crystalline silicon surface at the wavelengths indicated as a function of the fluence of 0.53 μm, 20 ps pump pulses, taken 15 ps after the peak of the pulse.

fluence indicates the heating of the lattice. No apparent effects due to the presence of the electron hole plasma are present at this wavelength. These are more evident at 1.06 μm as a decrease of reflectivity with pump fluence. They are due to the negative Drude contribution to the real part of the dielectric constant caused by an excess of carriers in the vicinity of the surface. At 2.8 μm the Drude contribution exceed the bulk value of the dielectric constant and its real part becomes negative. Plasma resonance conditions are reached at ~ 40 mJ/cm^2, i.e. at a fraction of the fluency required to melt the silicon surface[33].

A way to observe laser induced surface melting is by exploiting the nonlinear optical response of non-centrosymmetric crystals, such as GaAs[23]. Contrary to silicon, which has a centrosymmetric structure, the bulk second harmonic efficiency exceeds by several orders of magnitude the surface contribution arising from the symmetry breaking effect of the material-vacuum interface[34]. The second harmonic signal in reflection from GaAs <1 1 0> surfaces (see Fig. 4), is indeed a quadratic function of the input fundamental intensity. When the fluence of the latter reaches the critical value for surface melting, however, a saturation of the second harmonic signal is observed, which indicates a decrease of the total generation efficiency at the center of the illuminated area. A transition to a disordered, centrosymmetric phase has occurred. Taking into account the gaussian spatial profile of the beam, which acts both as excitation source and as fundamental frequency input, an upper limit of 2 ps to the onset of a disordered liquid phase is deduced as soon as the critical fluence for melting has been locally delivered to the surface. Thus, a single pulse experiments provides convincing evidence for ultrafast surface melting in the picosecond time scale.

Charge emission measurements from picosecond irradiated semiconductors are also sensitive to lattice heating. Due to the extremely small escape length of the electrons from the solid, interpretation of the data is non affected by the spatial gradients present inside the material. Electron emission from a material irradiated by intense laser pulses originates from several mechanisms. In the case of a semiconductor, such as silicon, irradiated by UV picosecond laser pulses[29], electrons can be emitted by single or multiple photon excitation processes, with intermediate levels being either virtual or real states, or by thermoionic emission. The latter effect originates when the electron energy distribution is so hot that the Boltzman tail exceeds the work function of the material. Experimental results such as the one illustrated in Fig. 5[35], set an upper limit to the electron temperature averaged over the laser pulse duration via the Richardson-Dushman relation, if one assumes that all the observed charges are indeed generated through thermoionic effect. At the critical fluence for melting, the measurement of Fig. 5 set this limit to about 3000 °K. During the laser pulse, therefore, most of the laser energy stored in the electron gas has already been dissipated to the lattice. Moreover, the sudden appearance at F_{th} of a positive ion signal shows the presence of a hot surface layer.

That ultrafast surface melting by picosecond laser irradiation is intimately related to a hot lattice has been further demonstrated by time

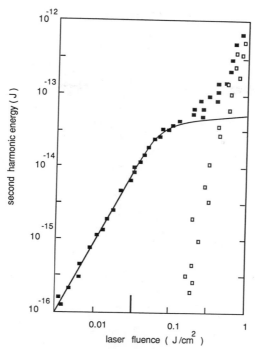

Fig. 4. Energy of the specular emission at 0.266 μm vs. incident 0.532 μm, 20 ps, laser fluence incident on a < 1 1 0 > GaAs surface. The collected second harmonic signal (full squares) refers to the central portion of the excited area. Open squares refer to spurious fluorescence radiation. The second harmonic energy (solid line) calculated assuming instantaneous surface melting is shown.

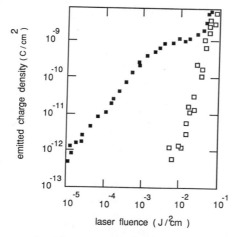

Fig. 5. Emitted charge density versus incident 0.266 μm, 12 ps, laser fluence on crystalline silicon. Full squares: negative charge. Open squares: positive charge. The threshold fluence for surface melting is at 0.03 J/cm².

resolved lattice temperature measurements in thin crystalline silicon on sapphire (S.O.S.) samples[36]. The variation of the complex index of refraction in the visible is deduced from simultaneous time resolved reflectivity and transmission measurements. At sufficiently long delay times after the exciting laser pulse, the electron–hole plasma contribution to the optical constants is negligible and optical measurements solely depend on the lattice temperature across the sample. The results are then compared with the measured optical constants of silicon versus temperature and surface and average temperatures across the sample versus incident laser fluence are deduced. Contrary to the results of early Raman scattering experiments[37], the lattice temperature obtained in this way are consistent with the attainment of the melting temperature of 1680 °K at the critical fluence value of 0.2 J/cm^2 found experimentally.

Optical detection of surface melting in metals is not as sensitive as in semiconductors, due to the very small decrease in reflectivity associated with the phase transition. This has however been observed with picosecond time resolution[38], confirming that heating and melting of metal surfaces is accomplished in picoseconds.

These experimental results provide direct evidence of:

1) melting of the surface in picoseconds after the proper fluence hase been delivered to the surface, and
2) heating of the lattice up to the melting point in the same scale.

This body of experimental evidence, nowaday widely accepted, points toward ultrafast energy transfer mechanisms from the photoexcited electron gas to the lattice, melting being achieved by purely thermal means even in the picosecond time domain.

HEATING OF METALS

Since thermalization occurs between electrons (carriers) and lattice in few picoseconds in metals (semiconductors), excitation of solids with longer laser pulses results in heating of the material in a depth d of the order of the absorption length $1/\alpha$. Melting may be achieved in this time scale, once the latent heat for fusion has been delivered to the material.

With picosecond and longer pulses, therefore, equilibrium conditions between electrons and lattice can be assumed in describing the evolution of the excitation at subsequent times. This approach, generally referred to as thermal model, has been highly successful in predicting experimental results in the picosecond and nanosecond laser irradiation experiments[39,40]. A one dimensional treatment of the problem is appropriate when the dimension of the illuminated area on the surface is much larger than the characteristic depths of the problem (absorption and diffusion lengths).

In metals, equation (6) for the lattice temperature $T_1(x,t)$ across the depth x of the illuminated sample is modified to take into account mechanisms occurring in pico- and nanosecond time scales

$$c \, \frac{\partial T_1}{\partial t} = S(x,t) + \frac{\partial}{\partial x} \, k \, \frac{\partial T_1}{\partial x} + Q_1(x,t) + Q_{ev}(x,t) \tag{9}$$

c is the specific heat per unit volume, k the thermal conductivity. The source term $S(x)=(1-R)\alpha \, I(t) \exp(-\alpha x)$, with R being the reflectivity of the surface and I the incoming laser intensity, implies an instantaneous energy transfer to the lattice, as opposed to the situation described by equation (5,6). The second term of equation (9) describes the effects of thermal diffusion toward the bulk of the sample. $Q_1(x,t)$ describes the rate of latent heat accumulated or released during the solid-liquid phase transformation. The kinetics of the interface can be thus deduced. It is important to note that in the interaction with picosecond laser pulses, the deviation from the melting temperature T_m of the interface temperature T_i can be several hundred degrees. The amount of everheating $(T_i > T_m)$ of the solid and subsequent undercooling of the liquid $(T_i < T_m)$ determine the velocity of the interface[41]. Resolidification velocities up to 100 m/s have been observed[38]. These are essential in determining the final structure and composition of the irradiated solid. Amorphous resolidification from crystalline substrates may occur when steep thermal gradients drive the interface at very high speeds (tens of m/s). Also, alloys may retain their liquid composition since there is no time for impurity segregation. The last term in equation (4) represents the heat loss at the surface of the sample by evaporation and describes the heat carried by atoms leaving the surface. Assuming equilibrium between the solid and the adjacent gas, evaporation can be evaluated through the Clausius Clapeyron equation. This term becomes significant only at long times and with high intensity, nanosecond heating pulses. Indeed, when the surface is heated above the melting point, evaporation during a picosecond laser pulse is likely to be negligible[42]. The recession of a hot surface due to evaporation at a temperature T during the laser pulse duration t_p is given by $d_{evap} = (\rho_g/\rho_s) \, v_{th} \, t_p$, with $v_{th} \sim (\kappa \, T/m)^{1/2}$ the component of the average thermal velocity of the evaporating particles normal to the structure and ρ_g, ρ_s the densities of the gas and solid phase, respectively. For a surface temperature of ~ 5000 °K, even with a density ratio $(\rho_g/\rho_s)=$ .01, the depth of evaporated solid matter amounts to a fraction of a monolayer in the picosecond regime. Essentially, thermally evaporating atoms cannot travel large distances in such short times. Therefore an optically thick gas layer cannot be formed in the picosecond time scale.

PLASMA KINETICS IN SEMICONDUCTORS

Contrary to metals, in semiconductors the energy transfer to the lattice is mediated by the electron hole plasma, where a substantial fraction of the total energy may be stored. It is therefore important to investigate the kinetics of the plasma and its characteristic recombination times. Within the framework of the thermal model, illustrated above for metals, the kinetics of the energy transfer is described in semiconductors by two coupled equations, carrier density $N(x,t)$ being the coupled variable to lattice temperature T_1. The rate of change of the carrier density $N(x,t)$ is expressed by source, ambipolar diffusion and recombination terms

$$\frac{\partial N(x,t)}{\partial t} = S(x,t) + D_a \frac{\partial^2 N(x,t)}{\partial x^2} - R(x,t) \tag{10}$$

where D_a is the ambipolar diffusion coefficient. Under the assumption of instantaneous relaxation of the electron hole plasma to equilibrium with the lattice, the energy $h\nu - E_g(T) - H \kappa T$ per absorption event is immediately transformed into heat. Here, $h\nu$ is the pump photon energy, $E_g(T)$ the lattice temperature-dependent gap energy and H ($H >= 3$) is the degeneracy factor which takes into account the occupation statistics of the bands in the presence of high carrier concentrations

$$H = F_{3/2}(\eta_e) / F_{1/2}(\eta_e) + F_{3/2}(\eta_h) / F_{1/2}(\eta_h) \tag{11}$$

H is derived in terms of Fermi integrals of order 1/2 and 3/2 evaluated at the reduced Fermi energy $\eta_{e,h} = E_F^{e,h} / k T$. Fermi energies are evaluated through the charge neutrality condition.

Further lattice heating occurs when the plasma ricombines. For each recombination event, the remaining fraction of the initial photon energy $E_g(T) + H \kappa T$ is dissipated into the lattice. Direct heating, which at low carrier densities is essentially driven by the laser pulse, is reduced at high concentrations by Pauli exclusion principle[43]. In a silicon electron hole plasma with 3×10^{20} carriers/cm^3 , for example, the quasi Fermi levels are pushed $\sim$ 230 and $\sim$ 150 meV above the bottom of the conduction and valence band, respectively. With visible laser excitation at 2.33 eV, only $\sim$ 34% of the photon energy results in direct heating. A substantial amount of energy is thus stored in the carrier system and is delivered to the lattice at the rate of the dominant recombination processes. At high carrier concentration direct heating of the lattice becomes progressively decoupled from the rate of input laser energy and delayed heating effects may become observable[44]. This is particularly true when the threshold laser fluence (J/cm^2) for surface melting is exceeded. The melting temperature will be approached in semiconductors at a rate characteristic of the plasma recombination processes and will not necessarily follows a fast excitation dynamics.

At high carrier densities Auger effect is known to play an essential role in limiting the carrier density. The recombination energy of an electron hole pair is transferred first to a third carrier. The assistance of a photon for momentum conservation is required in indirect recombinations. This has the effect of allowing transitions to a greater number of final states, thus partly overcoming rate reductions due to carrier degeneracy. This effect reduces the carrier density but preserve its energy content. The excited carrier then relaxes via phonon emission and instantaneously heats the lattice. The Auger relaxation time ($\tau_A \sim (\gamma N^2)^{-1}$) is strongly density dependent. Auger recombination is completely balanced by its inverse process, i.e. impact ionization, when the carrier density N_c reaches its equilibrium value $N_0(T_c)$ at the temperature T_c. However, the plasma looses energy faster than carriers, since the energy relaxation time towards the lattice is always shorter than the recombination rate. Therefore, the actual carrier density is always above its equilibrium value and Auger recombination dominates over impact ionization.

An example of calculated surface carrier density and lattice temperature versus time[42] is illustrated in Fig. 6 for a 20 ps, 0.1 J/cm^2 laser pulse illuminating a crystalline silicon sample. These results can be directly compared with experimental optical data by calculating the corresponding quantities and convoluting the results with the finite temporal profile of the probing laser pulse. Calculations also indicate for picosecond laser excitation maximum carrier densities of $\sim 10^{21}$ cm^{-3} for silicon and $\sim 9 \times 10^{20}$ cm^{-3} for germanium.

At the highest excitation regimes, a resonant condition may be reached when the plasmon energy $\hbar\omega_p$ equals the thermally reduced band gap

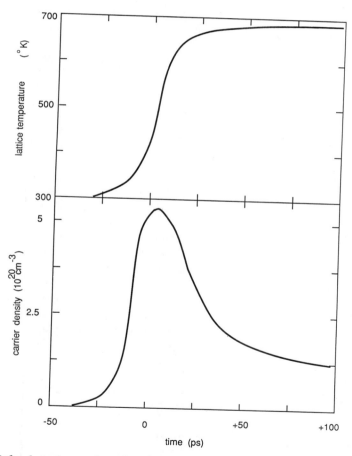

Fig. 6. Calculated carrier density and lattice temperature profiles at the surface of crystalline silicon irradiated by a 0.523 μm, 20 ps laser pulse at a 0.1 J/cm^2 incident fluence.

energy E_g [45]. In this case an electron and a hole recombine by emitting a plasmon to conserve energy. In addition, in indirect band gap semiconductors, a phonon is emitted to conserve momentum. The recombination energy is immediately transferred to the lattice providing a further delayed heating to the lattice, as explained above. This plasmon-phonon assisted recombination (PPAR) has been recently studied for the case of silicon[46]. Due to finite plasmon lifetime, this new recombination channel is active in a finite interval of plasma frequencies around the value $\hbar\omega_p = E_g$. With picosecond laser excitations, electron and lattice temperatures being equal, the effect is enhanced by thermal reduction of the band gap and therefore is expected to be dominant at the highest temperatures.

In silicon, calculated PPAR rates[47] dominate over Auger effects in the upper range of densities and temperature, as illustrated in Fig. 7. At a constant density of 5×10^{20} cm^{-3}, the Auger rate is $\tau_A \sim 5 \times 10^{10} - 10^{11}$ s^{-1}. This value is reached by PPAR at ~ 1000 °K. At the melting point of silicon, $T_m = 1680$ °K, PPAR rates for the same carrier density reach 10^{12} s^{-1}. The strong temperature dependence of this recombination channel is responsible for several features in picosecond irradiated silicon at high excitation levels which cannot be explained by temperature independent recombination mechanisms such as the Auger effect. During the first part of the heating pulse the lattice temperature is essentially low since most of the optical energy is being stored in the carrier gas. Plasmons are out of resonance with the band gap and recombination is governed by the Auger effect. As soon as recombination starts and the lattice is being heated up, PPAR rapidly sets in promoting further recombination and heating. This cumulative process, which lasts few tens of picoseconds after the laser pulse, stops when recombination has pushed the plasmon energy well below the band gap energy E_g. The recombination process is again driven by Auger effect alone. Just after the laser pulse higher excitation levels result in lower carrier densities, since higher initial carrier concentration 1) bring plasmon energy closer to resonance with the band gap and 2) provide more lattice heating which broadens the PPAR resonance. This is a very specific feature of PPAR in the picosecond regime which quantitatively explains the observed optical response of silicon at excitations close to the threshold for melting[48]. In agreement with picosecond experimental results[49], the maximum carrier density is kept around the values corresponding to the plasmon resonance with the band gap. Reduction of carrier density with input laser fluence, an effect peculiar of PPAR, has been documented by the disappearance, at high irradiation levels, of plasma resonances well developed at lower fluences[33]. Figg. 8 and 9 illustrate the situation at 2.8 μm for crystalline silicon illuminated by 0.53 μm, 20 ps laser pulses. Similar, magnified effects are expected to take place in germanium, where both coherent and incoherent routes for plasmon aided recombination can be envisaged[47].

With femtosecond laser irradiation, the role of PPAR is expected to be less important during the laser pulse, the reason being that high electron temperatures ($T_c \gg T_1$) leave a lower number of electrons in resonance with the plasmons. Only when thermalization with the lattice has occurred, I.e. ~ 1 ps after the excitation pulse, plasmon aided recombination may become again effective in reducing very quickly carrier density, as fem-

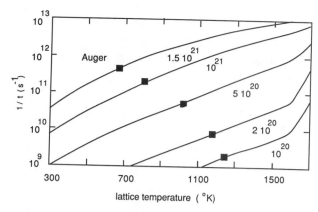

Fig. 7. Recombination rates $1/\tau(s^{-1})$ for plasmon-phonon assisted recombination in silicon in the picosecond case as a function of lattice temperature, for the indicated carrier densities (cm^{-3}). The squares indicate the Auger rates at the corresponding carrier densities.

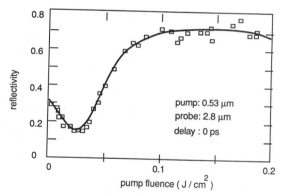

Fig. 8. Measured (squares) and calculated (full line) transient reflectivity of silicon excited by a 20 ps 0.53 μm laser pulse as a function of incident laser fluence. The pump and probe pulses are temporal coincident.

tosecond optical measurements in silicon seem to suggest[50].

CONCLUSIONS

The dissipation processes responsible for the energy relaxation of the photoexcited carriers occur in the femtosecond time domain. Therefore ultrashort laser pulses have proven unique tools for investigating such mechanisms. The role of carrier-carrier, -phonon, and -plasmon scattering phenomena has been elucidated for GaAs type compounds at low excitation levels. The extension of such investigations to higher exci-

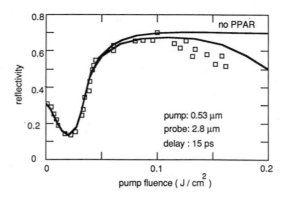

Fig. 9. Measured (squares) and calculated (full line) transient reflectivity of silicon excited by a 20 ps, 0.53 μm laser pulse as a
function of incident laser fluence. The upper solid line refers
to calculations without PPAR recombination.

tations and to elemental semiconductors will bring new basic information
on scaling laws and screening effects. In metals, observations of electron
temperatures far in excess of the one of the lattice clearly indicate the
limited coupling of electrons with phonons. Further investigations with
controlled phonon populations and in the area of hot electron energy transport are clearly needed.

Interaction of picosecond pulses with semiconductors results in
electron hole plasma formation in thermal equilibrium with the crystalline lattice. Experiments with picosecond excitation sources essentially probe the dynamics of the plasma and its energy exchange with the lattice. This is the fastest time scale for optically heating solid materials, alternative possibilities being the direct photon phonon coupling
in the far infrared. The role of plasma degeneracy in slowing the energy
transfer process to the lattice has been established. Plasma recombination and delayed lattice heating are driven by nonlinear density and
temperature dependent mechanisms such as Auger and plasmon phonon assisted recombination. The latter is instrumental in drastically reducing
the carrier density at the excitation levels close to the melting of the
surface. Experimental evidence of an upper limit to the carrier density
and a simultaneous ultrafast melting of the lattice clearly confirm the
validity of the thermal description in the picosecond time regime.

In strongly absorbing materials, steep temperature gradients are
generated and the heat diffuses inside the bulk with a diffusion coefficient $D_{th} = \kappa / c_v$ being κ the thermal conductivity and c_v the specific
heat per unit volume of the material. Picosecond irradiation regimes are
extremely attractive per investigating the high temperature properties
(0.5 - 1 eV) of matter, since both thermal diffusion and evaporation are
simultaneously minimized in this time scale. Surfaces at very high temperatures are obtained and may be studied before evaporation sets in and
complicates the observations.

In the spatial domain, local deposition of heat within short distances from the surface leads to ultrafast heat diffusion and melting. Its relevance for material applications in semiconductor processing and metallurgy justifies the technological effort in this fast growing area of investigations. With nanosecond laser pulses of duration t_p the diffusion length $(D_{th} t_p)^{1/2}$ is, for most absorbing materials, much longer than the absorption length $1/\alpha$. Thermal gradients are thus smoothed by heat diffusion into the bulk. The relevance of heat transport phenomena for material applications is well established.

ACKNOWLEDGEMENTS

I wish to thank Professor N. Bloembergen, Prof. H. Kurz and Dr. M. Rasolt for many enlightening discussions. This work is supported by the U.S. Office of Naval Research under contract no. N00014-85-K-0684.

REFERENCES

1. "Laser Solid Interactions and Laser Processing", S. O. Ferris, H. J. Lemay and J. M. Poate, American Institute of Physics, New York (1979)
2. "Laser and Electron Beam Processing of Materials", C. W. White and P. S. Peercy, Academic Press, New York (1980)
3. "Laser and Electron Beam Solid Interactions and Materials Processing J. F. Gibbons, L. D. Hess and T. W. Sigman, Elsevier North-Holland, New York, 1981, Mat. Res. Soc. Symp. Proc. , 1 (1981)
4. "Laser and Electron Beam Interactions with Solids", B. R. Appleton and G. K. Celler, Elsevier Nort-Holland , New York, 1982, Mat. Res. Soc. Symp. Proc., 4 (1982)
5. "Laser - Solid Interactions and Transient Thermal Processing of Materials", J. Narayan, W. L. Brown and R. A. Lemons, Elsevier North-Holland, New York, 1983, Mat. Res. Soc. Symp. Proc., 13 (1983)
6. "Laser Annealing of Semiconductors, J. M. Poate and J. W. Mayer, Academic Press, New York (1982)
7. "Energy Beam-Solid Interactions and Transient Thermal Processing", J. C. C. Fan and N. H. Johnson, Elsevier North-Holland, New York, 1985, Mat. Res. Soc. Symp. Proc., 23 (1984)
8. "Energy Beam-Solid Interactions and Transient Thermal Processing", D. K. Biegelsen and C. V. Shank, Elsevier North-Holland, New York, 1985, Mat. Res. Soc. Symp. Proc., 35 (1985)
9. "Beam-Solid Interactions and Phase Transformations", H. Kurz, G. L. Olson, J. M. Poate, Material Research Society, Pittsburgh, 1986, Mat. Res. Soc. Symp. Proc., 51 (1986)
10. N. W. Ashcroft, N. D. Mermin, "Solid State Physics", Saunders College, Philadelphia (1976)
11. F. Wooten, "Optical Properties of Solids", Academic Press, New York, (1972)
12. M. Born and E. Wolf, "Principles of Optics", Pergamon Press, Oxford (1980)

13. M. I. Kaganov, I. M. Lifshitz, L. V. Tanatarov, Sov. Phys. JETP 4, (1957)
14. A. Wilson,"The theory of metals"
15. J. G. Fujimoto, J. M. Liu, e. Ippen, N. Bloembergen, Phys. Rev. Letters, 53 :1837 (1984)
16. G. L. Eesley, Phys. Rev. Letters, 51:2140 (1983)
17. H. Elsayed-All, M. Pessot, T. Norris and G. Mourou,in:"Ultrafast Phenomena V",edited by G. R. Fleming and A. E. Siegman (Springer-Verlag, Berlin), p. 264 (1986)
18. R. W. Schoenlein, W. Z. Lin, J. G. Fujimoto and G. L. Eesley, in:"Ultrafast Phenomena V", edited by G. R. Fleming and A. E. Siegman (Springer-Verlag, Berlin) p.260, (1986)
19. D. Pines and D. Bohm, Phys. Rev. 85:338 (1952)
20. E. M. Conwell and M. O. Vassel, IEE Trans. Electron. Devices, 13:22 (1966)
21. D. von der Linde, J. Khul and H. Klingenberg, Phys. Rev. Letters, 44:1505 (1980)
22. J. A. Kash and J. C. Tsang, Phys. Rev. Letters, 54:2151 (1985)
23. A. M. Malvezzi, J. M. Liu and N. Bloembergen, Appl. Phys. Lett., 45:1019 (1984)
24. A. J. Taylor, D. J. Erskine and C. L. Tang, Appl. Phys. Lett., 43:989 (1983)
25. C. L. Tang and D. J. Erskine, Phys. Rev. Letters, 51:840 (1983)
26. W. Z. Lin, J. G. Fujimoto, E. P. Ippen and R. A. Logan, in: "Ultrafast Phenomena V", edited by G. R. Fleming and A. E. Siegman (Springer-Verlag, Berlin), p. 193 (1986)
27. J. L. Oudar, D. Hulin, A. Migus, A. Antonetti, F. Alexandre, Phys. Rev. Letters, 55:2074 (1985)
28. E. Yoffa, Phys. Rev., B 21:2415 (1980)
29. A. M. Malvezzi, H. Kurz and N. Bloembergen, Appl. Phys., A 36:143 (1985)
30. R. W. Schoenlein, W. Z. Lin, S. D. Brorson, E. P. Ippen, J. G. Fujimoto and G. L. Eesley, XV International Conference on Quantum Electronics, Baltimore, Md April 27, May 1 (1987) paper TuDD4
31. D. H. Auston, C. M. Surko, T. N. C. Venkatesan, R. E. Slusher and J. A. Golovchenko, Appl. Phys. Lett., 33:437 (1978)
32. F. Spaepen, Science Magazine, february 1987
33. H. M. van Driel, L. A. Lomprè and N. Bloembergen, Appl. Phys. Lett., 44:285 (1984)
34. N. Bloembergen, R. K. Change, S. S. Jha and C. H. Lee, Phys. Rev., 174:813, (1968)
35. A. M. Malvezzi, H. Kurz and N. Bloembergen, ref. 8, p. 75
36. L. A. Lomprè, J. M. Liu, H. Kurz and N. Bloembergen, Appl. Phys.Lett., 43:168 (1983)
37. M. C. Lee, H. W. Loo, A. Aydinil, G. J. Trott, A. Compaan and E. B. Hale, Solid State Comm., 46:677 (1983)
38. C. A. McDonald, A. M. Malvezzi and F. Spaepen, rf. 9, p. 271
39. P. Baerl, J. Appl. Phys., 50:788 (1979)
40. A. Lietola, F. Gibbons, Appl. Phys. Lett.,46:624 (1982)
41. F. Spaepen and D. Turnbull in ref. 6, p. 15
42. N. Bloembergen, ref. 9, p. 3
43. H. Kurz and N. Bloembergen, ref. 8, p.332

44. C. V. Shank and M. C. Downer, ref. 9, p.15
45. P. A. Wolff, Phys. Rev. Letters, 24:266 (1970)
46. M. Rasolt, Phys. Rev., B2:1166 (1986)
47. M. Rasolt and F. Perot, to be published
48. M. Rasolt and H. Kurz, Phys. Rev. Letters, 54:722 (1986)
49. A. M. Malvezzi, C. Y. Huang, H. Kurz and N. Bloembergen, ref. 9, p.201
50. C. V. Shank, R. Yen and C. Hirliman, Phys. Rev. Letters, 50:454 (1983)

OPTICAL DESIGN CONSIDERATIONS FOR CO_2 LASER INDUSTRIAL SYSTEMS

G. Alessandretti and P. Perlo

Centro Ricerche Fiat
Orbassano, Torino, Italy

INTRODUCTION

The coupling of laser sources and robotized systems for beam delivery and manipulation is one of the recent innovative aspects of laser technology for material processing[1]. By introducing robot machines with moving optical components, the laser beam can be directed to the workpiece, under automatic control, with variable trajectories, speeds, angles or relative positions. This allows to control more precisely the manufacturing process and to take full advantage of the laser benefits, in terms of productivity and flexibility.

With the evolution of "laser robotics", some particular trends have been emphasized, as in these instances:

- greater optical paths to the working point;
- processing at variable distances from the laser source;
- coupling of one laser to various workstations;
- switching optics to employ more lasers for the same operation (in case of maintenance or repairing).

As a consequence, the design of beam delivery systems has assumed a fundamental role, including aspects like beam propagation, size and position of optical components, beam shaping and beam focusing.

Some of these subjects are briefly discussed here, with particular reference to CO_2 lasers in the kW power range which are the most commonly employed sources for industrial applications on cutting, welding and heat treating of metals .

A general procedure for tracing the characteristics of a beam through a train of mirrors or lenses is presented and some particular cases, referred to practical problems, are illustrated.

PROCESS SPECIFICATIONS

Two examples of CO_2 laser industrial systems, representing some of the preferred approaches, are shown in Fig. 1. The first scheme is a gantry robot with flying optics: a reflective mirror is mounted on a precision sliding mechanism, on each axis, according to a cartesian geometry; two or three more degrees of freedom may be added by tilting the final focusing head, which generally includes a ZnSe lens.

The second scheme refers to a more integrated laser/robot system, with through-the-arm beam delivery optics. With this geometry, the operating field is a circular sector, and constant distance from the source to the workpiece is obtained.

The starting point for an overall analysis of an industrial system is to consider process requirements. To a great extent, these will depend on the specific application, but, in general terms, a given power density on the workpiece should be assured, with values from $10^3 - 10^4$ W/cm^2 for heat treating, to $10^5 - 10^7$ W/cm^2 for cutting and deep penetration welding of metals. In the latter case, spot sizes with dimensions of a fraction of a mm have to be obtained.

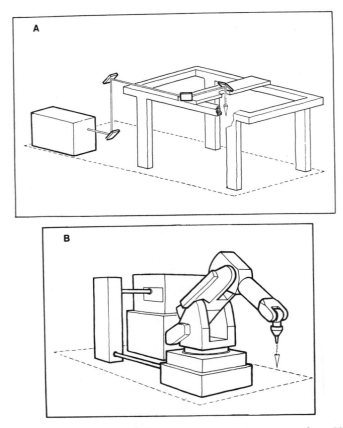

Fig. 1. Schematic design of laser robot systems using flying optics (A) or through-the-arm beam delivery (B)

The tolerance on the power density can be of about ±10% (or even less for some heat treating applications), which sets quite stringent limits on the admissible variations of spot size at the surface of the component. A typical depth of focus for a 127 mm focal length lens can be of the order of ±1 mm.

Thus the two principal subsystems, namely laser device and beam delivery optics, should be chosen and coupled in order to guarantee, among other requirements, adequate focusing within the whole working volume. Moreover, the mecanichal systems should possess the necessary positioning accuracies.

BEAM PROPAGATION

A convenient description of the laser beam is in term of Hermite-Gaussian or Laguerre-Gaussian modes[3].

In general the laser wave Ψ is a superposition of transverse modes Ψ mm in the form

$$\Psi = \Sigma \, Cmn \, \Psi \, mn$$

Experimental data at our laboratory[4], and by other groups[5], have shown that typical industrial CO_2 lasers with stable resonators of the transverse gas flow type operate on a single, or nearly single, high order mode with rectangular symmetry. Figs. 2 and 3 show representative acrylic burn patterns and intensity profiles for a 5 kW laser of this type and two different sets of cavity optics (for welding and for heat treating applications, respectively). The first beam is well represented by a mode of order six along the Y-axis (or maximum diameter) while the heat treating resonator is described by modes of order eight or nine, depending on the power and the alignment of the cavity.

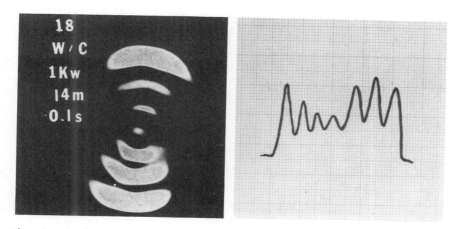

Fig. 2. Typical burn pattern and intensity profile in arbitrary units for a 5 kW transverse flow laser (cavity optics for welding applications)

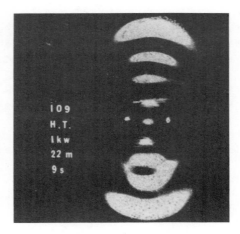

Fig. 3. Burn pattern for the same laser as Fig. 2 with cavity optics
for heat treating applications

Lasers of axial flow type show a cylindrical symmetry: a near-gaussian,
or low-order multimode corresponds to 1-2 kW sources, higher order modes
being observed for more powerful lasers. Fig. 4 refers to an almost single
mode of order two, typical for a fast axial flow laser.

Propagation of all these beams in free space is described by introducing three parameters[4]: the minimum beam waist radius W_o for the fundamental mode (at $1/e^2$ of the intensity on axis), the distance Z_o from output
window to beam waist, and a constant K, characteristic of the mode (or superposition of modes) which is essentially the ratio between radii of the
actual beam and the fundamental mode. (See W.H. Carter[6] for a definition
of this parameter).

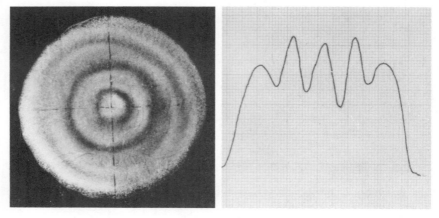

Fig. 4. Typical burn pattern and intensity profile for an axial flow laser

The beam radius W at position Z is then given for one transverse axis (X or Y) by

$$W(Z) = KW_o \left(1 + ((Z-Z_o)/Z_r)^2\right)^{1/2} \tag{1}$$

where $Z_r = \Pi W_o^2 / \lambda$ is the Rayleigh range, independent of mode order.

In the case of lenses, or spherical mirrors, the radius of curvature of the wave front is modified, so that the beam will have a waist with a new position and dimension (the parameter K remaining unchanged). As shown by S.A. Self[7] for a focal length f, the input beam waist KW_o, at distance s from the lens, is transformed into a waist KW_o' , at distance s', according to

$$s'/f = 1 + (s/f-1)/((s/f-1)^2 + (Z_r/f)^2)$$

$$W_o' = W_o /((s/f-1)^2 + (Z_r/f)^2)^{1/2} \tag{2}$$

Recursive application of equations (1) and (2) can be utilized for an optical train.

Equation (2) can be finally employed to compute the focal spot size on the workpiece. A good approximation, assuming Z_r/f and $s/f \gg 1$, is given by

$$d = k^2 \cdot \frac{4 \lambda f}{\Pi D} \tag{3}$$

(d is the spot diameter, D the laser diameter at the lens).

A depth of focus can be defined as the range over which d is changed no more than a given factor g. It is given by the expression

$$\Lambda = \pm \frac{\Pi d^2}{k^2 \lambda} \cdot (g^2 - 1)^{1/2} \tag{4}$$

PRACTICAL CONSIDERATIONS

Although effects like thermal blooming in the environment, diffraction due to apertures and gain in the laser medium are not considered by the procedure outlined in the proceeding paragraph, a good correspondence with actual results has been found in many practical situations, at least for distances up to around 30 m. An example[4] is shown in Fig. 5. A simpler approach based on geometrical optics proved to be generally insufficient.

An experimental determination of the three parameters W_o, Z_o, K has been done, for a number of industrial sources, by a least squares fit of beam diameters measured at various positions.

Results in Fig. 5 refer to a 5 kW laser with transverse flow design. Similar curves have been drawn for various types of industrial CO_2 lasers. The mode coefficient K versus maximum power are plotted in Fig. 6. The dif-

181

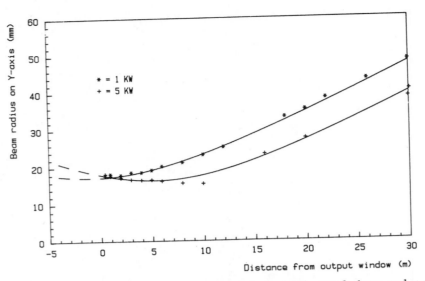

Fig. 5. Beam propagation at two power levels experimental data and computed result. (Fitted parameters at 1 kW : K = 2.68, W_o = 6.34 mm, Z_o = -1050 mm; at 5 kW : K = 2.63, W_o = 6.14 mm, Z_o = 5200 mm)

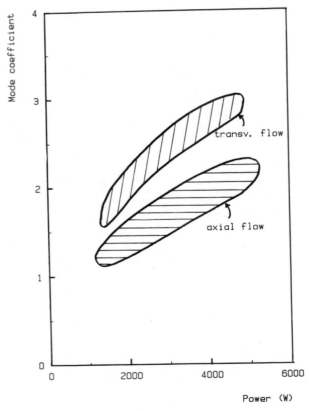

Fig. 6. Mode coefficient versus maximum power for various industrial CO_2 laser sources

ferences between axial flow and transverse flow lasers, and the increase of K with power are clearly related to the geometry of the various resonators (the Fresnel number being typically 1.5÷2 for axial flow, 5 or more for transverse flow design, and normally increasing with power).

In principle, the beam parameters could be obtained from geometrical data of the laser resonator such as its length, radii of curvature of the mirrors, and diameter of mode selective diaphragms. But the advantage of an experimental evaluation is that it takes into account any lens effect at the output window, a well known phenomenon caused by absorption of laser radiation[8].

The propagation of the beam, particularly the position of the waist, has been found to depend on laser power (see Fig. 5). Increased coating absorption with ageing of the output window has been observed, and implies similar effects. This aspect can be particularly important for systems of the type shown in Fig. 1B, where apertures and beam bending mirrors of limited size are present.

A general rule for dimensioning the mirrors or shaping the laser beam is to have a ratio typically 1.5-2.5 between their diameters. The free propagation equations still remain representative in the far field[9], as the beam is weakly diffracted, and cumulative effect are not too strong. In the case of lenses, a ratio around 1.5 may be preferred to reduce thermal stresses, since only edge cooling is possible.

To take a practical case, center-to-edge temperature differentials of up to 12° C with 1 kW power have been measured[10] on ZnSe lenses after prolonged use in an industrial environment. This corresponds, according to reference 8, to a 1/20 diopter additional power; by using equation (2), for f = 127 mm, a 25% reduction is found for the intensity at the target position corresponding to the unstressed lens.

Beam propagation analysis has also been applied with useful results to variable path optical systems; this is the typical situation of cutting machines with cartesian robots (Fig. 1A).

When the focusing head is moved, changes in the diameter and radius of curvature of the wave front at the lens may originate unacceptable variations in focused spot size and waist location.

Figs. 7 and 8 give a plot of these quantities calculated by using equation (1) and (2) with f = 127 mm and assuming the parameters of Fig. 5 for the beam. To describe a representative situation, the lens has been supposed to cover distances in the range 4 to 12 m from the laser source. Waist location is seen to shift of about 0.8 mm which is comparable with the depth of focus. Spot size changes by 15%, causing a 30% variation in power density. Software corrections on the control system could be introduced to reduce the problem of spot shift. The minimization of spot size variations requires attention in the optical design, one possible solution being to introduce corrective mirrors to bring the laser waist to a median position in the working volume.

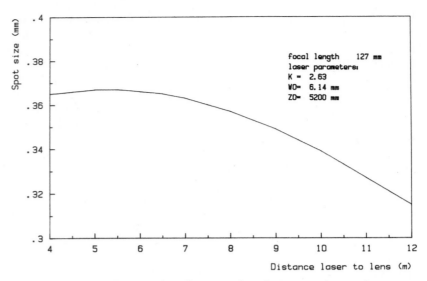

Fig. 7. Variation of focal spot size laser-to-lens distance

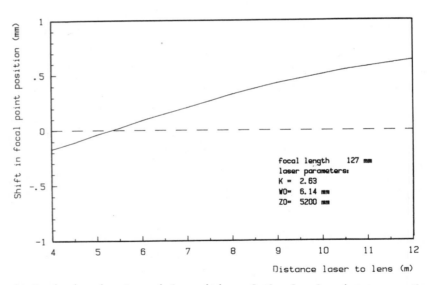

Fig. 8. Variation in the axial position of the focal point versus laser-
-to-lens distance

CONCLUSIONS

Industrial CO_2 lasers are conveniently described by Hermite-Gaussian or Laguerre-Gaussian modes. A generalization of Gaussian beam theory can be used to predict beam propagation through a series of mirrors or lenses. An experimental procedure has been described to obtain an evaluation of beam parameters.

The laser beam characteristics cannot be predicted from the resonator geometry alone, due to a lens-like effect cause by absorption at the output window. With regard to this point, power dependent variations of beam characteristics have to be considered in the optical systems design.

For the case of variable optical path systems it is important to account for changes in spot size and shifts of the focal point along the optical axis, when the lens scans the working volume.

REFERENCES

1. Laserobotics 2, Proceedings 2nd International Conference Combining Laser and Robot Technologies, Southfield, SME (1987)
2. D. Belforte, M. Levitt editors, "The Industrial Laser Annual Handbook", Pennwell Books (1986)
3. H. Kogelnik and T. Li, Laser Beams and Resonators, Proc. IEEE, 54:1312 (1966)
4. P. Perlo, Propagation of a multikilowatt laser beam: experimental characterization, in: "High Power Lasers and their Industrial Applications", SPIE Proc. 650:178 (1986)
5. J. T. Luxon and D. E. Parker, Practical spot size definition for single higher-order rectangular-mode laser beams, Appl. Opt.,20:1728 (1981)
6. W. H. Carter, Spot size and divergence for hermite-gaussian beams of any order, Appl. Opt., 19:1027 (1980)
7. S. A. Self, Focusing of spherical gaussian beams, Appl. Opt., 22:658 (1983)
8. M. Sparks, Optical distortion by heated window in high power laser systems, Journ. Appl. Phys., 42:5029 (1971)
9. P. Belland, J. P. Creen, Changes in the characteristics of a gaussian beam weakly diffracted by a circulàre aperture, Appl. Opt., 21:522 (1982)
10. G. Alessandretti and P. Gay, Characterization of coated CO_2-laser optical components employed in an industrial plant, in: "Gas Flow and Chemical Lasers", 1984, Adam Hilger Ltd., Bristol (1985)

LASER DIAGNOSTICS OF INDUSTRIAL CHEMICAL PROCESSES

J. Wolfrum

Physikalisch-Chemisches Institute der Ruprecht-Karls-

Universität, Heidelberg, West Germany

Industrial chemical processes take place under conditions where the many elementary chemical reactions interact strongly with the various transport processes (diffusion, heat conduction, convection, turbulence, radiational energy transfer). Detailed information on these interactions is necessary to improve the efficiency of industrial processes and to reduce the formation of unwanted side products and pollutants. Interesting new possibilities in this direction come from mathematical modeling of complex chemical processes in laminar or turbulent flows including heat and species transport and the influence of walls[1]. Realistic mathematical models require information on temperature, concentrations and velocities of reacting species with high spatial and temporal resolution. The development of powerful and tunable laser light sources from the infrared to the ultraviolet has greatly improved the possibilities for non-invasive spatially and temporally resolved measurements of species concentrations, mole fraction, temperature, density, velocity and pressure. In this lecture various industrial applications of linear and non-linear laser spectroscopic techniques are described.

LASER-INDUCED FLUORESCENCE (LIF)

In LIF, a tunable narrow band laser is used to excite a specific electronic transition in the species of interest. Either the total fluorescence (excitation spectrum) or the spectrally resolved fluorescence (fluorescence spectrum) is collected at right angles to the laser beam. A quantitative interpretation of the signal requires a detailed study of the collisional quenching and radiative and non-radiative energy transfer processes in the observed particle[2]. Thus, the fluorescence yield of OH radicals at atmospheric pressure is only one part in 10^3. One way to obtain a simple relation between fluorescence intensity and radical concentration is to work at high laser power so that the observed electronic transition can be saturated[3]. However, complete saturation is difficult to achieve since the temporal and spatial distribution of the laser light always contains regions of lower than saturation intensity. The degree of saturation in the beam must therefore be carefully measured[4]. Another possibility is to use laser pulses of duration shorter than the quenching and radiative times of the particle investigated. In this case, the maximum fluorescence power of the laser pulse is independent of the quenching and rotation relaxation rate[5]. Absolute concentrations can also be obtained by a combination of LIF and absorption measurements[6].

As Fig. 1. shows the application of powerful tunable UV laser sources such as excimer lasers provide the capability for simultaneous, spatially resolved measurements at a large number of points within a plane. Fluorescence light from the laser light sheet illumination is collected at right angles and imaged either directly onto an array detector or onto an image intensifier. Proximity-focused microchannel plate intensifiers are the best choice for imaging applications[7-9]. Due to the high sensitivity of LIF atoms and radicals (H, O, N, C, OH, CH, CN, NH_2, NCO, CH_3O etc.) with short lifetimes can be detected using multiphoton excitation. Atoms and stable molecules that absorb in the VUV region can be detected using visible or near infrared fluorescence[10].

Excitation of molecules directly with excimer laser wavelengths to pre-dissociating electronic states leads to a reduction in the fluorescence yield but removes the dependence of the fluorescence signal on quenching due to the very short lifetimes of such states[11-13].

Laser induced fluorescence excitation of organic ecxiplexes allows the simultaneous imaging of vapor and liquid distributions, which is important in many technical applications. These exciplexes (excited state complexes) are formed by a reversible reaction involving the liquid and an added organic compound. The transition between liquid and vapor phase is observed from the shift in the liquid-phase fluorescence emission by 100-200 nm from the peak of the gas phase emission. The fluorescence from both phases can be visualized simultaneously using a single wavelength laser source[14].

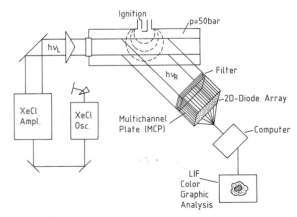

Fig. 1. Two-dimensional imaging of concentrations and temperature in high pressure combustion engines by planar LIF spectroscopy using tunable UV excimer laser

Temperature is an important parameter in many technical processes and laser induced fluorescence has also been used extensively for temperature measurements. One technique is to scan the laser frequency across different transitions followed by broadband fluorescence detection or excitation of molecules into different rotational levels in the ground state to the same upper level. By rationing the fluorescence intensities the temperature can be deduced[15]. The two-line excitation scheme has also been applied successfully to obtain 2D-images of temperature fields[16]. Fig. 2 shows an arrangement[17] in which a constant mole fraction of a stable tracer molecule (NO) is seeded into the flow. In this case the LIF signal is proportional to a simple function of temperature.

Velocity measurements by LIF are based on Doppler-shifted absorption of a narrow-linewidth laser source at separate locations within the absorption profile of the tracer species[18].

RAMAN, RAYLEIGH AND MIE SCATTERING

The Raman effect, predicted theoretically in 1923 and first observed by C. V. Raman in 1928, became, with the invention of the laser, a very important tool for temperature and concentration measurements in industrial devices. Compared to LIF, Raman scattering has the advantage of no quenching effect, no need for a tunable laser source, the feasibility of measuring several species simultaneously, of readily obtaining vibrational temperatures, and of investigating atoms, radicals, and molecules that absorb down in the VUV region at more suitable wavelengths. The only drawback is that the Raman effect is quite weak in intensity so that very intense laser sources are required. However, high power, narrow-band UV-excimer lasers are well suited to single-shot Raman measurements of temperature and concentrations in industrial devices (Fig. 3.).

Multiplex detection allows simultaneous measurements of several different species as well as 2-D imaging[20,21]. Compared to the inelastic scattering of photons in the Raman effect, the elastic scattering by molecules (Rayleigh) or particles (Mie) shows order of magnitude greater scattering cross-sections. Since in elastic scattering the wavelength of the light is unshifted from the illumination beam, other elastic scatterers such as particles or surfaces must carefully be avoided. However, gas mixtures in jets and flames can be studied under controlled laboratory conditions[21].

Due to the large cross-section for Mie scattering simple laser sources and photographic films or unintensified array cameras can be used as recording medium. As shown in Fig. 4, processes such as flame front movement can be recorded by seeding $TiCl_4$ into combustion gases. At the flame front, water molecules formed in the combustion process react with $TiCl_4$ forming submicron sized particles of TiO_2. The light scattered by these particles can be used to image the region where $TiCl_4$ contacts the water vapor[22].

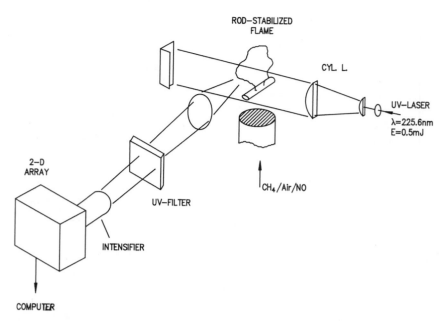

Fig. 2. Temperature imaging on a rod-stabilized CH_4 air flame using planar LIF of NO[17]

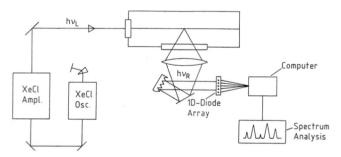

Fig. 3. Single-shot multi-component analysis in industrial devices using spontaneous Raman-scattering with narrowband excimer laser excitation

COHERENT ANTI-STOKES RAMAN SCATTERING (CARS)

The optical non-linear effect CARS was discovered in 1965[23]. CARS is observed when two properly phased-matched laser light beams traverse a sample with a Raman active mode. The non-linear interaction results in the generation of a coherent anti-Stokes laser beam. This beam can be used for non-intrusive measurements with high spectral, spatial and temporal resolution. Since the spectroscopic signal generated is coherent, local concentration and temperature measurements can be carried out in

190

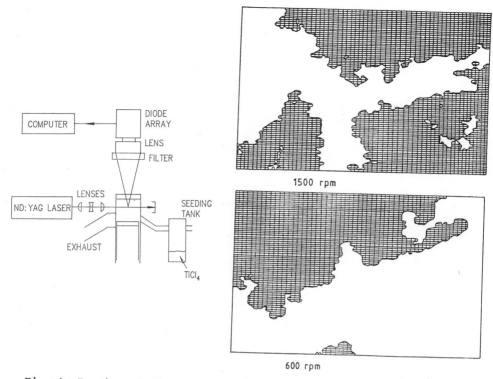

1500 rpm

600 rpm

Fig. 4. Imaging of flame propagation in a spark-inition engine using
Mie scattering at TiO$_2$ particles

the presence of other intense light sources such as electric discharges
chemiluminescence, thermal radiation of soot, etc., making this method
especially attractive for measurements in industrial devices[24,25].

The simplest type of phase-matching involves the colinear approach
of the two laser beams which results, however, in a poor spatial resolution. This problem can be circumvented by using the so-called folded
BOXCARS configuration where all the input laser beams are spatially separated[26]. For quantitative CARS temperature and concentration measurements, the CARS spectrum has to be simulated from the spectroscopic constants of the molecule investigated. Fig. 5 shows a comparison of simulated and experimental spectrum of N$_2$ from a counterflow diffusion flame[27].

Under the pressure conditions of many industrial devices (up to 2
bar) the collisional temperature dependent pressure broadening[28] and at
higher pressures, collisional narrowing[29] has to be taken into account.
Also, for the convolution of spectra with laser line shapes, the effect
of multi-mode lasers must be taken into account when significant inter-
ferences with a non-resonant background occur[30]. Reduction of non-resonant
signals can be achieved by taking advantage of the different temporal behavior of the resonant and non-resonant signals. This technique requires
pico-second laser pulses and has been shown to reduce the background by
two orders of magnitude[31]. The non-resonant background can also be rejected

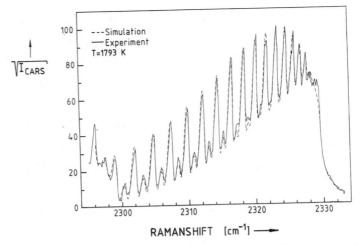

Fig. 5. Measured and simulated CARS-spectrum of N_2 in a CH_4/air
counterflow diffusion flame

by polarization rotation of the interacting laser beams which results,
however in a large reduction in signal strength[32]. When using rotational
transitions, several species may be detected simultaneously using a broadband dye-laser[33]. When using CARS with high-power laser beams, one must
carefully avoid saturation effects which come from stimulated Raman gain
where the Stokes beam is amplified and changes in the population of the
probed levels by the laser beam[34].

RESONANT MULTIPHOTON IONIZATION (REMPI)

The REMPI technique involves n-photon resonant excitation to an intermediate vibronic state followed by m-photon excitation to the ionization continuum. This allows very sensitive detection of atoms, radicals,
and molecules in technical processes as well as single-shot time-of flight
mass spectrometry with high repetition rates. Fig. 6. shows, as an example,
the detection of nitric-oxide (NO) by REMPI spectroscopy with optogalvanic
detection[35]. The optogalvanic technique is based on the measurement of the
current caused by the laser ionization of molecules or atoms. Metal atoms
such as Sr[36], Li[37], Th, and Pb[38] as well as hydrogen and oxygen atoms[39]
and methyl radicals[40] can be detected in flames directly. One must however
ensure that the electrodes and the applied voltage do not disturb the flow
conditions.

SPECTROSCOPY WITH TUNABLE IR-LASERS

Tunable infrared diode of molecular lasers can be applied in process
monitoring in a wide temperature range. As an application, the control of
NOx reduction in power plants is considered. Nitric oxide formation in combustion processes (power plants, car engines, and industrial processes

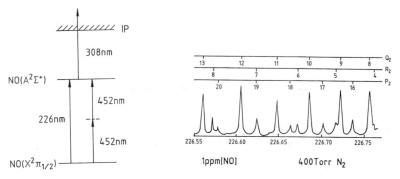

Fig. 6. Sensitive detection of NO in combustion processes using REMPI
spectroscopy

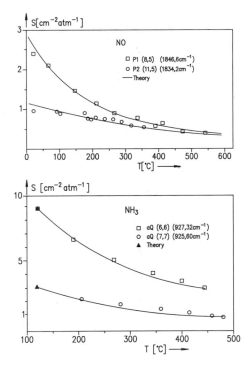

Fig. 7. Temperature dependence of NO and NH_3 absorption coefficients
measured with tunable diode-lasers

using hot air such as glass and ceramic production) involves many 300
elementary chemical reactions. At present detailed simulations of the
NO forming processes are restricted to zero- or one-dimensional configurations. For a more detailed understanding of NO formation and reduction,
the coupling of chemistry with mixing and transport processes must be investigated in detail[41]. This requires non-intrusive measurements of free

atoms (H, O, C, N) radicals (OH, CH, C_2, NH, NH_2, CN) and stable molecules (CO, NO, HCN, NH_3) by laser spectroscopy. Tunable diode-laser systems can cover many of these species[42]. Fig. 7 shows recent measurements on the temperature dependence of absorption coefficients of NO and NH_3. Selective homogeneous reduction of NO is possible in the post-combustion gases using the elementary reaction

$$NH_2 + NO \; ---> \; N_2 + H_2O$$

$$---> \; N_2H + OH$$
$$\downarrow$$
$$N_2 + H$$

This effective formation of an N-N bond leads rapidly to molecular nitrogen and provides an economic method for pollutant removal[44]. The selective reduction works in a temperature window around 1200 K. Complete reduction of NO is possible using a large excess of NH_3. However, the NH_3 remaining after the process can cause ammonium sulfate formation which is very harmful to the power plant installations[45]. To obtain complete NH_3 consumption, mixing of NH_3 to the flue gases must be carefully controlled and monitored. As shown in Fig. 8 this can be done using frequency ᴜmodulation of CO_2-waveguide lasers for tomographic reconstructions of concentration distributions.

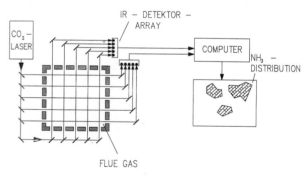

Fig. 8. Two-dimensional measurements of NH_3 concentrations in power plant flue gases using tomography with CO_2 lasers

REFERENCES

1. W. C. Gardiner Jr. Ed , Combustion Chemistry, Springer, New York (1984)

    J. Wolfrum, 20th Symp. Int. on Combustion, p. 559, Pittsburgh (1985)

    B. Rogg, F. Behrendt, J. Warnatz, Ber. Bunsenges. Phys. Chem., 90:1005 (1986)

    J. Wolfrum, VDI Berichte, 617:301 (1986)

2. J. B. Jeffries, R. A. Copeland, D. R. Crosley, J. Chem. Phys., 85:1898
    (1986)

3. E. H. Piepmeier, Spectrochim. Acta, 27 B:431, (1972)
    J. W. Daily, Appl. Opt., 16:568 (1977)
    Baronavski, J. R. Mc Donald, J. Chem. Phys., 66:3300 (1977)

4. K. Kohse-Höinghaus, R. Heidenreich, Th. Just., 20th Symp. Int. on
    Combustion, p. 1177, Pittsburgh (1985)
    M. J. Cottereau, Appl. Opt., 25:744 (1986)

5. D. Stepowski, M. J. Cottereau, Comb. and Flame, 40:65 (1981)

6. D. Stepowski, A. Garo, Appl. Opt., 24:2478 (1985)
    M. H. Hertz, M. Aldén, Appl. Phys., B 42:97 (1987)

7. M. Aldén, H. Edner, G. Holmsted, S. Svanberg, T. Högberg, Appl. Opt.,
    21:1236 (1982)

8. G. Kychakoff, R. D. Howe, R. K. Hanson, Appl. Opt., 23:704 (1984)

9. R. J. Cattolica, S. R. Vosen, 20th Symp. Int. on Combustion, p.1273,
    Pittsburgh (1985)

10. J. E. M. Goldsmith, R. J. M. Anderson, Appl. Opt., 24:607 (1985)
    J. E. M. Goldsmith, N. M. Larendeau, Appl. Opt., 25:276 (1986)
    M. Aldén, S. Wallin, W. Wendt, Appl. Phys., B 33:205 (1984)
    J. M. Seitzman, G. Kychakoff, R. K. Hanson, Opt. Lett., 10:439
    (1985)

    J. Haumann, J. M. Seitzman, R. K. Hanson, Opt. Lett., 11 (1986)

11. G. A. Massey, C. J. Lemon, IEEE J. Quant. Elec., QE 20:454 (1984)
    F. Itoh, G. Kychakoff, R. K. Hanson, J. Vac. Sci. Technol.,
    B 3(6):1600 (1985)

12. K. Kleinermanns, P. Monkhouse, R. Suntz, J. Wolfrum, Non-nuclear
    Energy R+D Contract no. EN3E-0080-D(B) Periodic Report 1986

13. K. Fujiwara, N. Omenetto, J. B. Bradshaw, J. N. Bower, Wirefordner,
    Appl. Spectr., 34:85 (1980)

14. L. A.Melton, Appl. Opt. 22:2224 (1983)
    L. A. Melton, J. F. Verdieck, Comb. Sci. and Tech., 42:217 (1985)
    A. M. Murray, L. A. Melton, Appl. Opt., 24:2783 (1985)
    G. M. Brown, J. C. Kent, in: Flow Visualization III, Ed. W. J.
    Wang, p. 118, Hemisphere Pub. Corp. (1985)

15. R. P. Lucht, N. M. Laurendeau, D. W. Sweeney, Appl. Opt., 21:3729
    (1982)

16. R. J. Cattolica, D. A. Stephenson, in: Progress in Astronautics and
    Aeronautics, 95:714 (1985)

17. M. P. Lee, P. H. Paul, R. K. Hanson, Opt. Lett., 11 (1986)

18. B. Hiller, R. K. Hanson, Opt. Lett., 10:206 (1985)
    B. Hiller, L. M. Cohen, R. K. Hanson, AIAA Reprint 860/61 (1986)

19. D. L. Hartley, in: Laser Raman Gas Diagnostics, Eds. M. Lapp and
    C. M. Penney, p. 311, Plenum (1974)
    K. F. Knoche, H. J. Daams, R. Bahnen, E. Hassel, VDI-Berichte,
    498:215 (1983)

20. F. Pischinger, E. Scheid, K. F. Knoche, H. J. Daams, T. Heinze, U.
    Reute, SAE Paper no. 861121 (1986)
    R. W. Dibble, A. R. Masri, R. W. Bilger, Combustion and Flame
    (1986)

21. B. Yip, M. B. Long, Opt. Lett., 11:64 (1986)
    M. B. Long, P. S. Levin, D. C. Fourguette, Opt. Lett., 10:267
    (1985)

22. A. O. Zur Loye, F. V. Bracco, D. A. Santavicca, Proc. Int. Symp. on Diagnostics and Modeling of Combustion in Reciprocating Engines, p. 249, Tokyo (1985)

23. P. P. Maker, R. W. Terhune, Phys. Rev., A 137:801 (1965)

24. M. Péalet, J. P. Taran, J. Taillet, M. Bacal, A. M. Bruneteau, Combust. Flame, 54:149 (1983)

25. R. M. Green, R. P. Lucht, Proc. Int. Symp. on Diagnostics and Modeling of Combustion in Reciprocating Engines, Tokyo (1985)

26. A. C. Eckbreth, Appl. Phys. Lett., 32:421 (1978)
    Y. Prior, Appl. Opt., 19:1741 (1980)

27. T. Dreier, B. Lange, J. Wolfrum, M. Zahn, F. Behrendt, Ber. Bunsen-ges. Phys. Chem., 90:1010 (1986)

28. R. J. Hall, Appl. Spectrosc., 34:700 (1980)

29. R. L. Farrow, R. P. Lucht, G. L. Clark, R. E. Palmer, Appl. Opt., 24:2241 (1985)

30. R. E. Teets, Opt. Lett., 9:226 (1984)
    R. L. Farrow, L. A. Rahn, J. Opt. Soc. Am., B2:909 (1985)
    J. P. Sala, J. Bonamy, D. Robert, B. Lavorel, G. Millot, H. Berger, Chem. Phys., 106:427 (1986)

31. F. M. Kamga, M. G. Sceats, Opt. Lett., 5:126 (1980)

32. L. A. Rahn, L. J. Zych, P. L. Mattern, Opt. Comm., 30:249 (1979)
    A. C. Eckbeth, R. J. Hall, Comb. Scie. Techn., 25:175 (1981)

33. J. B. Zheng, J. B. Snow, D. V. Murphy, A. Leipertz, R. K. Chang, R. L. Farrow, Opt. Lett.,9:341 (1984)
    B. Dick, A. Gierulski, Appl. Phys., B40:1 (1986)

34. A. Gierulski, M. Noda, T. Yamamoto, G. Marowsky, A. Sclenzka, "Pumping-induced Population Cganges in Broadband CARS-Spectroscopy", to be published in Opt. Lett. (1987)

35. T. Cool, Appl. Opt., 23 (10): 1559 (1984)

36. L. P. Hart, B. W. Smith, N. Omenetto, Spectrochim. Acta, 40B:1637 (1985)
    K. C. Smyth, W. G. Mallard, J. Chem. Phys., 77:1779 (1982)

37. B. W. Smith, L. P. Hart, and N. Omenetto, Anal. Chem., 58:2147-51 (1986)

38. N. Omenetto, T. Berthoud, P. Cavalli, G. Rossi, Appl. Spectrosc., 39:500 (1985)

39. J. E. M. Goldsmith, 20th Symp. Int. on Combustion, p. 1331, Pittsburgh (1985)

40. K. C. Smyth, P. H. Taylor, Chem. Phys. Lett., 122:518-22 (1985)

41. J. Wolfrum, "Schadstoffe in Verbrennungsvorgängen", p. 7, TEFCLAM, Stuttgart (1985)

42. S. M. Schoenung, R. K. Hanson, Comb. Sci. Techn., 24:227 (1981)
    J. J. Harris, A. M. Weiner, Opt. Lett.,6:142 (1981)

43. H. Neckel, J. Wolfrum, Appl. Phys. B (1987)

44. M. Gehring, K. Hoyermann, H. Schacke, J. Wolfrum, 14th Symp. Int. on Combustion, p. 99, Pittsburgh (1973)
    T. Dreier, J. Wolfrum, 20th Symp. Int. on Combustion, p. 695, Pittsburgh (1985)
    R. K. Lyon, U. S. Patent 3900544

45. G. Mittelbach, H. Voje, VGB-Handbuch, "NO_x-Bildung un NO_x-Minderung für fossile Brennstoffe" (1986)

LASERS IN INDUSTRIAL CHEMICAL PROCESSES

J. Wolfrum

Physikalisch-Chemisches Institut der Ruprecht-Karls-

Universität Heidelbergh, West Germany

The use of the laser in chemical synthesis offers a number of advantages over conventional light sources:

1) extension of the available wavelength range
2) with the narrow spectral bandwith of the laser light, a selective excitation of one kind of molecule, e.g. for isotope separation or ultra purification
3) because of the possibilities of pulsed operation and strong collimation, spatially and temporally controllable, homogeneous excitation of the reaction volume
4) due to the high power density, the possibility of multiphoton processes.

Nevertheless, in most cases Hg or Xe-lamps are to be preferred as light sources for industrial photochemistry, both for investment and operational cost as well as for maintenance expenditure, long term power, and lifetime. For economic application of lasers in this area, the effective cost of the photons produced must lie considerably below that for the desired product. Since the product price is often determined only to a small extent by the photochemical process steps, techniques employing lasers have to compete not only with a conventional photochemical process, but also with various other synthesis routes. As is generally the case with the introduction of new methods, the use of lasers should result in several improvements simultaneously, such as cheaper starting material, less or more valuable side products, fewer or cheaper process steps, improvement in product quality etc... The application of lasers for the production of cheap mass-produced chemicals is only economically viable at very high quantum yields (i.e., very large numbers of product molecules produced per photon in the reactor under industrial conditions), as can be achieved in radical chain reactions (see Fig. 1).

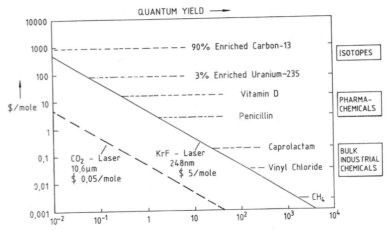

Fig. 1. Influence of quantum yield on chemical product cost.

1. LASER INDUCED RADICAL CHAIN REACTIONS

Vinyl chloride (VC), the monomer of PVC, is produced industrially mainly by thermal cleavage of HCl from 1,2-dichloroethane (DCE) by a chain reaction. With a worldwide production volume of over 25×10^6 tons per annum, VC is one of the leading products of the chemical industry as far as quantity is concerned. The advantage of photolytic over thermal initiation of the chain reaction is that the unimolecular process

$$C_2H_3Cl_2 \xrightarrow{+M} C_2H_3Cl + Cl$$

is rate-determining with a low energy threshold. This leads to a low activation energy for the total reaction and hence low reactor temperatures, higher conversions, and fewer side products. The use of laser radiation to initiate the chain reaction also permits detailed investigation of the reaction kinetics. Modern UV-lasers can be employed to produce radicals in a wide concentration range and to follow the reactions as a function of time. The experimental data can be compared with a kinetic simulation model. This is done by simulating the entire course of the reaction initiated by the laser using a system of coupled differential equations which describe the elementary chemical steps under the chosen temperature and pressure conditions, cfr. Fig. 2. From a comparison of the experimental data with the predictions of the model, missing rate constants for selected elementary steps can be determined.

A pilot plant presently under construction in Italy is shown schematically in Fig. 3. Thermal cleavage of DCE takes place in a tubular reactor. After heating the DCE, a segment of the reactor can be irradiated with a powerful excimer laser. The laser is tuned to a wavelength at which the absorption in the medium is as small as possible. Due to the high collimation of the laser beam, a very large volume can be irradiated, producing a steady, small concentration of active chlorine atoms and hence long chains of VC.

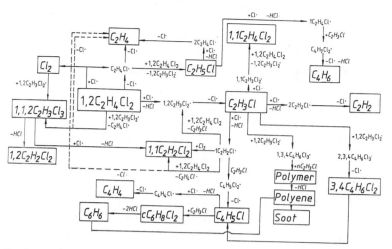

Fig. 2. Reaction paths for the cleavage of hydrogen chloride from dichloroethane to produce vinyl chloride in a laser-induced chain reaction.

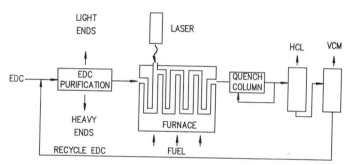

Fig. 3. Pilot plant for the production of vinyl chloride using an excimer laser

At the same time, irradiation of the wall and hence the initiation of heterogeneous processes is to a large extent avoided. With laser radiation, faster conversion at lower temperatures can be achieved compared with the conventional process. Thus we obtain higher total conversions with simultaneously reduced formation of side products.

As with the thermally initiated reaction, laser-induced production of vinyl chloride consumes energy. Therefore the conversion rate depends strongly on heat transport from the walls into the gas. This presents, in addition to a constraint on radiation intensity, a limitation on the shortening of the reaction times. In addition, interactions between the laser radiation and the reaction medium have to be investigated in detail. Density and concentration gradients arising in flow-through reactors can lead to undesired beam divergence and focussing.

A further problem in the industrial application of laser photolysis is the possible contamination of the windows through which the laser is coupled to the chemical reactor. One possibility is to flush the window with inert gas or cold DCE, which has a much lower absorption due to the strong temperature dependence of the absorption. Under certain conditions, the laser itself can "burn" the windows clean by virtue of its high intensity. Compared to conventional photolysis arrangements, the required transmission surface is considerably smaller here.

Overall, the use of high-power (1-10 kW) UV-laser for initiation of chemical processes is a very interesting new technological area. Compared to the discharge lamps employed previously, the good spatial and spectral collimation of the laser radiation can be used to good advantage. This avoids too high radical concentrations in the vicinity of the light source and allows the required production of reactive radicals under extreme conditions of pressure and temperature, as occurs in numerous large-scale industrial processes.

2. LASER-INDUCED POLYMERIZATION

Conventional light sources are used in industry for photopolymerization of thin films for reproduction technology as well as for annealing synthetic insulating materials. Because only the starting molecule needs to be excited for photopolymerization, large quantum yields can, in principle, be achieved.

As an example, laser-induced, high-pressure polymerization of ethylene is considered in more detail. Besides a number of global investigations, information on chemical and phase equilibria as well as kinetic parameters are readily available for the system[1-3]. Thermally induced radical polymerization of ethylene has a number of disadvantages. Long induction periods result from the need to heat the monomer/initiator mixture. Also, the use of a multicomponent mixture (monomer/initiator/regulator) is necessary. Polymerization is initiated by the need for heat transfer, initially to the walls. The polymer coating so formed raises the heat transfer resistance of the reactor walls considerably. In the exothermic polymerization process, however, the removal of heat from the reactor is an important parameter. Upon exceeding a critical temperature limit, explosive decomposition of the monomers can occur. In contrast, providing the necessary starting energy with the aid of laser radiation allows controlled and homogeneous initiation of the polymerization process. As with the UV laser irradiation of DCE, one can work in a region of low absorption and hence very large penetration depth in the starting material. However, the weak ethylene absorption (2×10^4 m^2/mol at 100°C) at 248 nm is strongly influenced by impurities such as O_2. The absorption of the reaction product polyethylene at these wavelengths can be neglected. Quantum yields of 10^4 can be achieved and rate constants for the propagation and termination reactions of polymerization can be measured directly as a function of pressure and temperature.

Polymerization by photochemical activation also succeeds with infrared lasers[4,5] and continuously operating ion lasers, which work more than two orders of magnitude more efficiently in copolymerizing cyclohexene oxide with maleic anhydride than the irradiation from conventional UV lamps[6]. Upon introducing a suitable photoinitiator, very rapid cross-linking in epoxy acrylate layers can be obtained by irradiation with an N_2-laser at 337 nm. Up to 450 crosslink points are produced per absorbed photon. Polymerization can also be initiated by laser photoionization. This can be done using either suitable salts for photochemical production of cations[8] or direct photoionization of the monomer via multiphoton processes[9]. The high temporal resolution feasible with laser photolysis also allows detailed investigation of the elementary processes in radical polymerization[10,11]. By introducing photochemically active catalysts, polymerization of acetylene and ethylene can be achieved in regions $3\mu m$ wide within the irradiation zone of an argon ion laser at 257.2 nm[12]. Interesting insight into the kinetics of the polymerization process can also be obtained using holographic techniques[13,14].

The irradiation of coating layers with excimer and dye lasers allows, even at very low laser power, very rapid polymerization without the use of solvents or heat. In this way, transparent coatings of $200\mu m$ and pigmented coatings of $20\mu m$ can be produced more than an order of magnitude faster than the processing rates obtained previously[15].

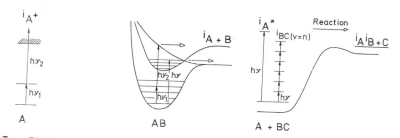

Fig. 4. Possibilities for photochemical separation of isotopes with lasers

3. SEPARATION OF ISOTOPES WITH LASERS

The monochromaticity of the electromagnetic radiation obtainable with the laser has given new impetus to attempts at separating isotopes photochemically, which started over half a century ago[16,17]. Fig. 4 shows that in laser isotope separation, one kind of isotope in a mixture with other isotopes is excited selectively by narrow band laser light and converted into an ion, isomer, or dissociation or reaction product that can be more readily separated chemically or physically. The focus of interest in laser isotope separation was initially the uranium isotope (^{235}U)[18].

The most developed process for this so far, AVLIS (Atomic Vapor Laser Isotope Separation), employs selective multiphoton ionization of uranium atoms as shown schematically in Fig. 4. The frequencies required for excitation are produced by dye lasers, pumped by powerful copper vapor lasers with high repetition frequency (10 kHz)[19]. Besides uranium and plutonium, a number of other elements can be isotope-selectively enriched by this method[20].

Following the observation of isotope-selective, infrared multiphoton dissociation of SF_6 with the aid of CO_2 lasers[21], analogous experiments were performed with UF_6. The radiation required for isotope-selective excitation of fundamental vibrations of UF_6 at 16 m can be obtained by rotational Raman scattering of CO_2 laser radiation on para-hydrogen[22]. However, the UF_6 gas beam has to be strongly cooled to obtain the necessary differential absorption between $^{238}UF_6$ and $^{235}UF_6$. By further selective vibrational excitation of UF_6, clear differences are obtained in the UV absorption spectra[23] of $^{238}UF_6$ and $^{235}UF_6$, which can be used for the isotope-selective dissociation shown in Fig. 4. The UV laser powers at 308 nm required for industrial application are in the 100 kW range[24]. Besides UF_6, other uranium compounds, such as $UO_2(OCH_3)_6$[25] and suitable uranium complexes[26], were investigated. The technique was also applied successfully for the enrichment of plutonium isotopes by selective excitation of PuF_6[27].

In nuclear technology, zirconium is used as a cover for fuel rods. In particular, the isotope ^{91}Zr has a neutron capture cross section more than an order of magnitude larger than the other isotopes. The selective ionization of ^{91}Zr was reported recently[28,29]. However, because the extension of the economic use of nuclear fission as a whole is developing more slowly and with more difficulty than was originally assumed, the requirement for additional capacity for isotope separation is considerably reduced. This is similarly true for the production of larger amounts of deuterium[30] for operating reactors with natural uranium. Tritium can be enriched, e.g. for experiments in controlled nuclear fusion, by IR multiphoton excitation of $CTCl_3$[31] and CTF_3[32,33].

The isotope-selective dissociation by one-photon excitation (Fig. 4) can also be achieved by selective excitation of predissociative states, as for isotopes of carbon in formaldehyde[34,36]. Interesting applications of ^{13}C-labelled compounds arise in various areas of medical diagnostics, for example, in metabolic investigations, in place of the more risky use of radioactive ^{14}C-isotopes and NMR tomography. For these purposes, an annual requirement of several hundred kilograms ^{13}C is to be expected. An industrially relatively well-developed process for laser-induced production of carbon isotopes[37] is the laser-induced multiphoton dissociation of Freon 22 (CF_2HCl). This results in the formation of C_2H_4, from which ^{13}C-enriched CF_2HCl can be obtained by thermal addition of HCl[38]. The cost of the CO_2 laser photons for this process can be reduced even more by employing a long-life discharge CO_2 laser[39]. Production of ^{13}C-isotopes via multiphoton dissociation of $^{13}CF_3Br$[40] is more expensive, due principally to the considerably higher cost of the starting material. Isotopes can also be enriched by selective excitation of bimolecular exchange[41,42] or addition reactions[43].

4. SYNTHESIS OF DEFINED SOLIDS AND CATALYSTS

Pyrolysis of gas mixtures with laser radiation can be used to produce catalytically active solids of variable composition (Fig. 5). After complete mixing of the gaseous starting materials and rapid heating in the laser beam, very small solid particles with homogeneous structure and large surface area are obtained, whose composition can be varied over a large range. In the Fischer-Tropsch synthesis with laser-synthesized Fe/Si/C catalysts, higher selectivity and preferential formation of the valuable light olefins ($C_2 - C_4$)[44] can be achieved.

By a similar process to that described in Fig. 5, powders of defined composition can be obtained for production of specialty ceramics[45]. CO_2 laser irradiation of $Fe(CO)_5SiH_4$[46], NH_3SiH_4[47], $(Me_3Si)_2NH$[48], or of $Si_4NH_3CH_4$[49-51], gas mixtured leads to particularly fine, pure, spherical, relatively nonclumping, and almost identically sized Si/Fe, Si_3N_4 and Si/C/N grains for production of ceramics.

Metallic glasses and other metastable phases can be produced by irradiation with intense laser pulses. Short laser pulses lead here to extremely rapid heating and melting, followed by very rapid cooling and solidification. Cooling rates up to 10^{12} K/s have been obtained for picosecond laser-heated alloy films. This allows the production of metallic glasses with unusual composition and properties[52-54].

5. LASER-INDUCED DEPOSITION OF MATERIALS ON SURFACES

In conventional thermally induced chemical vapor deposition of materials from the gas phase (CVD), relatively high temperature in the range 600-900°C must be used. The process temperature can be reduced to 350-450°C by employing electrical discharges in plasma-CVD. However, un-

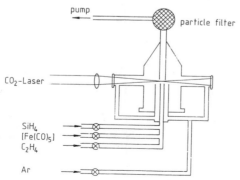

Fig. 5. Synthesis of Fe/Si/C catalysts by CO_2 laser pyrolysis of gaseous starting materials

desidered particles formed in wall reactions interfere with the process, radiation damage can be induced in the deposited films, the pressure range is restricted by the stability of the discharge, and the reaction volume can only be controlled to a very limited extent. The LICVD method (Laser Induced Chemical Vapour Deposition) for controlled production of certain species offers many advantages: control of the irradiation area in the range from μm^2 to cm^2 and of the irradiation time from fs to continuous operation; the absence of wall effects; and independent control of the substrate temperature for production of layers with extremely high purities.

Fig. 6 shows schematically three different possibilities for LICVD[55]. In the pyrolytic process (I) laser light is generally not absorbed by the reaction gas. The laser radiation absorbed produces localized heating at the substrate surface with a strong temperature gradient. Deposition therefore proceeds in a narrowly defined region, so that transport from the gas phase can occur from all directions. The pyrolytic method can be applied in a wide pressure region and allows deposition rates on the order of 1 to 10^2 $\mu m/s$, a factor of 10^2 to 10^4 higher than with a conventional CVD process[56]. In process II, a better directed and more homogeneous beam of the material to be deposited is obtained than in the ion vaporization technique otherwise generally employed[57]. In the photolytic process, the small energy width of the laser radiation for selective excitation of dissociating states is used to advantage. The products of laser dissociation can be condensed on the nearby surface or react chemically with other particles present in the surface layer. As shown in Fig. 6, the laser can also be focussed directly on the surface and scanned across it.

With the use of a frequency-doubled Ar^+ laser at 257nm, Cd from $Cd(CH_3)_2$[58] can be deposited by the photolytic method. Due to the non-linearity of the grain formation process, spatial resolution of 0.2 to 0.5 μm is obtained. In conjuction with computer control of the laser beam, these types of microscopic structures, which are written directly onto a substrate, allow "direct writing" of complex electronic circuits (VLSI microstructures)[59,60]. Besides metallic conducting layers, insulators and semiconductor layers can also be deposited. Not only continuously operating ion lasers but also CO_2 and UV excimer lasers can be employed,

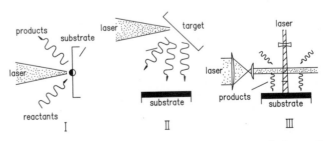

Fig. 6. Possibilities for the photochemical deposition of materials from the gas phase by means of IR or UV lasers

which permit large surface area production of thin layers. A principal application is the formation of amorphous Si layers for the production of low cost solar cells[61,62]. The application of UV-excimer lasers also offers interesting possibilities for economizing on valuable materials in the production of conducting compounds on various carriers (plastics, ceramics).

The rate of deposition can be increased considerably with the participation of the photoelectrons produced by the laser. Information on the primary processes occurring in applying UV excimer laser radiation for the deposition of metals from metal-organic compounds (MOCVD) can be obtained using multiphoton mass spectrometry with time-of-flight analysis. This technique allows the production of an entire mass spectrum at high sensitivity by using a single laser pulse[63].

Finally, the deposition and dissolution rate of metals on fluid/solid surfaces by laser radiation can be increased considerably ($> 10^3$). The laser beam is directed onto the cathode surface in an electrolyte solution and the laser wavelength chosen so that as far as possible no absorption occurs in the electrolyte[64,65]. Metal deposition without electric current is also feasible.

REFERENCES

1. P. Ehrlich, R. N. Pittilo, Hochpol., Forsch., 7:386 (1970)
2. G. Luft, Chemie Technik, 2:89 (1973)
3. M. Buback, H. Lendle, Makromol. Chem., 184:193 (1983)
4. S. G. Il'Yasov, I. N. Kalvina, G. A. Kyulyan, V. F. Moskalenko, E. P. Ostapchenko, Sov. J. Quantum Electron., 4:1287 (1975)
5. J. P. Fouassier, P. Jaques, D. J. Lougnot, T. Pilot, Polym. Photochem., 557 (1984)
6. R. K. Sadhir, J. D. B. Smith, P. M. Castle, J. Polym. Sci. Polym. Chem. Ed., 23:411 (1985)
7. C. Decker, J. Polym. Sci. Polym. Chem. Ed., 21:2451 (1963)
8. S. P. Pappas, B. C. Pappas, L. R. Gatechair, W. Schnabel, J. Polym. Sci. Polym. Chem. Ed., 22:69 (1984)
9. R. Bussas, K. L. Kompa, MPQ München Jahresbericht 1984
10. G. Amirzadeh, W. Schnabel, Makromol. Chem., 182:2821 (1981)
    T. Sumiyoski, W. Weber, W. Schnabel, Z. F. Naturforsch., 40a:541, (1985)
11. C. E. Hoyle, R. D. Hensel, M. B. Grubb, J. Polym. Sci. Polym. Chem. Ed., 22:1865 (1984)
12. D. J. Ehrlich, J. Y. Tsao, Appl. Phys. Lett., 46:198 (1985)
13. H. Niederwald, K. H. Richter, W. Güttler, M. Schwörer, Mol. Cryst. Liq. Cryst., 93:247 (1983)
14. C. Bräuchle, D. M. Burland, Angew. Chem., 95:612 (1983)
15. G. L. Paul, CLEO p. 48, San Francisco (1986)
16. W. Kuhn, H. Martin, Z. Phys. Chem., B 21:93 (1983)
17. Chemical and Biochemical Application of Laser, C. B. Moore Ed., Academic Press, New York (1977)
18. F. S. Becker, K. L. Kompa, Nuclear Technology, 58:329 (1982)

19. J. I. Davis, J. Z. Holtz, M. L. Spaeth, Laser Focus, 18.: 49 (1982), CLEO p. 89, San Francisco (1986)

20. R. W. Solarz, GRA, 85:24 (1985)
    R. L. Byer, IEEE J. Quant. Electr., QE 12:732 (1976)

21. V. S. Letokhov, Ann. Rev. Phys. Chem., 26:133 (1975)

22. P. Rabinowitz, A. Stein, R. Brickman, A. Kaldor, Appl. Phys. Lett., 35:793 (1979)

23. K. C. Kim, S. M. Freund, M. S. Sorem, D. F. Smith, J. Chem. Phys., 83:4344 (1985)

24. H. J. Circel, in: "Optoelectronics in Engineering, W. Waidelich. Ed., p. 706, Springer, Heidelberg (1986)

25. S. S. Miller, D. D. DeFord, T. J. Marks, E. Weitz, J. Am. Chem. Soc., 101:1036 (1979)

26. A. Kaldor, R. L. Woodin, Proc. of the IEEE, 70:565 (1982)

27. Laser and Appl., 4:11 (1985)

28. M. R. Humphries, O. L. Boarne, P. A. Hackett, Chem. Phys. Lett., 118:134 (1985)

29. A. V. Evseev, V. S. Letokhov, A. A. Puretzky, Appl. Phys., B 36:93 (1985)

30. Z. Linyang, Z. Yunwu, M. Xingxiao, Y. Peng, X. Yan, G. Mengxiong, W. Fuss, Appl. Phys., B 39:117 (1986)

31. F. Magnotta, I. P. Herman, J. Chem. Phys., 81:2363 (1984)

32. K. Takendu, S. Satooka, Y. Makide, Appl. Phys., B 33:83 (1984)

33. F. Magnotta, I. P. Hermann, Appl. Phys., B 36:207 (1985)

34. J. H. Clark, Y. Haas, P. L. Houston, C. B. Moore, Chem. Phys. Lett., 35:82 (1975)

35. R. E. M. Hedges, P. Ho, C. B. Moore, Appl. Phys., B 23:25 (1980)

36. L. Mannik, S. K. Brown, Appl. Phys.,B 37:75 (1985)

37. A. Outhouse, P. Lawrence, M. Gauthier, P. A. Heckett, Appl. Phys., 36:63 (1985)

38. J. Moser, P. Morand, R. Duperrex, H. V. D. Bergh, Chem. Phys., 79:277 (1983)

39. W. Fuss, W. E. Schmidt, K. L. Kompa, Verh. DPG, 20:1056 (1985)

40. O. N. Avatkow, A. B. Bakhtadze, V. Yu. Baranov, V. S. Dolyikov, I.G. Gverdsiteli, S. A. Kazakov, V. S. Letokhov,, V. D. Pismmenyi, E. A. Ryobov, V. M. Vetsko, Appl. Optic., 23:26 (1984)

41. D. Arnoldi, K. Kaufmann, J. Wolfrum, Phys. Rev. Lett., 34:1597 (1975)

42. T. J. Manuccia, M. D. Clark, E. R. Lory, J. Chem. Phys., 68:227 (1978)

43. M. Stuke, Spektrum der Wissenschaft, 4 (1982)

44. A. Gupta, J. T. Yardley, SPIE Vol. 458:131 (1984)

45. W. M. Shaub, S. H. Bauer, Int. J. Chem. Kin., 7:509 (1975)

46. D. J. Frurip, P. R. Staszak, M. Blander, J. Non-Cryst. Solids, 68:1 (1984)

47. Y. Kizaki, T. Kandori, Y. Fujitani, Jap. J. Appl. Phys., 24:806 (1985)

48. G. W. Rice, R. L. Woodin, SPIE Vol. 458:98 (1984)

49. W. R. Cannon, S. C. Danforth, J. H. Flint, J. S. Haggerty, R. A. Marra, J. Am. Ceram. Soc., 65:324 (1982)

50. W. R. Cannon, S. C. Danforth, J. S. Haggerty, R. A. Marra, J. Am. Ceram. Soc., 65:330 (1982)

51. J. H. Flint, J. S. Haggerty, SPIE Vol. 458:108 (1984)

52. E. Huber, M. von Allmen, Phys. Rev. , B 28:2979 (1983)

53. M. von Allmen, <u>Mat. Res. Soc. Symp. Proc.</u>, 13:691 (1983)
54. K. Attolter, M. von Allmen, Appl. Phys., A 33:93 (1984)
55. S. D. Allen <u>J. Appl. Phys.</u>, 52:6501 (1981)
    D. Bäuerle, P. Irsingler, G. Leyendecker, H. Noll, D. Wagner, <u>Appl. Phys. Lett.</u>, 40:819 (1982)
56. G. D. Davis, C. A. Moore, R. A. Goltscho, <u>J. Appl. Phys.</u>, 56:1808 (1984)
57. R. Solanski, W. H. Ritchie, G. J. Collins, <u>Appl. Phys. Lett.</u>, 43:454 (1983)
58. T. F. Deutsch, D. J. Ehrlich; R. M. Osgood, <u>Appl. Phys. Lett.</u>, 35:381 (1979)
59. D. J. Ehrlich, J. Y. Tsao, <u>J. Vac. Sci. Technol.</u>, B 1:969 (1983)
60. Laser Processing and Diagnostics, D. Bäuerle, Hrsg., Springer Series in Chemical Physics, 39 (1984)
61. R. Bilinchi, I. Hianinoni, M. Musci, R. Murri and S. Tacchetti, <u>Appl. Phys. Lett.</u>, 47:279 (1985)
62. H. M. Branz, S. Fan, J. H. Flint, D. Adler and J. S. Haggerty, <u>Appl. Phys. Lett.</u>, 48:171 (1986)
63. W. G. Hollingsworth, V. Vaida, <u>J. Phys. Chem.</u>, 90:1235 (1986)
64. R. J. von Gutfeld, E. E. Tynan, R. L. Nelcher, S. E. Blum, <u>Appl. Phys. Lett.</u>, 35:651 (1979)
65. H. Pummer, <u>Phys. Bl.</u>, 7:199 (1985)

LASERS IN THIN FILMS DEPOSITION

C. Calì

Dipartimento di Ingegneria Elettrica

Viale delle Scienze, 90128 Palermo (Italy)

1. INTRODUCTION

Laser sources are successfully utilized in thin film deposition: applications include thickness monitoring of dielectric thin films, deposition of compounds, fabrication of layers with spatially varying thickness, and generation of small patterns. In this paper I will describe some applications in this area, developed and used in our laboratory.

The first application employes laser light as a light source for interference measurements. In the second and third cases discussed, laser light acts as an energy source for materials evaporation. In the fourth case, laser light removes the adsorbed layer from a substrate surface, allowing material to be deposited on it.

2. THICKNESS MONITOR FOR DIELECTRIC THIN FILM DEPOSITION

The problem of controlling the thickness of a dielectric layer during its formation is of fundamental importance in fabricating antireflection coatings, dieletric mirrors, beam splitters, filters and other optical devices; although a wide variety of thickness-monitoring methods exist, optical monitoring techniques are preferred for dielectrics. These techniques are based on the relation between the intensity of a reflected or transmitted beam from a substrate-film combination, and the film thickness.

Assuming that a monochromatic beam propagates in a non-absorping substrate-film combination at near-normal incidence, equation (1) illustrates the relation between the reflectivity R and the physical thickness d of the film[1]

$$R = \left(\frac{n_o - 1}{n_o + 1}\right)^2 \frac{1+((n^2 - n_o)/(nn_o - n))^2 \tan^2 (\beta d)}{1+((n^2 + n_o)/(nn_o + n))^2 \tan^2 (\beta d)} \tag{1}$$

$$= R_o \frac{1+((n^2 - n_o)/(nn_o - n))^2 \tan^2 (\beta d)}{1+((n^2 + n_o)/(nn_o+ n))^2 \tan^2 (\beta d)} \quad ;$$

n_o denotes the refractive index of the substrate and R_o its reflectivity; n is the refractive index of the film; $\beta=2\pi/\lambda$, is the phase constant in the film, λ being the wavelength in the film material. R or R/R_o is a periodic function of d, so the thickness of a layer can be monitored by a reflectivity measurement.

Fig. 1 shows the reflectivity of a non-absorbing film-substrate combination, relative to the substrate reflectivity, as a function of the ratio d/λ. As the film of index 2.3 is deposited onto the substrate of index 1.5 the reflectivity of the substrate-film combination passes through maxima and minima at quarter-wave and half-wave optical thickness values, respectively.

Typical optical monitoring systems use a white light source and a monochromator to generate the probing beam[2,3,4]. The combination of white light and a monochromator allows an easy choice of wavelength; however, the low level and instability of the radiation intensity require complex electronics for detection.

Detection problems are simplified if a laser beam replaces the white lamp and the monochromator. The small size of the laser beam also allows the use of a simple technique to compensate for laser fluctuation effects[1].

The principle of operation is illustrated in Fig. 2. The monitor substrate (L) is in the form of a planoconvex lens, which is helpful in separating the reflection from the back side from that due to the coated surface. The beam from a laser is directed onto a plane mirror (M) mounted on the shaft of a motor. As the mirror axis is slightly misaligned from

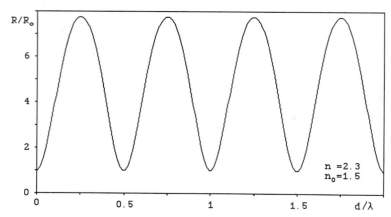

Fig. 1. Reflectivity of a non-absorbing film-substrate combination, relative to the substrate reflectivity, as a function of the ratio d/λ

the shaft, the reflected beam describes an elongated ellipse, which is focused at near normal incidence on the substrate on which the film is growing. Part of the elliptical path is shadowed from the incident material
being deposited, so the beam leaving the coated surface is periodically
modulated due to the difference in reflectivity of the two zones. This
beam is refocused on a detector (D) whose output is displayed on an oscilloscope. Also, a portion of the scanning beam path is blocked by a screen
(S) in order to display the zero level on the oscilloscope. The oscilloscope display can be directly used to calculate, by equation (1), both n
and d. Note that by measuring R/R_0 when $d=\lambda/4$ it is possible, using equation (1), to find n. Since the display is repeated at the line frequency,
slow fluctuations in laser beam intensity do not affect the measurement;
the rate of change in the display can be used for a rough estimate of the
evaporation rate. In its basic form, the technique has poor sensitivity
for films of thickness close to zero or a multiple of $\lambda/4$, due to the form
of the R versus d curve; also, the film index must be sufficiently different from n_0 to give a large variation in reflectivity between coated and
uncoated regions. All these shortcomings can be removed by depositing the
film to be monitored on a monitor substrate which has been previously coated with a pre-deposit of suitable index and thickness.

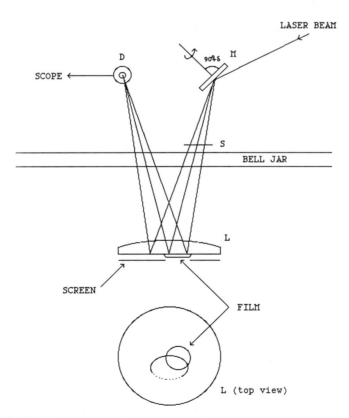

Fig. 2. Principle of operation of the film deposition monitoring method

3. COMPOUND DEPOSITION

The attractiveness of laser radiation as a means for inducing the evaporation of a substance lies in the possibility of selectively heating a small volume of evaporant and, for some lasers, in the ease with which the output can be pulsed with very high energy yields, leading to a precisely controlled technique of flash evaporation.

A convenient approach is to use a small CO_2 laser with fixed mirrors and a pulsed discharge. This technique provides sizable energy in a single pulse even with a conventional low-pressure laser, and useful speeds of deposition are readily obtained for many compounds at reasonable source-to-substrate distances. The pulses are fast enough to prevent decomposition of the evaporant during heating, allowing congruent deposition of several compounds that cannot be evaporated with the conventional thermal technique; further, even moderate output from the laser, when focused, is sufficient to ensure evaporation of highly refractory materials such as Al_2O_3 and MgO.

We have studied[5] the evaporation behavior of various dielectric and semiconducting materials, concentrating on those which are highly refractory or which react with the evaporant substrate.

We have investigated[5] the dynamics of the pulsed growth of SiO_2 and ZnS films. The thickness deposited in a single pulse is a fraction of 1 Å; this can be detected by our monitor when used in a nonscanning mode. A He-Ne laser beam is directed to a fixed spot on the exposed area of the monitor lens, and the intensity of the reflected beam is recorded as a function of time during laser irradiation of the evaporant sample. The monitor reflectivity can be made to approximate a linear function of film thickness with a suitable coating deposited prior to the film to be monitored[1]; further, the maximum change in reflectivity due to a given film thickness increment occurs when the preliminary coating is almost, but not quite, a perfect antireflection layer. With care in eliminating vibrations and by using two detectors in a differential arrangement to cancel fluctuations in the He-Ne laser beam intensity, a 0.2% change in the intensity of the reflected beam can be readily detected.

The pulsed growth of the thin film, which has been observed directly, offers interesting possibilities for multisource evaporation: for example, two synchronous trains of laser pulses could be directed onto two samples of different materials. The use of a shutter in front of the monitor would allow separate monitoring of the two rates of evaporation, which can be controlled individually by appropriate shaping of the two pulse trains. This technique would allow a precise control of composition during the growth of the film.

A different technique has been used for CdS thin film deposition, involving the photosublimation effect on the source material induced by Ar laser light at wavelengths above the CdS bandgap[6]. The method allows the formation of high quality films.

An Ar laser beam at λ=488 nm is focused on the surface of bulk CdS. Single CdS crystals are necessary to obtain films with high stoichiometry and purity. The evaporation proceeds very smoothly, compared with other techniques such as thermal or laser flash approaches. That is, no large particle ejection is observed and, as a consequence, very clean films are obtained. The absence of particle ejection can be explained by the concentration of radiation in the outer layers of the source crystal, which determines what is being evaporated. As a consequence, no deep heating occurs. Particle-free films are rather difficult to obtain using other techniques, because of non-uniform temperature distribution in the source material.

The supposed mechanism for the sublimation consists of a loosening of molecular bonds in the crystal by a concurrent action of heat and light, the latter creating a very high electron-hole concentration because of its above-bandgap energy.

4. VARIABLE THICKNESS LAYER DEPOSITION FOR GAUSSIAN MIRROR FABRICATION

Mirrors with a quasi-gaussian reflectivity profile as a function of radius are useful in unstable resonators for high-gain lasers[7]. Such mirrors can be built by stacking different index dielectric layers with spatially controlled thickness. Most applications require a reflectivity around 50% in the center. Only a multilayer structure can satisfy such a requirement. As an example the plot of the reflectivity versus the distance from the edge to the center of a multilayer mirror with spatially controlled thickness is plotted in Fig. 3. A schematic of the layer arrangement is represented in the left upper corner. A layer with thickness increasing linearly from 0 to λ/4 with index n=3 is superimposed another linearly increasing thickness layer (from λ/4 to λ/2) that is deposited on a glass substrate (n=1.5). By comparing a perfectly gaussian profile with that given by a linearly decreasing thickness film, it is possible to observe an acceptable agreement.

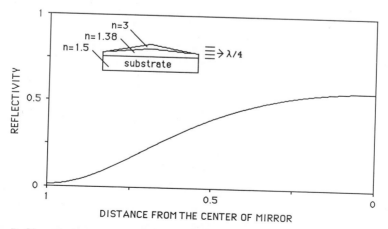

Fig. 3. Reflectivity versus the distance from the edge to the center of a spatially controlled thickness multilayer mirror.

213

Masking is the main difficulty in making a dielectric film of spatially variable thickness. For large area deposition, above a 1mm. diameter,
it is easy to use a properly shaped rotating mask. It is not easy, however,
to employ such a technique for substrates a few millimeters in diameter.
We have developed a method[8] based on the shadow effect given by non-contact
masking. The experimental setup is shown in figure 4. Films are deposited
by laser evaporation, creating a source much smaller than the mask aperture
d. A point source is essential to obtain a circularly symmetric deposit.
The thickness profile due to ideal mask shadowing can be easily calculated geometrically. The actual thickness profile, however, is smoothed with
respect to the calculated profile. A possible explanation of this phenomenon comes from molecular interactions in the released vapor stream. Non-uniform reflectivity mirrors have been realized for 633 and 1060 nm wavelengths. A laser scanning optical thickness monitor[1] has been used, for
"in situ" reflectivity profile measurements.

5. SMALL PATTERN GENERATION

Laser light affects the sticking coefficient of the impinging molecules of some materials. Investigations have been made when ZnS is evaporated in vacuo, on a CdS substrate[9,10,11].

In the initial experiments, a CdS film was first deposited on glass
by vacuum evaporation, using either a standard thermal source or, for
better stoichiometry, laser-induced vaporization from CdS pellets[5]. ZnS
was then similarly deposited; it was found that the initial sticking coefficient on the CdS film was very low, except in the area illuminated by
the Ar light. When a cylindrical lens was used to focus the Ar light, the
illumination was restricted to a line tens of microns wide. The power density on the sample surface was around 1 W/cm^2 and 4880 Å light was used

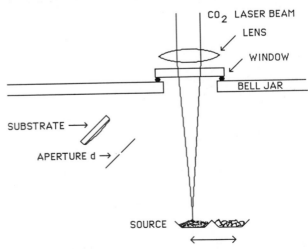

Fig. 4. Experimental setup for circular, variable thickness, layers
        deposition

for most experiments; shorter wavelengths gave the same results, whereas 5145 Å light (for which the CdS begins to show appreciable transmission) was much less effective.

The localized growth of ZnS can be directly observed in real time by monitoring the resulting change in reflectivity with a scanning He-Ne laser beam at 6328 Å using a plano-convex lens as the glass substrate[1]. If ZnS deposition is continued, growth starts occurring in the unilluminated regions of the CdS; the maximum light-induced thickness differential was found to be around 700 Å, which can be grown in about 20 seconds. The value of the ZnS thickness can be derived from the reflectivity change if the primary CdS layer optical constants are well known; an independent mechanical profilometer measurement was used to confirm the calculated values.

A memory effect is exhibited by the photo-induced growth: the preferential growth in the illuminated areas persists if ZnS deposition is performed after the Ar laser illumination is turned off, for a delay up to tens of minutes. Similar effects were observed by depositing, instead of ZnS, other materials as ZnO and pure Zn on CdS.

Adsorbed surface layers play a fundamental role in laser photochemistry. A developed optical diagnostic technique based on reflectometry[10] is particularly suited for real time studies of adsorbed layers, and has been applied to the investigation of the CdS-vacuum interface. With appropriate electronics and by choosing the optimum thickness for the CdS layer, an effective adsorbed layer variation of a fraction of monolayer can be detected differentially on the CdS surface, by comparing the reflectivities of two regions of the surface. In our experiment, 4880 Å radiation was directed on one of the reflectometer sampling regions and alternatively switched on and off with a slow period. The resulting change in reflectivity shows a fast and slow exponential response, with only the slow response remaining after the first laser illumination turn-on transient. The slow response can be explained as a thermally induced change in the refractive index of the film, which strongly absorbs the Ar laser radiation. The initial fast reflectivity change is due to the photo-induced desorption of an adsorbed layer form the CdS surface; with the optical parameters used in the experiment, the thickness of layer removed can be calculated from the measured $\Delta R/R$ and is of the order of 1 Å. This layer does not appear to form again for several minutes after the first illumination, in agreement with the memory effect mentioned before. The "fast" portion of the reflectometric signal disappears if the experiment is repeated at a sufficient high temperature, around 250 °C: this would indicate that the adsorbed layer is thermally desorbed at this temperature.

6. CONCLUSIONS

This paper has described a number of applications of laser radiation in the specific field of thin film deposition. It is evident that only a laser beam is suited for many of these applications.

REFERENCES

1. V. Daneu, Optical Thickness Monitor for Thin Film Deposition, Appl. Opt., 14, 962 (1975)

2. D. H. Harrison, Interference Coatings - Practical Considerations, Proc. Soc. Photo-Optical Instr. Eng., 1, Vol. 50, "Optical Coatings" (1974)

3. R. Hermann and A. Zöller, Automated optical coating processes with optical thickness monitoring, SPIE, 2, Vol.652, "Thin Film Technologies II" (1986)

4. M. Lardon and H. Selhofer, Thin film production with a new fully automated optical thickness monitoring system, SPIE, 10, Vol. 652, "Thin Film Technologies II" (1986)

5. C. Calì, V. Daneu, A. Orioli, and S. Riva-Sanseverino, Flash evaporation of compounds with a pulsed-discharge CO_2 laser, Appl. Opt., 15, 1327 (1976)

6. C. Arnone, C. Calì and S. Riva-Sanseverino, Laser evaporation technique for CdS thin film deposition, Technical Digest of 1984 IEEE International Workshop on "Integrated Optical and related technologies for signal processing", Sep. 10-11, 1984, Florence, Italy

7. P. Lavigne, N. McCarty and J. Demers, "Design and characterization of complementary Gaussian reflectivity mirrors", Appl. Opt., 24, 2581 (1985)

8. C. Zizzo, C. Arnone, C. Calì and S. Sciortino, "Fabrication and characterization of tuned Gaussian mirrors for visible and near infrared", to be published in Optics Letters.

9. C. Arnone, V. Daneu and S. Riva-Sanseverino, "Effect of laser light on the sticking coefficient in ZnS thin-film growth", Appl. Phys. Lett., 37, 1012 (1980)

10. C. Arnone, C. Calì, M. L. Rimicci and S. Riva-Sanseverino, "Laser Film thickness modulation and induced small pattern generation", Journ. Non-Cryst. Solids, 47, 263 (1982)

11. C. Arnone, C. Calì and S. Riva-Sanseverino, "Study of photo-induced thin film growth on CdS substrates", Proc. of MRS: Laser – Controlled Chemical Processing of Surfaces, Elsevier Science Publishing Co., 29 (1984)

FUNDAMENTALS OF LOW-POWER LASER PHOTOMEDICINE

T. Karu

Laser Technology Center USSR Academy of Sciences

Moscow Region, Troitzk, 142092 USSR

All biomedical laser application are based on the interaction of laser light with biological systems. Such interaction causes a broad spectrum of effects which can be divided into three principally different groups[1]. First, low-intensity laser light is absorbed, reflected or reradiated (as fluorescence) by the substance so that no changes occur within it. Such interactions form the basis for the laser diagnostics (spectral diagnostics of molecules, and macrodiagnostics on the tissue level). Second, low intensity UV and visible radiation can excite electronic states in molecules, and specific photobiological effects occur due to excitation of chromophores in cells (endogenous or exogenous). These processes occur under the light from incoherent sources as well, but the use of laser light can give several benefits from a practical point of view. This group of effects encompasses molecular photobiology and photomedicine. The third class of effects involves high intensity laser radiation which causes damage to tissues by thermal or hydrodynamical destruction. Such processes, rarely observed with incoherent light sources, form the basis for laser surgery. The principal methodologies based on the second and third type of light-biological system interactions are shown in Fig. 1.

My lecture deals with the area of molecular photobiology and photomedicine, and more specifically one of the methods of this group - low power laser therapy or, as it is sometimes called, laser biostimulation. The methods of molecular photomedicine are widely used in clinical practice, e.g. phototherapy of neonatal jaundice (hyperbilirubinemia), phototherapy and photochemotherapy of various skin diseases and photoradiation therapy of cancer with the aid of the haematopophyrine derivative (HPD)[2,3,4]. In some cases the photoacceptor molecules are artificially incorporated into cells (as in the case of psorales or HPD, which are called exogenous photoacceptors), and sometimes the light is absorbed by natural components of a cell (endogenous photoacceptors). Low-power laser therapy belongs to the group of methods of molecular photomedicine based on the action of light endogenous photoacceptors.

(LASER) MOLECULAR PHOTOMEDICINE	LASER SURGERY
Specific photobiological effects due to excitation of chromophores	Nonspecific photo-destruction
Photo(chemo)therapy of various dermatites	Photoablation
UV	
Phototherapy of hyperbilirubi-nemia	
BLUE, GREEN	
Photoradiation therapy of cancer with porphyrins	Thermal coagulation and surgery
RED	
Phototherapy with low intensity red (blue, far red) light	

Fig. 1. Methods of laser photomedicine and surgery

Even in ancient times, red light was used in medical treatment to cure disease, and treatment with red light was also among the methods used by N. R. Finsen, the father of contemporary phototherapy. Much experimental work was done in the last century and in the first third of this century. These data are reviewed in references 7, 8. Only in the last decades has laser light been used. It is clear that the stimulative effects caused by low-intensity blue and red light are not laser-specific, and the mechanism of light action does not depend upon the coherence and polarization of the light[8]. "Laser biostimulation" is a phenomenon of photobiological nature. In this case lasers are only handy tools, especially for clinical use. Table 1 illustrates possible fields of low-power laser therapy in contemporary medicine (data from references 5,6,9,10).

Despite the fact that low-intensity light (conventional and laser sources) has been used in clinical practice for a long time, the mechanism of its action is not clear. What is more, the method itself seems to be rather mysterious because of the very low intensities ($10^{-4} - 10^{-2}$ W/cm^2) of light used. This phenomenon is possible when we suppose that light acts as a trigger for metabolic rearrangements in the cell, as discussed in references 7,8.

THE ACTION OF MONOCHROMATIC VISIBLE LIGHT AT THE CELLULAR LEVEL

1. Irradiation enhances the rate of multiplication of microorganisms and the proliferation of mammalian cellular cultures

Fig. 2 shows the dynamic changes in the number of cells in irradiated and control cultures of E. coli and HeLa. The growth curves show that in case of eucaryotic cultures (T. spaerica[12], HeLa, Fig. 2b) the growth

Table 1. Photography with low-intensity (0.1 - 100 mW/cm^2) red (He-Ne laser, 632.8 nm), far red (diode lasers, 830-890 nm), and blue (He-Cd laser, 441.6 nm) light.

Direct irradiation	Irradiation of acupuncture points and reflexogene zones
– indolent wounds and trophic ulcers	– rheumatoid and metabolic-dystrophic polyarthritis
– superficial and deep burns	– bronchial asthma
– periodontosis and stomatitis	– pain relief
– osteochondrosis and arthritis	
– damage of peripheral nerves	

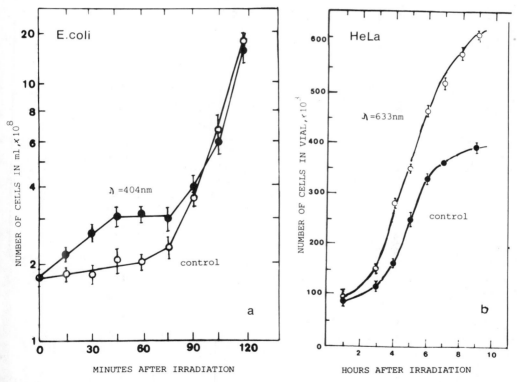

Fig. 2. Growth curves of control cultures and cultures irradiated with (a) blue or (b) red light for (a) E.coli or (b) HeLa. E.coli was irradiated in a buffer and then inoculated into growth medium[11,12]. HeLa cells were irradiated as a confluent monolayer, then trypsinized after 3 h and planted into fresh nutrient medium as described in reference 13.

is increased exponentially due to the decreasing generation time in the exposed culture (e.g., by 1,5 to 1,5 times in T. sphaerica[12]). In case of E. coli, the lag period of growth is substantially decreased or even absent in the exposed culture, whereas the difference in exposed and control cultures is minimal during the exponential period of growth (see Fig. 2a for blue light; for red light, see reference 15).

The stimulation of growth and metabolic processes in the cell depends on the dose and wavelength of the light used for irradiation. Some examples are shown in Figs. 3 and 4.

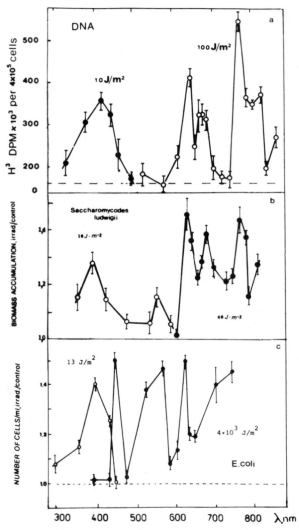

Fig. 3. Action of visible monochromatic light for (a) DNA synthesis stimulation in exponentially growing HeLa cells, (b) for protein synthesis stimulation of Saccharomycodes ludwigii, and (c) for growth stimulation of E. coli. The experimental conditions are described in references 14,12,15, respectively.

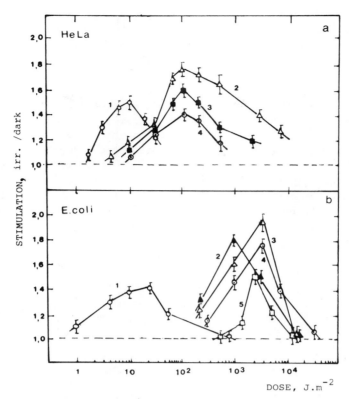

Fig. 4. The effect of irradiation dose on stimulation of (a) DNA synthesis rate in exponentially growing HeLa cells (1 − λ = 404 nm,
2 − λ = 760 nm, 3 − λ = 620 nm, 4 − λ = 680 nm) and (b) E. coli
culture growth (1 − λ = 404 nm, 2 − λ = 560 nm, 3 − λ = 750 nm,
4 − λ = 632,8 nm, 5 − λ = 454 nm)[28,29,34]. The irradiation conditions and measurements are described in references 14,15.

Fig. 3 shows the action spectra of visible light for stimulation of DNA
synthesis in HeLa cells and for growth stimulation of T. sphaerica and
E. coli. Some other examples are reviewed in reference 7. All these action
spectra are practically of the same type, having maxima in almost every
visible-light band (at about 400–450, 570, 620, 680 and 830 nm). As one
can see, the He−Cd (411,6), He−Ne (632,8) and diode (830) laser radiation
wavelength fall within these regions.

Fig. 4 presents data on the stimulation of (a) DNA synthesis in HeLa
cells and (b) the growth of E. coli culture for various irradiation doses
and wavelengths. It can be seen that there are two groups of active spectral regions. The first covers wavelengths of 365,404 and 434 nm (see Fig.
4 for λ = 404 nm), and has a stimulative action at doses lower by a factor
of 10 − 100 than those of the second group of wavelengths (454, 560, 633
and 750 nm). In other words, achieving a maximum effect with light in the
near UV and blue regions requires at least an order magnitude lower dose
then that required to achieve the same effect with red or far red light.

As we mentioned above, irradiation caused a decrease in the generation time of yeast in its exponential phase of growth (Fig. 1b)[12]. Autoradiographic studies indicated that in HeLa cells the irradiation with He-Ne laser caused the decrease in the cell generation cycle by shortening its G_1 phase (Fig. 5).

To confirm the hypothesis that light can indeed serve as a stimulus for proliferation, the resting cells (G_o cells), human peripheral lymphocytes, were irradiated with a Ne-Ne laser and the resulting conformational changes of DNA were studied[18]. By measuring the degree of acridine orange binding to DNA, it was found that the changes occurring in the lymphocyte chromatine structure a few hours after irradiation were similar to those caused by phytohemagglutin in (PHA), a well known mitogen. The binding of the dye to DNA increased during 45-90 min. after the irradiation, then dropped to the control level in 3-4 hrs, and then increased again (Fig.6).

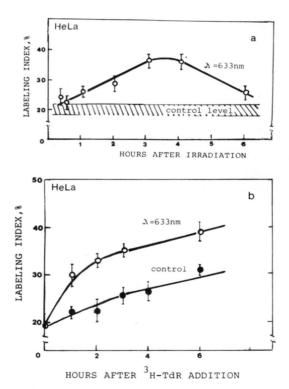

Fig. 5. Changes in (a) labeling index (per cent of cells in S phase of cellular cycle, measured by pulse labeling with H^3-thymidine at various times after the irradiation), and (b) variation of the per cent of the labeled cells during continuous labeling with H^3-thymidine after the irradiation of exponentially growing HeLa cells with a He-Ne laser (λ = 632,8 nm, D = 100 J/m^2), measured as described in reference 16

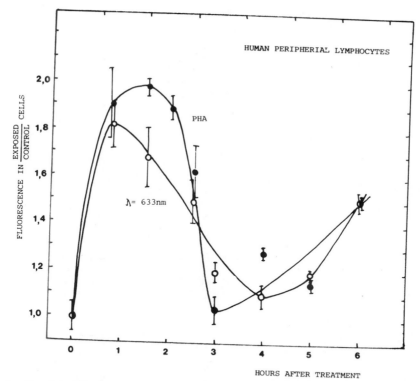

Fig. 6. The binding of acridine orange to DNA of human peripheral lymphocytes after irradiation of lymphocytes with a He-Ne laser
dose of 56 J/m^2, or after adding phytohemaggylutinine (4 μg/ml),
expressed as ratio of fluorescence intensity of treated and control cells as described in reference[18].

If the light is the proliferative stimulus, then it must have some
effect on the systems known to regulate cellular proliferation (or at
least taking part in this process). The cyclic adenosine monophosphate
(cAMP) system has been demonstrated to control both the biosynthesis of
DNA and RNA and the realization of the biologic activity of these macromolecules[21]. The irradiation of Chinese hamster fibroblasts with blue
(404 nm) or red (632,8 nm) light caused changes in the intracellular cAMP
level, whereas irradiation with light at 546 or 700 nm had no appreciable
effect[17] (see Fig. 7). There are some similarities between the wavelengths
which are effective for DNA an RNA synthesis stimulation (Fig. 3a)[14], and
those which cause variations in the cAMP level.

Our observation of changes in the intracellular cAMP concentration
following irradiation may help relate the growth stimulation effects to
the known regulatory mechanisms of the proliferation activity of cells.
These observations may also assist in further studies aimed at revealing
the mechanism of the bio-stimulating effect of red light, as it is well
known that there exist causal relations between variations in the concentrations of cAMP and Ca^{2+} on the one hand and the rate of synthesis

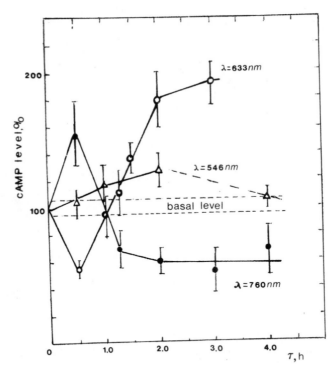

Fig. 7. Changes in intracellular cAMP level after irradiation of Chinese hamster fibroblasts with light (1) 632,8 nm, (2) 760 nm, or (3) 546 nm of dose 1 x 10^3 J/m^2, as described in reference 17

of DNA and RNA on the other at the early stages of regenerative processes[22].

In summary, we may conclude that irradiation with monochromatic visible light in the blue, red, and far red regions can enhance metabolic processes in the cell and activate its proliferation.

2. The primary photoacceptors are respiratory chain components, and regulation of the cellular metabolism occurs via changes in redox potential

The probability of respiratory chain components being the primary photoacceptors (i.e., compounds absorbing light at wavelengths effective in bringing about responses for irradiation) was discussed in references 7,8, using three approaches:

(1) - comparison between the action spectra for photoresponse and the absorption spectra of respiratory chain components, since an action spectrum should closely parallel the absorption spectrum of the photoreceptor compound;

(2) - use of substances which are known to act as quenchers of the
    excited state of the presumed photoacceptor molecules;
(3) - demonstration that the components or procedures known to in-
    fluence the probable mode of action of a potential photoacceptor
    do in fact influence the magnitude of the photoresponse.

The latter approach enables us to answer the next question: how does
the absorption of light quanta influence cellular metabolism? The answer
could be: via changes in cellular redox potential. Four series of experiments support this suggestion.

First, reducing agents such sodium dithionite ($Na_2S_2O_4$) modify the
light-growth response. Addition of $Na_2S_2O_4$ to the E. coli suspension reduces the stimulation level caused by light. The effect depends on the
$Na_2S_2O_4$ concentration, no matter what spectral band, blue, red or far red,
is used for irradiation[19]. When the irradiation dose is varied, with the
concentration of $Na_2S_2O_4$ being kept invariable at 5 mM, the influence of
$Na_2S_2O_4$ on the light-growth response displays a characteristic dependence
on the irradiation dose (Fig. 8). The reducing agent iself has non influence on the division of non-irradiated bacteria or on bacteria irradiated
in low doses. When $Na_2S_2O_4$ was added to cells that were subsequently irradiated with such doses as caused the maximum effect in the absence of
$Na_2S_2O_4$, cell division was observed to be inhibited. As the irradiation
dose was further increased, the inhibition of cell division decreased,
(see Fig. 8). The results of these experiments give us reason to believe
that blue, red, and far red light cause changes in redox potential of the
cell, this process being the starting mechanism for bacterial cell division.

Second, the decrease of oxygen concentration (and the associated
decrease of the intracellular redox potential) achieved by blowing the
cultural medium into the vials with cells using N_2,changes the magnitude
of the stimulative effect of light (632,8 nm)(Table 2) as well as the
shape of the action spectrum[20]. In this kind of experiment the redox potential of the medium and respectively the redox potential of the cells
are reduced, as evidenced by an increase of the NADH fluorescence[23]. This
result enables us to propose that irradiation causes a change of intracellular redox potential towards oxidation.

Third, the redox dye pethylene blue stimulates DNA synthesis in the
dark; in other words, this dye appears to mimic the effect of irradiation
(Table 2).

Fourth, the addition α-naphtylacetic acid (auxin) to the cells prior
to irradiation with red light modifies the effect of light in a concentra-
tion-dependent manner. Auxin was found to have no effect on the nonirradia-
ted cultures (Fig. 9). Auxin is believed to conduct protons through membranes in pH-dependent manner[27]. In as much as the bacterial respiratory
chain, when activated by light, acquires the ability to produce the necessary pH gradient to enable the cell to start dividing earlier[29], auxin
causes a decrease in this gradient.

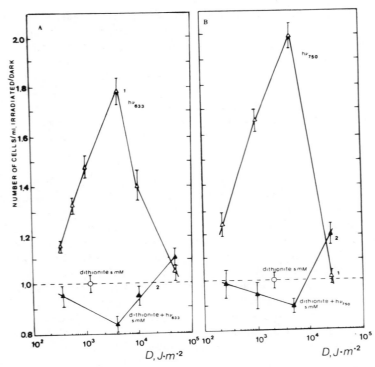

Fig. 8. Changes of E. coli growth as a function of irradiation dose by
irradiation with (a) red or (b) far red light with or without
$Na_2S_2O_4$ (5×10^{-3} M). Sodium dithionite in concentration 5 mM
did not influence the growth rate of non-irradiated cultures,
as denoted with dashed line. $Na_2S_2O_4$ was added to a bacterial
suspension just befor the irradiation, as described in reference 19.

Table 2. Action of red light = 632,8 nm on the DNA synthesis
rate in HeLa cells (measured as in ref.14) after decreasing pO_2 in nutrient medium by blowing through with
N_2 60 sec., or after 60 min. incubation with methylene
blue (MB, 1 mM).

Dose, J/m^2	0	6	30	10^2	10^3
Cells					
Intact	100%	–	–	–	–
Irradiated					
– in normal conditions	--	102±5	150±8	183±12	153±9
– decreased pO_2	--	106±8	163±12	222±11	188±7
– with MB	--	200±9	227±11	213±8	189±6
With MB in dark	202±6	–	–	–	–

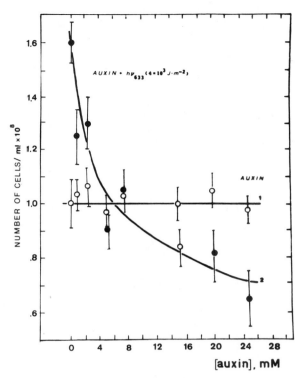

Fig. 9. Changes in E. coli growth rate as a function of auxin concentration when auxin was added (1) to non-irradiated cells or (2) to the culture before the irradiation its with a He-Ne laser (D = 4×10^3 J/m^2), as described in ref. 29.

The general conclusion from these experiments appears to be that in these cases when the treatment before irradiation lowers the redox potential in the cell, the magnitude of the effect from irradiation is stronger. And when the redox potential of cell is high enough before irradiation, the effect of light is weak, if any. This means that the photosensitivity of cells is not an all-or-nothing phenomenon, and the cell can respond to the light stimulus in various degrees, depending on the redox potential of the cell before irradiation.

Our hypothesis about redox regulation makes it possible to explain the relationship between such distant events as the absorption of light and the acceleration of growth or stimulation of proliferation (Fig. 10). The redox state of a cell appears to be controlled by the state of phophorylation of adenine nucleotides in the cytoplasm and mitochondria, and therefore by the operation of the respiratory chain[24]. The redox level of the cell may affect its metabolism by influencing the cAMP level[26]. Changes in cellular redox potential are connected with the electrophysiological parameters of cellular membrane (E_m, ion fluxes). Increasing the redox potential causes the influx of K^+ 30, and connected with it

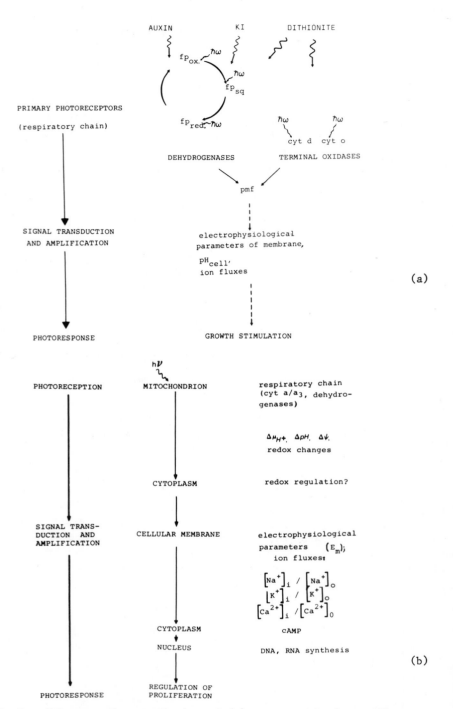

Fig. 10. Possible ways for regulation of (a) procaryotic (E. coli) and
(b) eucaryotic cell metabolism by irradiating the cells with
monochromatic visible light. fp_{ox}, fp_{red}, fp_{sq} - flavoproteins
in oxidized, reduced or semiquinone forms; E_m - electrical po-
tential; $\Delta\mu$(pmf) - proton motive force, ΔpH and $\Delta\psi$ - components
of pmf. In case (a) the possible action of chemicals in modifying
the light-growth response (auxin, KI, $Na_2S_2O_4$) is shown.

the efflux of Na^+ and H^+. This leads to the change in E_m and an increase in the cytoplasmic pH. The short-term alkalization of the cytoplasm[31] as well as the changes in E_m[26] are necessary components of mitotic stimulus transmission in a great variety of cells. Methylene blue increases the influx of K^+[30], and this is the reason why it mimics the action of light. The inhibition of the respiratory chain occurs in presence of external K^+, as well as by incubation with the proton conductors[28] such as auxin in our experiments.

ON THE MECHANISM OF LOW-POWER THERAPY

The "local" effects of phototherapy in treating trophic ulcers and indolent wounds with the He-Ne laser or with the He-Cd laser, or else with diode lasers operating in the far red region, may be explained by an action of low-intensity visible light on cell proliferation. In the area of such injuries conditions are created (low oxygen concentration and pH, lack of necessary nutrients) which prevent proliferation, so that the cells enter the G_0 phase or remain in the G_1 phase. For such cells, light may serve as a signal to increase proliferation.

When irradiating fresh wounds, the effect of irradiation can be minimal, if any. This happens in cases where proliferation is active and the regeneration of tissue integrity occurs at a more or less maximal (normal) rate. This may be the reason why no phototherapeutic effect is generally observed when irradiating fresh experimental wounds, and whereas light proves to be more effective in the case of "old' and "bad" wounds. Of course, it should be born in mind that proliferation control in the case of a whole organism is a much more complex process than in the case of a cell culture, because in the organism there is a hierarchy of proliferation control structures.

Many authors have described "systemic" effects of laser irradiation, meaning effects observed at sites distant from the treated areas[5,9,10]. It is quite possible that in these cases the irradiation acts on peripheral nerves. Indeed, the peripheral nerve effects of laser treatment have been reported (see review[10]). On the other hand, there data exist describing photosensitivity of neurons. Microbeam irradiation of the mitochondrial area of a crustacean neuron with blue light changed its impulse frequency[32]. It has been supposed that the photobioelectrical processes in neurons are mediated by changes in redox processes in membranes and cytoplasm[33]. This approach needs much more experimental data.

Finally, it is also quite possible that laser irradiation influences the immune systems. This approach seems to be supported by data concerning changes in chromatine structure of human peripheral lymphocytes after irradiation with a He-Ne laser[18].

REFERENCES

1. V. S. Letokhov, Laser Biology and Medicine , Nature, 316, 325:330 (1985)

2. R. Pratesi, C. A. Sacchi, "Lasers in Photomedicine and Photobiology", Springer, Berlin, Heidelberg, New York (1980)
3. D. Kessel and Th. J. Dougherty, "Porphyrin Photosensitization", Plenum, New York, London (1983)
4. R. V. Bensasson, G. Jori, E. L. Land and T. G. Truscott, "Primary Photo-Processes in Biology and Medicine", Plenum, New York, London (1985)
5. N. F. Gamaleya, Laser biomedical research in the USSR, in: "Laser Application in Medicine and Biology", M. L. Wolbarsht, ed., Plenum, New York, London (1977), vol. 3, 1:175
6. G. Galletti, "Laser", Monduzzi Editore, Bologna (1986)
7. T. I. Karu, Photobiological fundamentals of low-power laser therapy, IEEE J. Quant. Electr., QE-23, vol. 10 (1987)
8. T. I. Karu, Molecular mechanism of therapeutic effect of the low-intensity laser radiation, "Lasers in the Life Sciences", vol. 2 (1987) (in press)
9. A. S. Kryuk, V. A. Mostovnikov, I. V. Khokhlov and N. S. Serdyuchenko, "The Therapeutic Efficiency of Low-Intensity Laser Light", Science and Techn. Publishers, Minsk (1986) (in Russian)
10. J. R. Basford, Low Energy Laser Treatment of Pain and Wounds: Hype, Hope or Hokum?, Mayo Clin. Proc., vol. 61; 671:675 (1986)
11. O. A. Tiphlova, T. I. Karu, Effect of Ar laser radiation and noncoherent blue light on E. coli growth, Radiobiology, vol. 26, no. 6, 829:832 (1986) (in Russian)
12. G. E. Fedoseyeva, T. I. Karu, V. S. Letokhov, V. V. Lobko, N. A. Pomoshnikova, T. S. Lyapunova, M. N. Meissel, Effect of He-Ne laser radiation on the reproduction rate and protein synthesis in the yeast", Laser Chemistry, vol. 5, 27:33 (1984)
13. T. I. Karu, G. S. Kalendo, V. V. Lobko
14. T. I. Karu, G. S. Kalendo, V. S. Letokhov, V. V. Lobko, Biostimulation of HeLa cells by low intensity visible light, Parts I-IV, Il Nuovo Cimento D. , vol. 1, 1761:1767 (1982), vol.3, 309:325 (1984), vol. 5, 483:496 (1985)
15. T. I. Karu, O. A. Tiphlova, V. S. Letokhov, V. V. Lobko, Stimulation of E. coli growth by laser and incoherent red light, Il Nuovo Cimento D., vol. 2, 1138:1144 (1983)
16. T. I. Karu, G. S. Kalendo, L. V. Pyatibrat, Investigations into the effects of He-Ne laser irradiation on the proliferation of HeLa cells, "Laser in the Life Sciences"(in press).
17. T. I. Karu, G. G. Lukpanova, I. M. Parkhomenko, Yu. Yu. Chirkov, Changes in cAMP level in mammalian cells after irradiation with monochromatic visible light", Dokl. Akad. Nauk USSR (Proc. USSR Acad. Sci.), vol. 281, 1242:1244 (1985)
18. G. E. Fedoseyeva, N. K. Smolyaninova, T. I. Karu, A. V. Zelenin, Human lymphocyte chromatin changes following irradiation with He-Ne laser, Radiobiology, vol. 27 (1987) (in press) (in Russian)
19. O. A. Tiphlova, T. I. Karu, Action of low-intensity red and far red on growth of E. coli, Microbiology, vol. 56, no. 3 (1987)
20. T. I. Karu, Biological action of low-intensity visible monochromatic light and some of its medical application, in: "Laser", Proc. Int. Congress on Lasers in Medicine and Surger, Bologna, 25:29, (1985), G. Galletti, ed., Monduzzi Editore, Bologna (1986)

21. A. L. Boynton, J. F. Whitfield, The role of cyclic AMP in cell proliferation: a critical assessment of the evidence, in: "Advances in Cyclic Nucleotide Research", P. Greengard, G. A. Robinson, eds., vol. 15, Raven, New York, 192:294 (1983)

22. I. Martelly, R. Franquinet, Planarian Regeneration as a model for cellular activation studies, Trends. Biochem. Sci., vol. 9, 468:471 (1984)

23. B. Chance, P. Cohen, Fr. Jobsis and B. Schoener, Intracellular oxidation-reduction states in vivo, Science, vol. 137, 499:508 (1962)

24. M. A. Krebs, Veech, Regulation of the redox state of the pyridine nucleotides in rat liver, in: "Pyridine nucleotide dependent dehydrogenases", H. Smid, ed., Springer, Berlin, Heidelberg, New York, (1979), pp. 413:434

25. F. L. Crane, H. Goldenberg, D. J. Morre, H. Löw, Dehydrogenases of the plasma membrane, in: "Subcellular Biochemistry", O. B. Roodyn, ed., vol. 6, Plenum, New York, London (1979), pp.345:399

26. C. Cone, Unified theory on the basic mechanism of normal mitotic control and oncogenesis, J. Theor. Biol., vol. 30, pp.151:181 (1971)

27. J. Gutkneeht, A. Walter, Trasport of auxin (indoleacetic acid) through lipid bilayer membranes, J. Membrane Biol., vol. 56, 65:72 (1980)

28. P. C. Maloney, E. R. Kashket, T. H. Wilson, A proton-motive force drivesATP synthesis in bacteria, Proc. Nat. Acad. Sci., USA, vol. 71, 3896:3900 (1974)

29. O. Tiphlova, T. I. Karu, The action of low-intensity laser light on the transient metabolic processes in E. coli, Dokl. Akad. Nauk, USSR, Proc. USSR Acad. Sci.,(1987) (in press)

30. N. G. Aleksidze, The influence of redox potential of the medium on the acetylcholine sensitivity of muscula, Biophysics, vol. 7, 602:608 (1962) (in Russian)

31. J. Poussegur, A. Franchi, G. L. Allemain and S. Paris, Cytoplasmic pH, a key determinant of growth factor-induced DNA synthesis in quiescent fibroblasts, FEBS Lett., vol. 190, 115:119 (1985)

32. A. B. Uzdenskii, Action of laser microbeam irradiation on the isolated crustacean neuron, Biological Sciences, no.3, 20:28 (1980) (in Russian)

33. R. G. Ludkowaskaya, Yu. M. Burmistrov, Light action on the processes of electrogenesis in pigmented crustacean neurons, Dokl. Akad. Nauk., USSR, Proc. USSR Acad. Sci, vol. 230, 1462:1465 (1976)

LASER APPLICATIONS IN BIO-MEDICINE:

TUMOR THERAPY AND LOCALIZATION USING PHOTOSENSITIZING DRUGS

R. Cubeddu, R. Ramponi, C. A. Sacchi and O. Svelto

Centro di Elettronica Quantistica e Strumentazione Elettronica, CNR, Istituto di Fisica del Politecnico, Milano, Italy

INTRODUCTION

Among the various applications of lasers in Biomedicine, a particularly interesting one makes use of a laser beam and photosensitizing drugs, for therapeutic and diagnostic purposes. Out of the several possibilities offered by this combination of laser light and drugs, we shall consider in this paper only the case of tumor therapy and tumor localization. The basic principles of these applications rely on the fact that a wide class of drugs present a high accumulation or retention in a tumor, or in a highly proliferative tissue, as compared with the surrounding normal tissue. For therapy the drug must also be able to induce a cytocidal reaction when activated by (laser) light of suitable wavelength. On the other side, for localization, when bound to the tissue, the drug must exhibit a reasonably high fluorescence quantum yield so that its emission can be detected by a suitable system. Thus it may happen that a drug suitable for tumor therapy is not suitable for tumor localization. In both these cases, however, two further conditions must be fulfilled:

(i) - the drug must be nontoxic (in the dark) and non-mutagenic;
(ii) - the drug must absorb and emit at wavelengths that can permit a reasonably good penetration into the tissues.

PHOTODYNAMIC THERAPY

With almost all the drugs this therapy is based on the photodynamic action, a photochemical reaction well known since the beginning of this century. In this "action" a suitable drug, henceforth called "photosensitizer", is first raised by a photon of appropriate energy to its excited singled state S_1. A fraction of the excited molecules then undergoes intersystem crossing thus decaying to the triplet state T_1. Once in the triplet state the drug cannot decay radiatively to the ground state, since the triplet to singlet transition is forbidden by the selection rules. The decay of the triplet state can therefore occur only by non-radiative decay. For a photosensitizer, this decay predominantly occurs by collision with oxygen, which is normally present in the tissue and which, in the ground state, is in a triplet level. After the collision, an energy exchange process may occur wherein the photosensitizer returns to its ground state and the oxygen is raised to one of its excited states. Since the total spin of the two interacting molecules must be conserved upon collision (Wigner's selection rule), the oxygen needs to be excited to its excited singlet state. Once in this state, the oxygen cannot decay radiatively to its ground state (singlet-to-triplet transition is forbidden) and, furthermore, it has sufficient energy to overcome the potential barrier for most chemical processes. This means that singlet oxygen can react with many cellular components, mostly at the membrane sites of a cell, to create photo-oxidation products which eventually lead to the cell death. The "photo-dynamic action" is thus the principle upon which the kind of tumor therapy described here is based: this therapy is, accordingly, called photodynamic therapy. Note that the energy transfer can occur provided that the triplet energy of the sensitizer is larger than or about equal to the energy of the excited singlet state of the oxygen. Since the corresponding wavelength of the latter energy is about λ = 800 nm, this requires that the wavelength of the singlet to triplet transition ($S_o \to T_1$) and hence that of the singlet to singlet transition, $S_o \to S_1$, of the sensitizer be larger than this wavelength.

The most common drugs, currently used for the photodynamic therapy of tumors, are Hematoporphyrin Derivative (HpD) and the so-called DHE (Di-Hematoporphyrin Ether or Ester). They are commercially available under the trade names of Photofrin I and Photofrin II, respectively. These drugs are derived from Hematoporphyrin (Hp), a well known dye belonging to the wide class of porphyrins. Its chemical structure consists of four phenol rings which form the chromophoric group. HpD is formed from Hp through a standard chemical procedure which consists in acidification with sulfuric and acetic acid, followed by neutralization. DHE is prepared from HpD by a suitable procedure using gel filtration, thus obtaining a higher percentage of aggregate fractions. The absorption spectra of DHE and HpD in buffer solution, besides a strong absorption peak at about λ = 365 nm (the Soret band), present weaker absorption bands extended to the red (the last peak occurs at $\lambda \cong 615$ nm in a buffer solution and at $\lambda \cong 630$ nm when the drugs are bound to the cell components).

HpD and DHE are complex and still not completely unravelled mixtures of many compounds. This can be evidenced by high-pressure liquid chromatography. Chromatograms, where the optical density of the solution at λ = 400 nm (approximately correspond to the peak of the Soret band) at a given position in the distillation column is plotted vs distillation time, consist of three well defined peaks occurring in the early stage of the distillation, and a broad continuoum occurring later. The peaks have been identified as due to Hematoporphyrin, (Hp), Hydrossiethylvinil deuteroporphyrin, (HVD), and Protoporphyrin, (PP), i.e. deydrated Hp in monomeric form. The broad continuum likely consists of aggregates of the previus monomers, in dimeric or oligomeric forms. The monomers and the various aggregates occupy different positions along the distillation column and can thus be physically separated. Dividing the column in three parts, one thus gets three different drugs. The fraction which is first distilled consists mostly of monomers (Hp, HVD and PP). It amounts to about 45% of HpD and has been found to be inactive for the photodynamic action both "in vivo" and "in vitro". The intermediate fraction, amounting to about 20% of HpD, shows still the presence of a good fraction of monomers plus aggregates. It has been shown to provide for best fluorescence localization of tumors. The last fraction, amounting to about 25% of HpD, is the one which best localizes in the tumor tissue, giving the most effective photodynamic action. This is the fraction which is used commercially under the name of Photofrin II. Recent works indicate that this fraction mostly consists of oligomers (an average of 8 molecules of Hp or HVD and PP), linked by either an ether or ester bond. The term DHE actually stands for Di-Hematoporphyrin Ester or Ether, since initially the structure was thought to be dimeric. Depending on the environment, DHE may be either in the planar form or in a folded structure. Extensive research on all these subjects has already been performed in several qualified institutions and discussed in specialized topical conferences[1-4].

As a second important aspect, in order to choose the best excitation wavelength, let us consider first the absorption spectra of three of the main tissue constituents namely water, melanin and haemoglobin. If we consider only the absorption, the visible to near IR region and, in particular, the interval from 600 nm to ~ 1.5 μm appears to be the most suitable for the lowest absorption of the aforementioned compounds. To obtain a complete picture, however, one must also take into account the scattering phenomenon, which occurs at large extent in the tissue. If Mie scattering is considered, the corresponding cross-section is known to vary with the wavelength as λ^{-2} and longer wavelength are thus favoured. On account of both scattering and absorption, the penetration depth in the tissue then turns out to be maximum at about $\lambda \cong 800$ nm. Note that the best 1/e penetration depth, in this case, is equal to about 4 mm. If now the absorption spectrum of HpD or DHE is taken into account, one then reaches the conclusion that the best excitation wavelength corresponds to the peak of HpD in the red, at $\lambda = 630$ nm. Note that with a dye absorbing at longer wavelength, a better penetration depth could be obtained. For the photodynamic action to occur, however, we must require that the excitation wavelength be shorter than $\cong 800$ nm, as explained before. Possibly, more convenient excitation wavelengths with new drugs may thus range between 600 and 800 nm.

In the clinical procedure commonly used for therapy, an intravenous injection of 2 ÷ 5 mg/Kg of body weight of HpD or DHE in saline solution is used. After 48-72 hours from the injection, a maximum contrast ratio between the drug concentration in the tumor and in the surrounding tissue is achieved. Depending on the histological type of tissue involved, this ratio may range between 4 and 10. At the time of maximum contrast ratio, the tumor mass is irradiated at 630 nm. If irradiation is made at the tumor surface, the used irradiance ranges between 100-200 mW/cm^2 and the total therapeutic energy fluence ranges between 60-500 J/cm^2. If the irradiation is performed with the help of one or more optical fibers inserted into appropriate needles, the power in each fiber is limited to 100 mW, to avoid carbonization problems at the tip of the fiber. If necessary, the

treatment can be repeated within 10 days after the injection. The main
problem with this treatment is that the skin and the retina of the patient
become photosensitive for about 4 weeks. During this time the patient must
be kept away from sun light and must possibly remain in dim light.

Three laser are most commonly used for the therapy:

(i) Rhodamine B dye laser pumped by a cw Argon laser. Starting with
an Ar^+ laser with a power of 10-18 W (in all lines) a power of
1 ÷ 4 W with a wavelength tunable between 610 and 680 nm can be
obtained from the Rhodamine B laser. The principal limitation
of this laser system is due to its high investment and running
cost.

(ii) Gold vapour laser. It is a pulsed laser with a pulse repetition
rate of ~ 10 kHz, average power 1-9 W, and it oscillates at
627.8 nm. As main disadvantages, this laser operates at a fixed
wavelength and requires a long warm up time (~ 45 minutes).

(iii) Rhodamine B laser pumped by a Copper vapour laser oscillating
at the two wavelengths of 510.6 and 578.2 nm. Since the average
power of the Copper vapour laser is in the 10 ÷ 50 W range and
since the conversion efficiency to the Rhodamine B laser wavelength may be up to 30%, the average power in the red is in this
case comparable to that of the previous case. However the tunability is here obtained with a somewhat more complicated and
expensive system.

Using the previously described clinical procedure and laser systems,
clinical studies have been performed in the las few years in several institutions[5,6]. According to the results there referred, one can say that
the photodynamic therapy has been found to be effective (although to a
different extent) in all the treated tumor types. The average percentage
of complete tumor remission is about 50% and it raises to more than 70%
in early stage cancers. It has been found that the main reason of tumor
necrosis arises from damage in the tumor blood vessels where HpD and DHE
appear to be primarily localized. However, single cell death due to the
binding of HpD and DHE cell membranes also occurs. The possible failure
of PDT may be due to:

(i) Insufficient penetration of the light in the tissue (in particular in heavily pigmented tissue).

(ii) Low oxygen content in the tumor (mostly in the so-called anoxic
regions of the tumor).

(iii) Insufficient technical skill of the operator.

As a conclusion of this discussion, PDT offers a new and interesting
possibility of tumor treatment with two main limitations:

(i) Low penetration depth of the laser light. Only superficial
tumors with a dimension smaller than about 1 cm can therefore
be effectively treated.

(ii) Likewise to other radiant treatments (e.g. X-rays or γ-rays), only
local regions containing the tumor mass can be treated. This

leaves up the problem of treating methastasis. As a consequence, the field of applications is presently limited to:

(i) Inoperable tumors, in particular for palliative treatment.
(ii) Early stage tumors not yet spread out at other distant sites (e.g. limphonodes). It must be noted, however, that PDT is compatible with all other treatments currently used (surgery, therapy by ionizing radiation, chemotherapy). Its most interesting future may therefore rely on its use in connection with other kinds of therapy. For instance, PDT can be used, as it is now initially done, to sterilize the tumor bed after a surgical removal of the main mass.

Before ending this section on PDT we like to point out that the research activity in this field continues at high rate and new directions are being explored. A particularly interesting possibility arises from the use of new drugs with stronger absorption bands in the red-to near IR. To this purpose some chlorin derivatives of Hp such as the hematoporphyrin-mesochlorin ester[7] and the hematoporphyrin-bonellin ester[8] appear to be particularly interesting. The absorption spectrum of both these dimers presents in fact a strong absorption peak centered at $\sim \lambda = 640$ nm. Even more attractive appear[9] phthalocyanines whose absorption spectrum has a strong peak in the red extending up to $\sim$ 750 nm.

TUMOR LOCALIZATION

As previously mentioned HpD and DHE in solution exhibit two main fluorescence peaks centered at 630 nm, and 670 nm respectively. Since the drug's concentration is higher in the tumor compared with the surrounding tissue, its fluorescence can be used for tumor localization. This is particularly useful for small tumors (with a diameter smaller than one mm) which cannot be localized with other techniques (such as X-ray analysis or endoscopy). Through endoscopic techniques[10], in this case, a laser light of appropriate wavelength, sent through an optical fiber inserted in an ordinary endoscope excites the fluorescence of the suspected region in the patient treated with HpD. The fibers normally used in the endoscope for the direct vision can now be used to collect the fluorescent emission of the irradiated region that is imaged at the entrance of an image intensifier. A band-pass filter in the system allows for the observation only in the desired spectral region. The fluorescence of HpD bound to tumor tissue can be distinguished from tissue autofluorescence and an unambiguous signal indicating an anomalous HpD concentration can thus be obtained. As the excitation source, a Kr ion laser oscillating at the wavelength of $\lambda = 405$ nm with a power of 0.1 - 0.5 W is commonly used. This wavelenght is preferred since it practically coincides with the peak of HpD fluorescence excitation, and on account of the large discrimination with the fluorescence wavelength.

This technique is very promising, with application possibilities not only limited to the field of endoscopy, but in its present form suffers from some limitations:

(i)   On account of the low penetration depth at $\lambda \cong 400$ nm, only
       superficial tumors can be localized.
(ii) The fluorescence intensity of HpD in the tumor is comparable
       to tissue autofluorescence so that low signal to noise discrim-
       ination is achieved.
(iii) Since HpD is also a photosensitizing drug, the patient undergoing
       the clinical test must be kept in dim light for several days.
       It should also be noted that, partly due to the previous points,
       the problem of false-positive (strongly fluorescent sites not
       corresponding, according to biopsy, to a tumor tissue) or false-
       negative (tumors not showing a stronger fluorescence) still needs
       to be solved.

RESEARCH ACTIVITIES AT CEQSE

As a final part of this paper we briefly mention the research activities
undertaken at our Center (CEQSE) in this field, during the last seven years.
These researches are the result of a vast interdisciplinary collaboration
promoted by our group with several other institutions, namely:

(i)   Centro di Istochimica del CNR, Pavia, for studies of HpD in
       solution and cells;
(ii) Centro di Oncologia Sperimentale, Naples, for studies of HpD
       in solution and in cells;
(iii) Istituto di Farmacologia of Milan, for studies on experimental
       tumors on mice;
(iv) Veterinary Clinic, Milan, for treatment of spontaneous tumors
       in animals;
(v)   Istituto Tumori of Milan, for general purpose aspects;
(vi) Department of Chemistry of the Paisley College, Scotland, for
       studies of new drugs.

This is perhaps a unique example of most extended interdisciplinary
collaboration in this field. The results obtained are so numerous that it
is not possible to summarize them here in some detail. We will therefore
limit ourselves to point out two research fields in which our contributions have been of particular significance, namely:

(i)   New type of photochemical action;
(ii) Photophysical studies of HpD and DHE bound to different bio-
       molecules.

We proposed a new scheme based on a two step excitation of the dye
(e.g. HpD) bound to a given biomolecule[11]. The first step with a beam at
wavelength λ_1 provides for the excitation of the dye to its first excited
singlet state. The second step, with a beam at wavelength λ_2, provides for
further excitation of the dye from its first excited singlet state (or from
the first excited triplet state) to some higher lying excited state. As
experimentally shown, such a high excited dye molecule can dissociate in
(charged) radicals which then react with the surrounding tissue, thus
leading to cell death. Note that this new kind of two-step photochemical

action, is essentially different from the photodynamic action and can find many applications in photochemistry and photobiology. In particular, oxygen may not necessarily be involved in this process. Note also that the rate of this two-step process increases as the square of the irradiance of the incident radiation, in contrast to the photodynamic action, the rate of which is linear with the irradiance. Higher laser irradiances are thus required for two-step photo-action and these are obtained by using dye lasers pumped by a high peak power (2 MW) excimer laser.

The second main field of research wherein many contributions have been given, is the photophysical study of HpD and DHE in different environments[12]. The experiments have been performed with the technique of time-resolved fluorescence spectroscopy with high resolution in time. To this aim, a general purpose instrumentation has been developed[13]. As the excitation source, a mode-locked dye laser (pulse duration of ~ 5 ps) synchronously pumped by a mode-locked Ar-ion laser is used. The HpD fluorescent signal is passed through a monochromator and then sent to a micro-channel-plate photomultiplier working in the single photon counting regime. The time resolution of the detection apparatus is ~ 50 ps. When microscopic observations have been required (for studies in single cells), a fluorescence microscope has been used to focus the excitation beam on the sample, and to detect the resulting fluorescence. As a meaningful case of application of this instrument, we mention the results obtained with DHE bound to the surfactant CTAB (Cetyltrimethylammonium bromide). This surfactant, at sufficiently high concentration, aggregates in the form of micelles and provides perhaps for the simplest model system that reproduces the hydrophobic environment of cell membranes. With the mentioned experimental apparatus (without the microscope, in this case) a decay curve of the fluorescence emitted at the wavelength set by the monochromator is obtained. This curve is then fitted with one or more exponential curves of given amplitudes. Several examples of experimental curves are shown in the literature[12]. With measurements at different emission wavelengths (the excitation wavelength was in this case set at λ = 364 nm) and at different CTAB concentrations, we then obtained the decay times and amplitudes of three exponentials which give the best fit of the experimental curves. Although a detailed discussion of these results is beyond the scope of this paper, we limit ourselves to mention that, according to this detailed discussion, the three exponential curves are attributed to molecular configurations in the polymeric chains that are the constituents of DHE, namely[12]:

(i)   The component with a lifetime of ~ 14 ns is considered to arise from "impurity" monomers present in the DHE (~ 4.5%) and from "totally free chromophores", i.e. some of the porphyrins rings belonging to DHE polymer in an unfolded configuration.

(ii) The component with a lifetime of ~ 0.7 ns is due to the folded polymer DHE.

(iii) The component with a lifetime of ~ 3.3 ns is attributed to a "free chromophore" interacting with (or being encapsulated by) the remainder of the polymer or polymer/surfactant mixture.

The general conclusion of this work is that different aggregational states are detectable by time resolved fluorescence spectroscopy. This technique can thus turn out to be very useful to unravel the complex modifications that DHE undergoes once injected into a living body, to eventually lead, in the tissue, to the formation of monomers, which are thought to be the main agents of the photodynamic action at tissue level.

REFERENCES

1. A. Andreoni and R. Cubeddu (ed.s), "Porphyrins in Tumor Phototherapy", Plenum Press, New York and London (1984)
2. G. Jori and C. Perria (ed.s), "Photodynamic Therapy of Tumors and Other Diseases", Edizioni Libreria Progetto, Padova (1985)
3. D. Kessel (ed.), "Methods in Porphyrin Photosensitization", Plenum Press, New York and London (1985)
4. Proc.s Clayton Foundation Conference on Photodynamic Therapy, Los Angeles, Ca, Feb. 15-19 (1987), Photochem. Photobiol., 46:561-952 (1987)
5. T. J. Dougherty, Photodynamic Therapy, in "Photodynamic Therapy of Tumors and Other Diseases", G. Jori and C. Perria, ed.s, Edizioni Libreria Progetto, Padova (1985)
6. Y. Hayata, Photodynamic Therapy in Japan, in Proc.s 1st Int. Conference on the Clinical Applications of Photosensitization for Diagnosis and Treatment, Tokyo, Japan, April 30 - May 2, 1986
7. R. Cubeddu, W. F. Keir, R. Ramponi, T. G. Truscott, Photophysical Properties of Porphyrin-Chlorin Systems in the Presence of Surfactants, Photochem. Photobiol., 46:633 (1987)
8. D. Kessel and C. J. Dutton, Photodynamic effects: Porphyrin vs chlorine Photochem. Photobiol.; 40:403 (1984)
9. E. Ben-Hur and I. Rosenthal, Photosensitized inactivation of Chinese Hamster cells by phtalocyanines, Photochem. Photobiol., 42:129 (1985)
10. A. E. Profio, M. J. Carvlin, J. Sarnaik and L. R. Wudl, Fluorescence of Hematoporphyrin-Derivative for Detection and Characterization of Tumors in "Porphyrins and Tumor Phototherapy", A. Andreoni and R. Cubeddu, ed.s, Plenum Press, New York and London (1984) J. P. A. Marijinissen, W. M. Star, J. L. van Delft and N. A. P. Franken, Light intensity measurements in optical phantoms and in vivo during HpD-Photoradiation treatment, using a miniature light detector with isotropic response in "Photodynamic Therapy of Tumors and Other Diseases", G. Jori and C. Perria, ed.s, Edizioni Libreria Progetto, Padova (1985) A. Goetz, J. Feyh, P. Conzen and W. Brendel, In-vivo Detection of Hematoporphyrin Derivative (HpD) Fluorescence by Digital Subtraction of Videomicroscopic Images, Photochem. Photobiol., to be published.
11. A. Andreoni, R. Cubeddu and O. Svelto, Pulsed-Laser photoactivation of porphyrins in "Photodynamic Therapy of Tumors and Other Diseases", G. Jori and C. Perria, ed.s, Libreria Progetto, Padova (1985)
12. R. Cubeddu, R. Ramponi and G. Bottiroli, Time-resolved fluorescence spectroscopy of hematoporphyrin derivative in micelles, Chem. Phys. Lett., 128:439 (1986)

13. R. Ramponi and M. A. J. Rodgers, An instrument for simultaneous acquisition of fluorescence spectra and fluorescence lifetimes from single cells, Photochem. Photobiol., 45:161 (1987)

FUNDAMENTALS OF LASER ANGIOPLASTY:

PHOTOABLATION AND SPECTRAL DIAGNOSTICS

N.P. Furzikov, V.S. Letokhov, and A.A. Oraevsky

INTRODUCTION

Laser angioplasty is a new and very promising method for treating arterial obstructions resulting from atherosclerotic degeneration of the blood vessels[1-3]. There are several reasons for current interest in this technique. The first is the wide spread of atherosclerosis among the population of developed countries and the high mortality (about 50% of the total death rate) from this disease, especially from occlusions of the major arteries. The second reason is the inadequacy of existing methods of treating atherosclerotic lesions in blood vessels. Balloon dilatation yields positive results in a limited number of cases only, and prothesis and aortic-coronary shunting are very traumatic and frequently suffer from complications. And finally, the third reason for the advent of angioplasty is the rapid progress in development of laser radiation sources and optical fiber technology.

The general scheme of laser angioplasty is the following. High intensity laser radiation is transmitted through a flexibile glass or quartz fiber to the site of irradiation in an artery, and destroys an occlusion impeding normal blood circulation (Fig. 1). It is possible to introduce such a fiber by a catheter in a suitable place into a vascular lumen, to move the distal fiber end to the zone of treatment, and to deliver the radiation to this zone instead of penetrating through surrounding tissues. The method is potentially simple and general and it can probably cost less than current surgical operations.

The first suggestions and experimental modeling appeared in the early eighties, and some time later the first clinical operations were begun. In some cases, blood circulation improved after such operations[4,5], but sometimes a thrombosis of the neolumen occurred[6]. These results excited considerable interest and stimulated great activity in this field. Particular emphasis has been placed on prevention of vascular wall perforation, precise control of irradiation dose, and design of mutifunctional catheters having modified tips, diagnostic information, etc.[7].

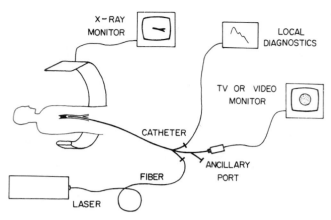

Fig. 1. A simplified laser angioplasty scheme: laser radiation is introduced into a flexible optical fiber and delivered with a
multifunctional catheter to the target area within an atherosclerotic blood vessel. The passage of the catheter through
the blood vessel system is monitored and the treatment results
checked by X-ray angiography. To make a local diagnosis, use is
made of spectral and fluorescence techniques or of some other
method allowing atherosclerotic plaques to be distinguished
from normal tissue. If necessary, one may resort to visual
angioscopy using a television monitor. The target area is
illuminated by means of either a laser or a standard incoherent
light source used in endoscopy. The ancillary port of the catheter
is employed to deliver contrast substances and saline or to remove
plaque destruction products.

It is evident that technical realizations and medical applications
must be based on a background of physical and biological investigation.
This lecture is devoted mainly to the latter subjects. The first problem
is the selection of the laser and of its mode of operation. Earlier models
and clinical operations have predominantly used CW visible, near-IR or
mid-IR lasers[4-6,8,9]. Some laboratory experiments have used pulsed radiation
of visible, near-Uv or IR lasers[10,13]. This wide variety of wavelengths and
operating modes demonstrated a lack of definite criteria for selecting these
parameters. Such criteria must be founded on quantitative estimation of the
radiation-tissue interaction. The reason is that in laser angioplasty the
range of permissible conditions is usually narrow, so that qualitative
comparison between the various lasers acceptable in general laser surgery
is insufficient in this case. In this lecture, quantitative estimates have
been carried out for various conditions of laser angioplasty[14].

The next problem - conditions for plaque of clot destruction - is
closely connected with the first one. Experimentally, tissue destruction
is characterized by a threshold. From a general point of view the destruction efficiency would be better at higher radiation intensities. However,
too high an intensity can cause injury in the vessel wall by means of direct
absorption of light or thermal diffusion. This result is inacceptable, so
the intensity used is an important parameter. We have done estimations of

the intensity needed and of the quantity of heat which can diffuse from the irradiated tissue volume into surrounding layers.

The third problem addressed here is the composition of tissue destruction products. This composition is important from the viewpoint of both the near-term and long-term results of laser angioplasty operations, as these products would stay within the patient's body if no special measures were undertaken.

Large fragments of the destroyed tissue may obstruct fine blood vessels in other organs and in other parts of the body. The same result may occur from embolisms due to gaseous products insoluble in the blood. Also, we cannot rule out the possibility that laser destruction of tissues may yield toxic and carcinogenic products. We report here the results of spectroscopic analyses of the main products of destruction[15].

Laser angioplasty operations are essentially percutaneous, and so the success of these techniques will largely depend on the possibility of differentiating between the desired ablation zone and other tissue areas and controlling the actual course of ablation (Fig. 1). The existing analysis is based on X-ray angiography, intravessel endoscopy (angioscopy), and spectral techniques. We describe here some experimental results relating to the latter method of diagnosis.

MODELLING OF PLAQUE DESTRUCTION BY LASER RADIATION

A thermal model with negligible relaxation time will be used for estimation. It assumed that destruction starts when the tissue temperature reaches the boiling point of water, $T_b=100°C$. Radiation is assumed to fall normally on the plaque surface and penetrate it to a depth equal to the inverse of the attenuation coefficient K for a given wavelength (Fig. 2). Energy absorption and tissue heating are believed to be uniform in depth down to K^{-1} and throughout a spot of a radius equal to the standard fiber radius r=0,2 nm. Radiation divergence, surface reflection, and heat diffusion to the surrounding medium are neglected. Radiation scattering is taken into account of the tissue thermal and optical constants are assumed to be negligible. The heat capacity of the tissue is taken to be equal to that of water. The plaque thermal conductivity x and the termal diffusity χ used have the average experimental values x=5x10^{-3} W/cm °C and χ=1.3x10^{-3} cm^3/s, respectively[16]. Attenuation coefficients were obtained from experimental data or were assumed as given in Table 1[17-20,3].

Tissue heating and destruction by laser radiation may be continuous (steady-state) or pulsed (transient). The boundary between these cases is determined by the thermal diffusion time[23]

$$\tau = \ell^2/4\chi \tag{1}$$

where ℓ is the minimum linear dimension of the volume being heated. For visible and near IR wavelengths the coefficients K are small, ℓ is equal to the spot radius, $\ell = \gamma = 0,2$ nm, and the time τ is constant: $\tau \cong 0,08$ s.

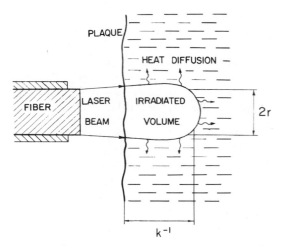

Fig. 2. Ideal laser angioplasty geometry used for making thermal model
estimates: laser radiation emerges from the optical fiber and
strikes the plaque surface at normal incidence. The irradiated
volume is assumed to be a cylinder whose diameter is the same as
the fiber diameter and whose height is equal to the inverse of
the attenuation coefficient of the plaque tissue for the given
laser wavelength. The absorbed energy is converted to heat which
is consumed in destroying the irradiated tissue volume and in
heating the adjacent tissue layers by diffusion.

For UV and IR radiation ℓ is determined to be $\ell = K^{-1}$, so in these cases
τ varies approximately from 10^{-5} to 10^{-2}s (see Table 1). The heating will
be pulsed if the laser pulse duration t_p is much shorter than the time τ,
$t_p \ll \tau$, and quasistationary if $t_p \gg \tau$. Thus microsecond pulse durations
provide pulsed plaque heating for all optical wavelengths; moreover, milli-
second durations are sufficient for the visible range and some adjacent
wavelengths. We must note that an analogous condition limits the maximum
pulse repetition rate f if one wishes to obtain a purely pulsed mode for
each successive pulse, i.e. $f \ll \tau^{-1}$. Data on τ^{-1} are also included in
Table 1.

The continuous mode is characterized by a destruction threshold for
the laser power. The threshold can be determined from the balance of the
power supply rate and the thermal diffusion rate. For small K this conditions has a form

$$P/_{\rho q V} \geq 4\chi/_{\ell^2} \tag{2}$$

where P is the power, ρ is the tissue density, V is the irradiated volume,
and q is the specific destruction energy, which is assumed to be the sum
of the specific energy of tissue heating from $T_n = 37°C$ to T_b and of the
latent vaporization energy of water, i.e. $q \cong 2.6kJ/g$. For small K, $\ell = \gamma$,
$V = \pi\gamma^2/K$ and

$$P \geq P_{th} = 4\pi\chi\rho q/K. \tag{3}$$

Table 1. A list of data for laser angioplasty

Laser	KrF	XeCl	dye	dye	Ar$^+$	Nd:YAG	Er^{3+}	CO$_2$
Wavelength, nm	249	308	465	465	514.5	1064	2940	10600
Attenuation coefficient, cm^{-1}	650	200	54 (plaque)	26 (wall)	32	7.2	5000	500
Thermal diffusion time, s	4.5×10^{-4}	4.8×10^{-3}	7×10^{-2}	8×10^{-2}	8×10^{-2}	8×10^{-2}	7.7×10^{-6}	7.7×10^{-4}
Limiting repetition rate, Hz	2.2×10^3	2.1×10^2	14	12	12	12	1.3×10^5	1.3×10^3
CW threshold, W (estimated/measured)	0.02 / –	0.02 / –	0.8 / –	1.6 / –	1.3 / 2[a]	5.9 / 9[a]	0.02 / –	0.02 / –
Pulsed threshold, J/cm^2 (estimated/measured)	0.40 / 0.35[b]	1.3 / 1.4[b]	5.0 / 6.8[c]	10 / 15.9[c]	–	37 / 30–40[d]	0.05 / –	0.52 / –
Ablation threshold, J/cm^2 (estimated/measured)	1.5 / –	5.0 / –	19 / –	38 / –	–	–	0.52	5.2 / 6[e]; 5.2[g]
Pulse energy at threshold, mJ (2r= µm)	1.8	6	23	46	–	–	0.62	6.2

(a) Measured thresholds for plaque destruction in saline from ref. 8; (b) from ref. 21; (c) from ref. 18; (d) from ref. 15; (e) this value was calculated using an experimental intensity of 60 kW/nm^2; (g) at this fluence a pronounced decrease of tissue sample mass is initiated[22].

For the argon ion laser the formula (3) gives $P_{th} = 1.3$ W and $P_{th} = 5.9$ W for the Nd:YAG laser. These estimates are given in Table 1 together with experimental results and other data. For large K the irradiated volume appears as a flat disk and the threshold is determined by the condition of stationary temperature difference $\Delta T_b = T_b - T_n$ on the tissue surface[23]

$$P \geq P_{th} = \pi \, x\Delta T_b. \tag{4}$$

Thus, for CW UV and IR lasers the threshold power is constant ($P_{th} \cong 0.02$ W for r = 0.2 nm) and depends only on the spot radius.

For the pulsed mode the destruction threshold is determined simply by the condition of tissue heating up to T_b

$$\Phi_{th} = C\rho\Delta T/K \tag{5}$$

where Φ is the input energy fluence and C is the tissue heat capacity. The estimated threshold fluences vary from a fraction of a J/cm^2 up to tens of J/cm^2 at optical wavelengths (see Table 1).

Experimentally we investigated in vitro the dependance of the yield of plaque destruction products on the fluence of the excimer XeCl laser radiation used for tissue destruction[24]. A flowing cell with distilled water was used, and the optical density A of soluted organic products was measured at a wavelength of 206 nm. The product yield (A per unit of the spot area) depending on the fluence is shown in Fig. 3. At small fluences (0.3 – 1.5 J/cm^2) the yield was small and nonlinearly increased to higher fluences. At some point this increase appeared to be linear up to the highest fluence used of 8 J/cm^2. Qualitatively, the dependences were analogous both for fibrous and calcified plaques, but the extrapolation of their linear parts to zero fluence gave different values: 1.5 ± 0.2 J/cm^2 and 3.3 ± 0.4 J/cm^2, respectively. The first value agrees relatively well with the estimate from the formula (5) for $\Delta T = 80°C$ and $K = 200$ cm^{-1}: 1.7 J/cm^2, and also with the gaseous product yield extrapolation: 1.4 J/cm^2 [21]. Therefore the thermal model explains both the threshold existence and its value for 308 nm laser radiation. The calcified plaques have much more inorganic substance, so the temperature corresponding to their distruction threshold must probably be higher. For our results, $\Delta T \cong 150$–$200°C$ would give better agreement. For the Krf laser the thermal model predicts $\Phi_{th} = 0.40$ J/cm^2, which is also close to the experimental extrapolated value of 0.35 J/cm^2 [21].

The threshold intensity corresponds only to the initiation of the plaque destruction; higher fluences are needed for complete destruction of the irradiated volume. These fluences were determined in two steps. Initially, we measured the specific destruction energy for three pulsed lasers: the excimer KrF and XeCl lasers, and the TEA CO_2 laser[15]. At their wavelengths the plaque attenuation coefficients have similar values and their pulse durations satisfy the condition for pulsed heating. Moreover, the energy fluences used were approximately the same (5 J/cm^2) for all three lasers. The specific destruction energy was determined by simple division of the delivered radiation energy by the plaque mass decrease. The sample mass was measured by a torsion balance with an accuracy of 0.25 mg before

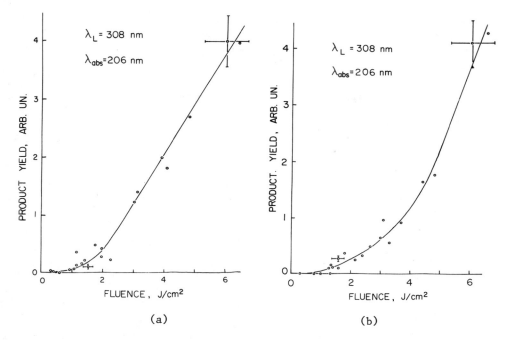

Fig. 3. Yield of soluble plaque destruction products as a function of
XeCl laser energy density for (a) fibrous and (b) calcified
plaques and an observation wavelength of 206 nm.

and after an irradiation of 10 to 100 pulses. The obtained values were
0.97±0.28 kJ/g, 1.1±0.3 kJ/g, and 2.65±0.86 kJ/g for the KrF, XeCl and
CO_2 lasers, respectively (Table 2). The last value is the same as needed
for tissue vaporization, which confirms that the thermal path applies for
tissue destruction by the TEA CO_2 laser. The excimer lasers are 2.5 times
more effective, but this most probably means that UV ablation does not
require complete vaporization, and thus a portion of the tissue may be
ablated in the condensed state.

For the second step, we used the specific destruction energy values
to define the required energy fluences and pulse energies for the standard
fiber diameter (2r = 0.4 nm) at other wavelengths[14]. The UV and visible
lasers were believed analogous to the KrF and XeCl lasers, and the IR
lasers, to the CO_2 laser, respectively. The values obtained are summarized
in Table 1. They give an estimate for selecting the optimal laser.

THE PROBLEM OF HEATING OF ADJACENT TISSUE LAYERS

Let us introduce a boundary heating coefficient K_γ which is equal to
the ratio of the heat diffused through the side surface of the irradiated
tissue volume to the laser energy absorbed in this volume[14]. For CW angioplasty,

$$K_\gamma \cong x \frac{\partial T}{\partial n} S/P \tag{6}$$

Table 2. Specific photodestruction energy

Laser	Wavelength, nm	Pulse duration, ns	$\bar{\varepsilon}$ kJ/g
KrF	249	20	0.97 ± 0.28
XeCl	308	15	1.1 ± 0.3
CO_2	9470	300	2.65 ± 0.86

where $\partial T/\partial n$ is the temperature gradient normal to the side surface, and S is the area of this surface. The characteristic width of the thermally injured tissue is of 0.5 nm, so $\partial T/\partial n \cong 10^3$ °C/cm. The surface S is typically $10^{-2} - 10^{-1}$ cm^2 (for 2r = 0.4 nm) and for P = 5 W the coefficient K_r is of the order of 10^{-1} and 10^{-2}. This small amount of energy is sufficient to injure surrounding layers, as such injury results merely from heating to 60–70°C, without any energetically expensive vaporization. An acceptable solution to this problem may be to reduce the irradiation time to the thermal diffusion time τ , but this can reduce the destruction efficiency.

For pulsed laser angioplasty the coefficient K_γ takes the form

$$K_\gamma \cong x \frac{\partial T}{\partial n} t_\rho \, S/E \tag{7}$$

where t_ρ is the effective lifetime of the irradiated tissue volume and E is the pulse energy absorbed in this volume. The absolute quantity of heat transferred to the adjacent layers, $K_\gamma E$, and therefore the extent of their thermal injury, depends only on the lifetime t_ρ . This lifetime is controlled by exceeding the threshold fluence, and varies from infinity near the threshold to the pulse duration t_p at the highest fluences. An increase of the input fluence Φ decreases the lifetime t_ρ and the heat diffusione to the adjacent tissue layers. At some fluences practically all the absorbed energy goes into tissue vaporization and ejection of products. This mode is called ablation. From the practical point of view ablation is close to the effective destruction discussed above. We must note that ablation requires both short laser pulse duration and intensity well above the threshold. Otherwise, the irradiated volume survives for a long time, and too much heat can diffuse into adjacent layers.

The thermal model predicts that ablation can be realized at every pulsed laser wavelength. This was illustrated by the work of Deckelbaum et al. using a TEA CO_2 laser with $t_p \cong 10^{-6}$ s [13]. They have observed that thermal injury to the tissue near the destruction crater appeared to be absent above an intensity of 60 kW/nm^2 ($x\tau_\rho$ = 6 J/cm^2). Let us estimate the boundary tissue heating assuming the lifetime t_ρ to be equal to 10 t_p and the heated layer width $\delta\Gamma$ to be 2 µm. In this case, $\frac{\partial T}{\partial n} \cong 3 \times 10^5$ °C/εm, and with $E = 0.2$ J and $S \cong 1.2 \times 10^{-3}$ cm^2, we obtain $K_\gamma \cong 2 \times 10^{-4}$. The temperature rise in the layer δr

$$\Delta T = K_\gamma E / C \delta r S = x \frac{\partial T}{\partial n} t_\rho / C \delta r \cong 18 \; °C \qquad (8)$$

is actually within the acceptable limits. It is interesting to note that the effective destruction of tissue by TEA CO_2 laser radiation as determined by sample mass decrease begins at a fluence of 5.2 J/cm^2, which is close to the corresponding value above of 6 J/cm^2 [22].

Analogous estimates were made for excimer lasers. They show that tissue near the irradiated volume does not undergo thermal injury if the lifetime of this volume does not exceed 10^{-6}–10^{-5} s. These estimates help to select operational fluence range for laser angioplasty.

PRINCIPAL PRODUCTS OF PLAQUE DESTRUCTION

Analysis of products was based on spectral data[15]. The near UV and visible absorption spectrum for organic products of plaque destruction by XeCl laser radiation in distilled water is presented in Fig. 4. The spectrum contained just one intense peak at about 200 nm, two shoulders of lesser intensity near 235 and 270 nm, and a weak band with its maximum at about 420 nm. Samples of various plaque types and of intact vessel walls only altered the relative intensities of these spectral features and, to some extent, the position of the intense peak.

A tentative interpretation of this spectrum follows: for lipid plaques or deeply penetrated fibrous plaques, the peak at 200 nm arises from lipids and partly from proteins[25]. Other spectral features are due to photooxidation products of these lipids and proteins by the laser radiation. These are, respectively, diene (235 nm) and triene (270 nm)[26]. For surface destruction of fibrous plaques, the two shoulders at 235 and 270 nm may be also due to sulfur-containing and aromatic amino acids, respectively[25]. We estimated an apparent quantum yield for these products assuming that the initial plaque consisted only of cholesterol and that UV destruction gave pure cholesterol in solution. Knowing the solution volume, the pulse energy, the number of pulses, and the estimated molar extinction coefficient of about 5×10^3 1/cm mole at 200 nm[27], the cholesterol quantum yield did not exceed 10^{-2}. This relative value showed that the yield of complex organic products and their photooxidation derivatives, which have potential toxicity, should be low. Similar absorption spectra may be obtained for plaque destruction by other laser wavelengths e.g. by the argon ion laser or by Cu laser radiation.

The IR absorption spectrum of gaseous products from UV plaque destruction in a vacuum is given in Fig. 5. The two most intense bands coincide with absorption bands of water molecules, and the position of the high frequency band maximum shows that water is highly aggregated or even liquid. The two remaining bands can be found in the liquid water spectrum. This seems to be due to a condensate on the salt cell windows. Thus, water is the principal gas-phase product of UV plaque destruction. This conclusion is

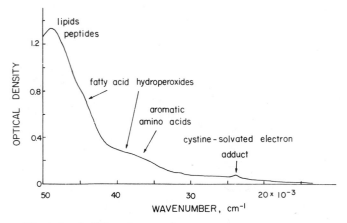

Fig. 4. Near-UV and visible absorption spectra of water-soluble products
of plaques destruction using XeCl laser radiation.

straightforward since water accounts for 70 to 80 percent of tissue.

The apparent quantum yield of gaseous products was determined from the
increase in pressure produced by sample irradiation in an evacuated cell of
known volume. For fresh samples this yield was 0.8±0.3, assuming that the
gas formed did not condense. Since spectral data suggest partial water con-
densation, the actual quantum yield may be higher. For sublimated samples,
the measured quantum yield is lower, equal to 0.1-0.2.

FLUORESCENCE AND REFLECTION SPECTROSCOPY OF VESSEL WALLS

In the angioplasty of blood vessels suffering from atherosclerotic
lesions, it its essential that pathological areas can be identified by
their optical properties. This would make it possible to use one and the
same fiber catheter both to locate the diseased vessel wall areas and to
control the laser ablation process on the basis of spectral analysis. Re-

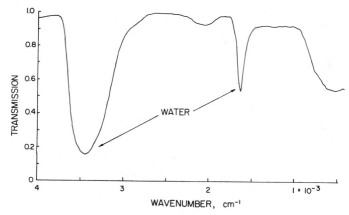

Fig. 5. IR absorption spectrum of gaseous products of plaques destruction
using XeCl laser radiation.

cently, reports have appeared in the literature about some differences between the fluorescence spectra of normal and pathologically altered zones (plaques) in human atherosclerotic vessels[28-30]: the spectrum of normal vessel wall features three peaks instead of the single wide fluorescence band typical of atherosclerotic plaques. In the above-mentioned works, optical excitation was effected at λ = 350, 476 or 480 nm. However, the data reported are not exhaustive and contain no explanation of the detected difference.

In the work reported in ref. 30, we measured the reflection and fluorescence spectra of human blood vessel walls in vitro. The results obtained enabled us to draw certain conclusions as to the causes of the difference in the spectral properties of plaques and normal wall areas of atherosclerotic vessels. Measurements performed using optical fibers to deliver and gather radiation give reason to believe that it will be possible in the future to realize direct spectroscopic diagnosis of atherosclerotic lesions in vivo.

We studied human cadaver thoracic and ventral aorta and femoral artery samples. The samples were extracted within 2-6 h of sudden death. Prior to experiment, the samples were stored in a normal saline solution at 4°C for not over 24 h after extraction. Special care was exercised prior to and during measurements to keep intact the endothelium covering the inside surface (intima) of the vessels. The vessels under study were subjected to a selective histologic analysis prior to and after the experiment. The plaques were from 0.8 to 1.5 mm in thickness. We compared the fluorescence and diffuse reflection spectra of fibrous plaques and normal wall areas of atherosclerotic vessels.

The fluorescence investigation results are illustrated in Fig. 6. Curve I of Fig. 6 is the fluorescence spectrum of a thoracic aorta wall area with a clearly expressed atherosclerotic lesion -- fibrous plaque. The maximum of the spectrum is in the region of 515 nm under excitation at a wavelength of 480 nm. This spectrum is due mainly to the emission of oxidized flavoproteins (FAD-components) of the oxidative metabolism system and,

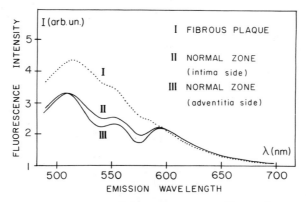

Fig. 6. Fluorescence spectra of human aorta walls under excitation at λ = 480 nm: I — fibrous plaque; II -- normal wall (taken from the intima side); III -- normal wall (taken from the adventitia side).

to a lesser degree, to that of reduced pyridine nucleotide (NAD) molecules[31].

Curve II of Fig. 6 represents the fluorescence spectrum of a normal wall area of the same vessel under excitation at λ = 480 nm. As can be seen, the spectrum shown in the figure differs substantially in its form, and the emission spectrum of the intact wall area featuring three characteristic peaks with maxima at λ = 515, 560 and 595 nm.

We also compared the fluorescence excitation spectra of fibrous plaques and normal wall areas and found them to be similar. The form of the spectra remained unchanged as the fluorescence detection frequency was varied over the range 500–650 nm. These measurements allowed us to conclude that the plaques contained no additional fluorophores in the visible region of the spectrum as compared with normal vessel wall areas. The key to the explanation of the fluorescence of blood vessel wall tissue was found when exciting and measuring fluorescence from the adventitia, i.e., the exterior wall side of the vessels. This spectrum (curve III of Fig. 6) features even more pronounced peaks than the emission spectrum measured from the intima side of the normal zone of the vessel wall (curve II of Fig. 6). It is well known that the microcapillary network (vasa vasorum) supplying major blood vessel wall is located much closer to the adventitia than to the intima, and we hypothesize that the difference between the fluorescence spectra is due to the presence of hemoglobin.

Figure 7 shows the absorption spectrum of normal saline containing blood. The peaks observed at 418, 544, and 578 nm are due to the Soret, α -, and β - bands of oxygenated hemoglobin, respectively[32]. By subtracting the α - and β - bands of the absorption spectrum of oxygenated hemoglobin from the fluorescence spectrum of fibrous plaques, we obtain a spectrum similar in form to the fluorescence spectrum characteristic of intact wall areas of atherosclerotic vessels. In the latter case, filtration of the luminescence emission occurs, and the dips in the fluorescence spectrum of normal vessel wall areas due to reabsorption of the emission in oxygenated hemoglobin are perceived as additional peaks at neighboring frequencies. On the other hand, a dense tissue of sufficiently thick ($h \cong 1$ nm) fibrous plaques infiltrated with lipids (primarily cholesterol and its esters) strongly absorbs the exciting radiation and screens the vasa vasorum layer containing blood oxygenated hemoglobin. Our measurements of the absorption and scattering factors of fibrous plaques and normal wall areas of atherosclerotic arteries[30] confirmed this inference.

We also noticed that in contrast to atherosclerotic plaques which are whitish or yellowish in color, the neighboring normal wall areas were frequently of a reddish-pink hue. Such a coloration is evidently due to the diffusion of hemoglobin inside the intima as a result of damage to the endothelium lining the interior surface of the vessels. It has been known that the development of atherosclerotis causes various damages to the endothelium in vivo[33]. Mechanical damage to vessel samples in vitro may also lead to staining of the vessel walls with hemoglobin in the course of prolonged storage. To find out which of these two factors is the main cause requires experimenting in vitro. At the present time, it can only be stated that the

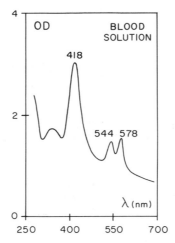

Fig. 7. Absorption spectrum of human blood solution.

fluorescence spectra of reddish normal wall areas of atherosclerotic vessels feature clearly observable dips at λ = 544 nm and λ = 578 nm similar to those in the spectra measured from the adventitia side of the vessels (curve III of Fig. 6).

Fig. 8 shows the diffuse reflection spectra of optically thick human aorta walls taken from the intima (lumen) side of the vessel. These spectra were obtained with a two-channel spectrophotometer equipped with a MgO integrating sphere. Note two specific features of the spectra of Fig. 8. First, the absorption spectrum of oxygenated hemoglobin shows itself in the reflection spectrum of normal vessel wall areas, both the strong Soret band and the α – and β – bands being seen as in the fluorescence spectra. In the reflection spectrum of fibrous plaques on the other hand, even the Soret band is expressed much more weakly. Secondly, in the wavelength range 440–540 nm, the reflection spectra presented differ significantly. The reason for this is the difference in absorption and scattering factors between the fibrous plaque and normal wall areas of atherosclerotic vessels.

LASER DIAGNOSIS OF ATHEROSCLEROSIS

The results obtained to date suggest that spectral analysis of blood vessels in vivo can be used as a method of diagnosing atherosclerotic lesions. To make a spectral diagnosis, it would be necessary to temporarily replace the blood in the vessel with a saline solution. In our opinion, the method of measuring fluorescence spectra is preferable, since excitation can be effected with an argon-ion laser at λ = 476.5 or 488 nm. Note that a fibrous plaque can be identified by its fluorescence intensity at the wavelengths corresponding to the fluorescence peaks (e.g., at 595 nm where no oxygenated hemoglobin absorption is present) and dips (e.g., at 578 nm where an oxygenated hemoglobin absorption band is present). The statistics of the investigations carried out show that the ratios between the fluorescence intensities at the above-indicated wavelengths differs substantially between the fibrous plaque and normal wall areas of one and the same vessel:

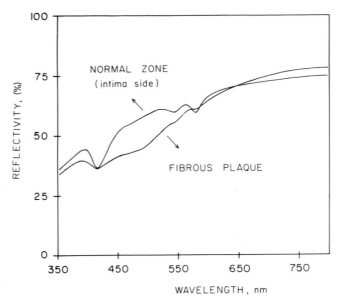

Fig. 8. Diffuse reflection spectra of human atherosclerotic vessel walls taken from the intima side.

$$(I_{f.p.}^{595} /I_{f.p.}^{578} /I_{n.w.}^{595} /I_{n.w.}^{578}) = 1.25 \quad 0.1$$

Despite the fact that measuring reflection spectra requires that white light be delivered to the target area (which is more difficult to do), the reflection technique is advantageous in that the intensity of the reflected signal is several orders of magnitude in excess of the fluorescence intensity, for the same excitation intensity.

ACKNOWLEDGEMENTS

We are very grateful to Drs. R. S. Akchurin, A. A. Belyaev, L. A. Vertepa, A. K. Dmitriev, T. I. Karu, V. G. Omel'yanenko, S. E. Ragimov, and B. V. Shekhonin for their useful cooperation.

REFERENCES

1. J. M. Hisher, R. H. Clarke, IEEE J. Quant. Electron., QZ-20, 1406 (1984)
2. J. J. Livesay, P. G. Hogan, H. A. McAllister, Herz, 10, 343 (1985)
3. M. J. C. Van Gemert, A. J. Welsh, J. J. M. Bonnier, J. W. Valvano, G. Yonn, S. Rastegar, Seminars Intervent. Radiol., 3, 27 (1986)
4. R. Ginsburg, D. S. Kim, D. Guthaner, J. Toth, R. S. Mitchell, Clin. Cardiology, 7, 54,(1984)
5. H. J. Geschwind, G. Boussignac, B. Teisseire, N. Benhaiem, R. Bittoun, D. Laurent, Brit. Heart J., 52, 484 (1984)
6. D. S. J. Choy, S. H. Stertzer, R. K. Myler, J. Marco, G. Fourkial, Clin. Cardiology, 7, 377 (1984)

7. See, for example: CLEO '86, June 9-13, 1986, San Francisco, CA,"Digest of Technical Papers", sections Lasers Angioplasty: 1 and 2

8. G. S. Abela, S. Normann, D. Cohen, R. L. Feldman, E. A. Geiser, G. R. Conti, Amer. J. Cardiol., 50, 119 (1982)

9. M. Eldar, A. Battler, H. N. Neufeld, E. Gaton, R. Arieli, S. Akselrod, A. Katzir, J. Amer. Coll. Cardiol., 3, 135 (1984)

10. W. S. Grundfest, F. Litvack, J. S. Forrester, I. Goldenberg, H. J. C. Swan, L. Morgenster, M. Fishbein, I. S. McDermid, D. M. Rider, T. J. Pacala, J. B. Laudenslager, J. Amer. Coll. Cardiol., 5, 929 (1985)

11. J. M. Isner, R. F. Donaldson, L. I. Dec elbaum, et al., Amer. J. Coll. Cardiol., 6, 1102 (1985)

12. G. A. Abil'sitov, A. A. Belyaev, M. A. Bragin, E. P. Velikhov, V. S. Zhdanov, T. I. Karu, V. S. Letokhov, S. E. Ragimov, M. Ya. Ruda, A. V. Trubetzkoj, N. P. Furzikov, E. I. Chazov, Kvatovaya Electronika (in Russian), 12, 1991 (1985)

13. L. I. Deckelbaum, J. M. Isner, R. F. Donaldson, S. M. Laliberte, R. H. Clarke, D. N. Salem, J. Amer. Coll. Cardiol., 7, 898 (1986)

14. N. P. Furzikov, IEEE J. Quantum Electron., QE-23, October 1987 (to be published)

15. N. P. Furzikov, T. I. Karu, V. S. Letokhov, A. A. Beljaev, S. E. Ragimov, Lasers in Life Sci., 4, 1 (1987)

16. A. J. Welsh, J. W. Valvano, J. A. Pearce, L. J. Nayes, M. Motamedi, Lasers Surg. Med., 5, 251 1985

17. W. Gallakher, J. Spenser, paper at the working session "Laser and Fiber Application in Cardiovascular Research", 26-27 Sept. 1985, Washington DC, USA

18. M. R. Prince, T. F. Deutsch, A. H. Shapiro, R. J. Margolis, A. R. Oseroff, J. T. Fallon, J. A. Parrish, R. R. Anderson, Proc. Natl. Acad. USA, 83, 7064 (1986)

19. M. J. C. van Gemert, R. Verdaasdonk, E. G. Stassen, G. A. C. M. Schets, G. H. M. Gijsbers, J. J. Bonnier, Lasers, Surg. Med., 5, 235 1985

20. M. L. Wolbarsht, IEEE J. Quant. Electron., QE-20, 1427 (1984)

21. D. L. Singleton, G. Paraskevopoulos, G. S. Jolly; R. S. Irwin, D. J. McKenney, W. S. Nip, E. M. Farrell, L. A. J. Higginson, Appl. Phys. Lett., 48, 878 (1986)

22. J. T. Walsh, T. F. Deutsch, T. Flotte, M. R. Prince, R. A. Anderson, CLEO '86, June 9-13, 1986, San Francisco CA,"Digest of Technical Papers", paper TUL2

23. J. F. Ready, "Effects of High-Power Laser Radiation", Academic Press New York-London (1971) Ch.3

24. I. A. Vertepa, A. K. Dmitriev, N. P. Furzikov, Kvatovaja Electronika (in Russian, to be published)

25. R. B. Setlow, E. S. Pollard, "Molecular Biophysics", Addison-Wesley Reading London (1962)

26. S. V. Konev, I. D. Volotovsky,"Photobyology" (in Russian) 2nd, ed. Belorussian University Edition Minsk 1979; pp.265, 272

27. J. P. Phillips, L. D. Freedman, J. Craig Cymerman (eds), "Organic Electronic Spectral Data", vol. VI, 1962-63, Wiley Interscience New York,(1970) p. 1148

28. C. Kittrell, R. L. Willett, C. De Los Santos-Pacheo, N. B. Ratlift, J. R. Kramer, E. G. Malk, M. S. Feld, Applied Optics (1985), vol. 24, no. 15, p.2280-2281

29. R. Richards-Kotrum, C. Kittrell, M. S. Feld,"Digest of Technical Papers", CLEO '86, San Francisco, CA, June 9-13 1986
30. A. A. Oraevsky, V. S. Letokhov, S. E. Ragimov, V. G. Omeljanenko, A. A. Belyaev, B. V. Shekhonin, R. S. Akchurin, Lasers in Life Sci., to be published
31. B. Chance, C. M. Connely, Nature, vol. 79, p. 1235 (1957)
32. V. N.Karnaukhov, V. G. Brailovskaya, L. G. Hashekov, V. P. Zinchenko, L. D. Lukianova, Proceed. of USSR Acad. of Sci. (Dokladi-Russ), 1971, v. 201, p. 227-229
33. V. N. Smirnov, V. S. Repin, Bulletin of USSR National Cardiological Scientific Centre, v. 8, n. 2, p. 12-13. (1985)

PHOTONS IN MEDICINE

R. Pratesi

Istituto di Elettronica Quantistica del CNR

Via Panciatichi 56/30, 50127 Firenze, Italy

HISTORICAL EVOLUTION OF THE THERAPEUTICAL USES OF LIGHT[1-10]

The conventional wisdom that the sun cures human illness comes to us from ancient times, and the ancient belief that a "robust" color was a sign of health persists to this day. The worship of the sun as a health-bringing deity is probably as old as man itself. Possibly the earliest medical association with sun light is the observation of Herodotus, who, in 525 BC, related the strenght of the skull to sunlight exposure. The social attributes of healthy glow and a tan have led to popular use of sunbathing and sunlamps.

Sunbathing

Sunbathing is not a new custom; it goes back to Greek and Roman times when enclosed patio solaria open to the sun were quite popular. Many early physicians prescribed sunbathing for such diverse conditions as epilepsy, jaundice, and obesity. The systematic use of sunbaths as a preventive and therapeutic measure in rickets and other diseases was suggested by Palm in 1890. Rickets may be caused by either the lack of vitamin D in the diet of humans or inadequate exposure to sunlight, and it may be considered the first air-pollution disease. In fact, rickets was prevalent among children in Northern European industrial towns, where the burning of soft coal produced heavy black smog along the narrow street and UV-B antirachitic radiation from sunlight was greatly reduced. The action of UV-B on skin o-curs through a photochemical precursors of vitamin D, present in living skin cells, which leads to the formation of vitamin D_3.

"Chromotherapy"

One hundred years ago "color therapies" based on claimed therapeutical properties of various colors or combination of colors received wide attention in U.S.A. The claim that exposure to sunlight filtered through blue glasses would cure certain diseases and increase the yield of crops and the fecundity of domestic beasts produced a blue-light mania, which lasted for over a decade. Not only blue, but also red light was soon after recommended for treating various diseases. The use of colored light had lost scientific credence by the turn of the century, when the importance of the UV component of sunlight was realized. Surprisingly, there are still "chromopaths" today, sometimes with tragic consequences for their credulous patients[9].

Heliotherapy and Actinotherapy (UV-phototherapy)

The foundations of modern-day UV photobiology began with the work of Niels Finsen, who promoted natural sunlight (heliotherapy) and later artificial UV radiation (actinotherapy) for the treatment of tuberculosis of the skin (lupus vulgaris). Following his pioneering work, heliotherapy expanded rapidly throughout Europe and USA, accompanied by an enormous literature on the subject. Most of the irradiation protocols for the countless number of diseases described in the literature are now of historical interest only. The advent of effective antibiotics and the realization that the success claimed in many of these diseases were little more than anecdotal have resulted today in a much reduced role of UV radiation in clinical medicine. Strangely, despite the lack of convincing proof on the role of sun's UV radiation in curing many diseases, not everyone has abandoned the view that the sun's rays have such curative value.

The use of UV radiation from solar or artificial sources to treat diseases is now usually confined to the therapy of certain skin diseases (acne vulgaris, mycosis fungoides, psoriasis, and some forms of chronic eczema).

Photosurgery

It has been known for thousands of years that the sun can cause ocular damage (eclipse blindness). In 1949 Meyer-Schwickerath reported the first beneficial use of sun light in surgery. Retinal detachment, close-angle glaucoma and other eye pathologies were successfully treated by focusing sunlight into the eye, thus avoiding direct surgical intervention. A few years later the famous eye photocoagulator employing a high-power xenon arc lamp was developed by Zeiss; it was replaced in current clinical practice only twenty years later by the introduction of the argon laser.

INTERACTIONS OF PHOTONS WITH BIOLOGICAL MATTER[7,11,12]

The sequence of changes that follow exposure of the body to nonionizing radiation is very complex and only partly understood for any given biological response.

The photobiological effects of exposure to nonionizing radiation involve the initial essential step of absorption of photons by some molecule, called a chromophore. Once radiation has been absorbed, photochemical changes may occur in the chromophore, and this leads eventually to an observed biological response. Changes in cell and tissue function, and probably participation of mediators, are steps that produce such a response.

When the chromophore is known, the wavelengths of radiation used to initiate a biological response can be tailored to the absorption spectrum of the molecule. More often, the chromophore for a given response is unknown, and the approach taken is to determine the action spectrum for the biological response being studied. This is determined by testing the magnitude of the response at various wavelengths; the observations then permit construction of a curve of exposure dose of radiation as a function of the size of the response. The peak of the action spectrum defines the most effective wavelengths. Figure 1 shows the possible sequence of events following exposure of skin to radiation leading to beneficial and/or adverse effects.

Optical radiation is presently utilized as a therapeutic agent in medicine and surgery. Different mechanisms of action are involved depending on the particular application and light source characteristics.

Thermal interaction

All surgical applications of light rely on the conversion of light energy into thermal energy.

At the microscopic level, the photothermal process originates from bulk absorption occurring in molecular rotational-vibrational bands, followed by subsequent rapid thermalization through nonradiative decay. The temperature rise can be described by a two-step reaction involving i) the excitation of the target molecule; ii) the deactivation of the excited molecule through an inelastic scattering with a collisional partner belonging to the sorrounding medium, which carries away as kinetic energy the internal energy released by the excited molecule.

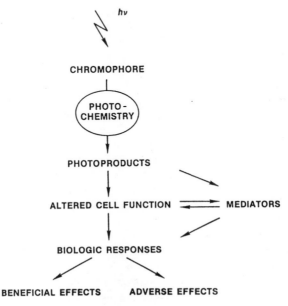

Fig. 1. Possible sequence of events following exposure to optical radiation leading to beneficial and/or adverse effects.

The biological effects of heating are largely controlled by molecular target absorption, essentially from free water, hemoproteins, pigments (such as melanin, charotenoids, flavins, etc.), and other macromolecules such as nucleic acids and aromatics. The photophysical parameter of interest is the absorption coefficient α, and its reciprocal, $1/\alpha$, which measures the characteristic absorption length of the biological medium. In low-absorbing media, the scattering properties of the medium may influence the wavelength selection and the depth of penetration.

When a soft tissue is heated from the normal body temperature (37 °C) to the hyperthermic range (43-45 °C) fundamental changes in the structure of macromolecules and cell components begin to occur (configurational changes, bond destruction, membrane alterations). A further increase of temperature makes these processes more dramatic and denaturation of proteins occurs, leading to the "coagulation" of the tissue. Of particular interest is the denaturation of collagen: above 60 °C the trihelical structure of collagen is disrupted and the coils become randomized. Macroscopically a blanching of the irradiated surface, due the increased scatter of illumi-nating light in the tissue, is observed together with retraction of the tissue. At around 100 °C free water in cells starts boiling. Due to the large volume expansion the cell walls are ruptured explosively. The large heat of vaporization of water has the beneficial effect to prevent any further temperature increase since the steam generated carries away any excess of heat above that required to raise the water to the boiling point. When the water has dried up, the temperature of the residual material in-creases rapidly. At 300-500 °C carbonization of the tissue occurs with decomposition of tissue constituents.

Photochemical interaction

Light may act as a "reactant" in a photochemical reaction. In this case the photon energy is used to excite a particular chromophore that in turn initiates a complex biological process whose final products have a therapeutical relevance. A chromophore capable of causing light-induced reactions in molecules that do not absorb light is called a "photosensi-tizer".

Many kinds of molecules can act as photosensitizers including both naturally-occurring (chlorophyll, iron-free porphyrins, flavins, etc.) and synthetic (acridines, anthraquinones, xanthenene dyes, etc.) organic com-pounds and some inorganic materials.

In general, these sensitizers are excited to a short-lived singlet excited state (^{1}S*) by the absorption of a photon, and then cross over into a long-lived triplet excited state (^{3}S*) which mediates the photobiological damage. The ^{3}S* state has generally the greates reactivity because of its favourable spin configuration and relatively long lifetime (1 ms - 1 s). The subsequent reactions involved can proceed by a variety of (competing) pathways depending on the chemical nature of the photosensitizer and sub-strate (molecule being photodegraded) as well as on the reaction conditions. In some of these process, the ^{3}S* sensitizer abstracts an electron from

the substrate to give free radicals which react in various ways to give
photooxydated substrate products and a regenerate photosensitizer (in this
case the photosensitizer behaves as a photocatalyst). In other cases,
sensitizer and substrate radicals interact to give sensitizer-substrate
and substrate-substrate covalent photoadducts.

Most of the photosensitized reactions in biological systems that have
been studied involve the participation of molecular oxygen, i.e. they are
sensitized photooxydation processes. Such reactions are often termed "photodynamic action" or "photodynamic reactions". In these reactions, $^3S^*$ sensitizers transfer excitation energy to ground state molecular oxygen, $^3O_2$,
to give singlet state oxygen, $^1O_2^*$, and ground state sensitizer. Singlet
oxygen, in turn, can react with many types of biomolecules not susceptible
to ground state oxygen. In fact, owing to its peculiar electronic orbital
and spin configuration, $^1O_2^*$ is higly electrophilic: it efficiently oxidizes
electron-rich sites in neighbouring biomolecular targets.

In many photosensitized processes, other reactive species such as
hydrogen peroxide, superoxide anion and dydroxyl radical are produced
which can alter biomolecules. Finally, in a few cases, excited photosensitizer is converted into a toxic photoproduct.

Photomechanical interaction

Extremely high-peak power light pulses may locally generate high
electric fields comparable to average atomic or intramolecular Coulomb
electric fields. At suche field dielectric breakdown of the target material
can occur with the production of a microplasma, i.e. an ionized volume
with a very large free electron density. The shock wave following the
plasma expansion can produce a mechanical rupture of the tissue.

In the initial phase, the optical pulse produces the separation of
bound electrons from individual molecules. With IR lasers the absorption
of one photon is not sufficient to ionize the molecule (ionization potential
7-10 eV), and multiphoton absorption must occur. In the field of molecules
or ions free electrons can acquire energy by absorbing incoming photons
according to the inverse Bremmstrahlung process. When the electrons have
gained sufficient energy, collisions with neighbouring molecules create
new free electrons. An electron avalanche growth is achieved: in a few
hundred picoseconds a very large plasma density (typically 10^{21} electrons/
cm^3) is thus created in the focal volume of the pulsed laser beams (laser-
induced breakdown).

Photoablative interaction

Ablative photodecomposition consists of the direct breakup of intramolecular bonds in polymeric chains, caused by the absorption of energetic
UV photons, followed by the expulsion of the fragments at supersonic
velocities. The microscopic mechanisms correspond to transitions of a
molecule AB that is promoted to a repulsive electronic state that yields
photoproducts A and B. Photoablative techniques have arisen much interest

in photosurgery in view of their ability to produce sharp incisions with minimal thermal damage to adjacent unirradiated tissues.

THERAPEUTIC PHOTOMEDICINE

We now briefly illustrate the most important applications of light in treating human diseases.

As pointed out in the introduction, the therapeutical use of light was gradually reduced as the understanding of the biological action of light progressed. Of the large collection of light therapies only two, namely the photochemotherapy of psoriasis and the phototherapy of hyper-bilirubinemia, both based on photochemical processes, are still applied with wide diffusion and good success. The advent of the laser greatly improved the photocoagulative treatment of ocular diseases and opened the entire field of surgery to optical techniques.

It is important to note that the introduction of lasers in Medicine and Surgery has represented only a technological improvement; in fact, up to date there are no indications for a specific effect of the coherence of laser photons in the therapeutical action of light. The peculiar charac-teristics of laser light, i.e. high directionality, high continuous wave and peak powers, and, to a lesser extent, high monochromaticity, make the laser light superior to standard thermal, incoherent light in terms of better control of light delivery, energy deposition, pulse duration, and wavelength selection.

Incoherent (thermal) photons

Photochemiotherapy of psoriasis[6,7,9]. Psoriasis is a common skin disease which affects between 1 to 5% of the world's population. This skin disorder is characterized by an accelerated cell cycle and rate of DNA synthesis: the turn-over rate of mitosis in the epidermis in 2-5 days compared to 25-28 days in normal skin. The increased epidermal prolifera-tion leads to the formations of lesions in the form of red, raised, scaling plaques. UV irradiation, with or without added photosensitizers, represents an effective treatment of psoriasis: the cytotoxic effect of UV photons produces a transient retardation of the growth of epidermal cells.

Psoralens. The furocouramins are three-ringed compounds found in a variety of plants. Popularly called "psoralens", they are composed of con-joined coumarine and furan rings (Fig. 2). Only certain furocoumarins are effective photosensitizers. Psoralens are phototoxic compounds that can interact with various components of cells and absorb photons to produce photochemical reactions that alter the function of the cellular constit-uents.

Cellular responses. Psoralens readily penetrate intact cells and intercalate between base pairs in the DNA. Absorption of photons by psoralens leads to covalent binding of the psoralens to DNA or RNA, with the formation of monofunctional adducts of psoralens with pyrimidine bases

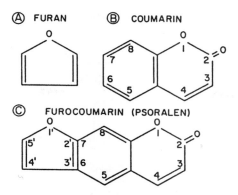

Fig. 2. A furocoumarin (C) is formed by combining furan (A) and coumarin
coumarin (B) rings.

(thymine, cytosine) (Fig. 3). This first photochemical reaction is usually
a conjugation of the 4', 5' double bond of the furan ring of the psoralen
molecule with the 5, 6 double bond of the pyrimidine base. This results in
the formation of a fluorescent adduct that absorbs at 360 nm. If two pyrim-
idines are adjacent at appropriate distances and on opposite strands, subsequent absorption of a second photon by the monoadduct leads to the formation of additional covalent linkages between the 3, 4 double bond of the
psoralen molecule and the 5, 6 double bond of the pyrimidine base of the
opposite strand, thus cross-linking the two strands of the double helix.
Such cross-links are also called "bifunctional adducts".

"PUVA therapy". Psoralens absorb mainly in the UV (Fig. 4). The psora-
lens/UVA reaction produces suppression of DNA and RNA synthesis in cells,
and this effect is usually considered to be its mechanism of action in
psoriasis. Suppression of macromolecular synthesis and cell proliferation
is a temporary phenomenon. PUVA treatment of psoriasis represents the first
beneficial effect of UVA-drug combination that hitherto had always been
considered deleterious; this new pharmacologic concept was termed "photochemiotherapy".

Risks of PUVA treatment. Photocarcinogenesis is one of the possible
long-term side effects that may occur during PUVA treatment. Which of the
two lesions, monofunctional adducts and cross-links, produced in DNA by
PUVA therapy is more carcinogenetic is still matter of debate. Cells have
more difficulty in repairing crosslinks so that this lesion tends to be
lethal. Monofunctional adducts tend to be repaired to a greater extent,
and, because repair of DNA is not error-free, mutations and transformations
can occur with a probability greater than for crosslinks. Because of this
concern great effort is presently devoted to the development of new compounds capable of producing monofunctional adducts with a higher degree of
lethality and with the possibility of excitation with longer wavelenght
light.

Phototherapy of neonatal jaundice[6,9,13]. Neonatal jaundice, or hyperbilirubinemia, is a common disorder found in about 10-30% of newborn babies.

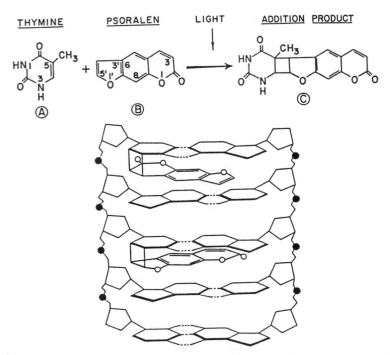

Fig. 3. Covalent linking of psoralens to thymine (single-strand linkage).

Bilirubin (BR-IXα) is derived principally from the catabolism of heme. It is a lipid-soluble, non-polar substance that binds to albumin in blood and is transported to the liver. There it is conjugated with glucuronic acid and excreted in bile. Newborn infants are temporarily deficient in their capacity to conjugate bilirubin, and after all albumin and tissue-binding

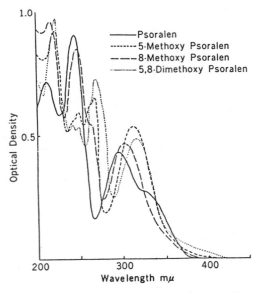

Fig. 4. Absorption spectra of psoralens.

sites are saturated, free bilirubin circulates and can enter neural tissues and produce irreversible brain damage (bilirubin encephalopathy).

The discovery that visible light can be used to influence bilirubin metabolism favourably in jaundiced infants was made in England in 1956. Phototherapy soon became an accepted treatment in England, but it did not become widely used until the results of more extensive controlled studies had been reported in USA in 1967. Phototherapy is now used almost routinely in the management of neonatal hyperbilirubinemia.

Photochemistry of bilirubin. Bilirubin-IXα is an unsymmetrically sustituted bichromophoric molecule containing two non-identical pyrro-methenone units. Each of these units has a Z configuration at the central exocyclic double bond. Figure 5 shows the preferred conformation of bili-rubin, with the polar NH and COOH groups linked together by hydrogen bonds. Bilirubin has a single broad absorption band in the visible extending from about 350 nm, to 550 nm, with a maximum near 450 nm. When exposed to light, the pigment undergoes several competing photochemical reactions.

Photooxydation. In the presence of O_2 and light BR is converted to monopyrrolic and dipyrrolic fragments that are colorless and mostly very polar. The quantum yield of this reaction in oxygenated solutions contain-ing albumin is very low, and so the process is probably inefficient "in vivo".

Configurational photoisomerization. The most quantum yield efficient photoreaction that BR is known to undergo is reversible Z-E configurational isomerization at the bridging C=C double bond at C-4 and C-15, that leads to the formation of three isomers 4E, 15Z; 4Z, 15E; and 4E, 15E – bilirubin. When BR is bound to human serum albumin (HSA) the predominant product is 4Z, 15E-BR. During Z-E isomerization a pyrrole ring at one end of the molecule becomes rotated through 180 degrees (Fig. 6). This partially opens ut the hydrogen bonding in one half of the molecule, exposing polar func-tions, and making the molecule less lipophilic than the native BR. The quantum yields for the Z-E and E-Z processes are 0.2 and 0.8, respectively.

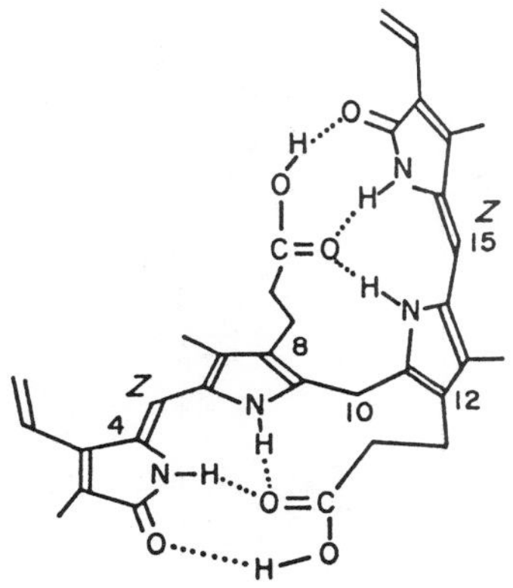

Fig. 5. Preferred conformation of bilirubin.

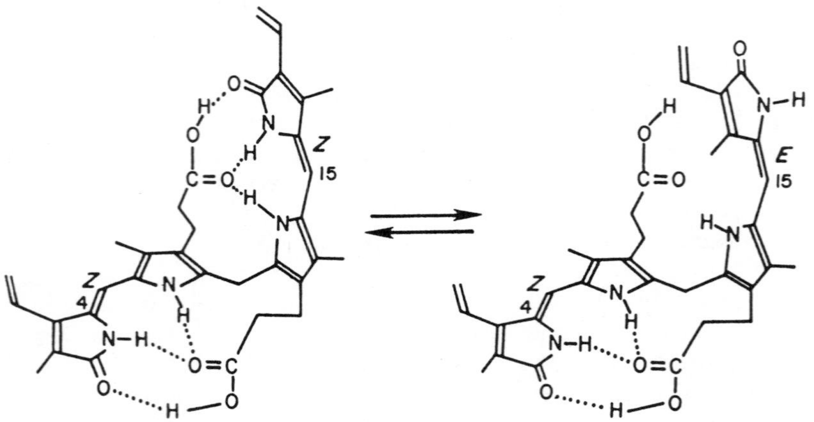

Fig. 6. Z-E configurational isomerization of bilirubin.

Due to the overlapping of the absorption spectra of BR and its isomers, irradiation of BR in a closed system leads to a photoequilibrium mixture with a proportion of isomers that depends on the solvent and on the excitation wavelength. The most efficient wavelength "in vitro" is 390 nm which produces more than 40% conversion of BR to Z,E-BR/HSA at equilibrium, while the corrisponding value with green light excitation (520 nm) is only 5%.

Structural photoisomerization. Configurational isomerization is accompanied by an irreversible reaction which involves intramolecular cyclization of the endovynil group at C-3 (Fig. 7). This "cyclobilirubin" is named "Z-lumirubin". The exposure of the polar groups of the molecule following cyclization makes the molecule more polar than BR, to such a degree that it no longer has to be metabolized to a glucuronide to be excreted. Although lumirubin formation is more quantum efficient than photooxydation, the reaction nevertheless has a low quantum yield ($\cong 0.001$).

Mechanism of phototherapy. The evolution of the understanding of the basic processes responsible for phototherapy in jaundiced babies has closely followed the progress of the photochemistry of BR "in vitro" and "in vivo".

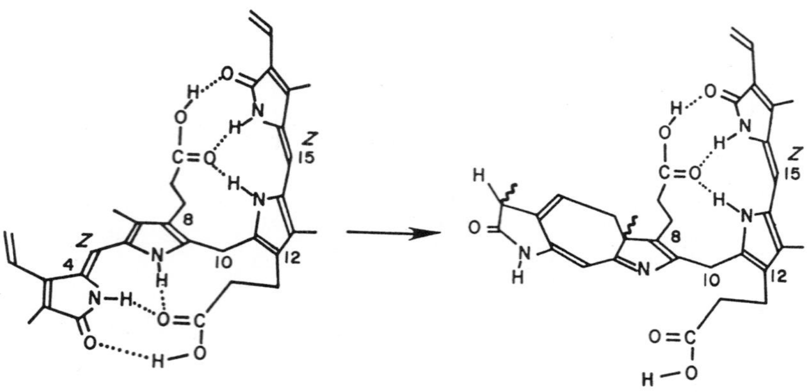

Fig. 7. Intramolecular cyclization of bilirubin.

The discover in 1978 that BR has high quantum yield configurational isomers and the observation that they are promptly excreted in Gunn rats (rats with a genetic deficiency of the enzyme responsible for conjugation) led to the abandonment of the previous photooxidative hypothesis. In 1983 two different observations that contrasted with the new mechanism based on configurational isomerization were reported[13]:

i) in babies exposed to phototherapy the Z,E-BR isomer was efficiently produced with an efficiency consistent with the high quantum yield value of the "in vitro" reaction, but its excretion was too slow to be responsible for the phototherapy;

ii) fluorescent green lamps with emission spectrum poorly overlapping the absorption spectrum of BR ("in vitro") were found to be efficient light sources for phototherapy.

On the contrary, Z-lumirubin, although formed at small percent, appeared to be quickly cleared by the organism. Successive studies demonstrated that:

(a) blue and green fluorescent lamps have comparable efficiencies in clinical phototherapy and that the cut-off of the blue components of the green lamps does not sensibly affect their efficiency;

(b) green fluorescent lamps produce more lumirubin and less Z,E-bilirubin than white fluorescent lamps in the serum of jaundiced infants;

(c) the proportion of isomers at photoequilibrium in the bile of Gunn rats is the same under green and blue fluorescent lamp irradiation. Moreover, the quantum yield for Z-lumirubin formation "in vitro" has been found to depend on the excitation wavelength, with a maximum in the green.

With all these new results at disposal it is, however, not yet possible to outline with precision the mechanism of phototherapy and the relative roles of the various photoprocesses. Lumirubin is presently hypothesized as the main route of the light-induced clearance of BR from the organism, and the fast Z-E reaction is supposed to represent a rapid detoxifying mechanism of phototherapy due to an expected lower toxicity of the Z,E-BR molecule compared with the Z,Z-BR one. More informations on the photochemistry of BR "in vitro", the modification of the photophysical properties of BR in extravascular tissues, and the effects of the filtering action of skin on the efficiency of the various spectral components of a polychromatic light are needed to establish the optimal protocol for phototherapy.

Coherent photons[14,15,16]

The advent of the laser has permitted the introduction and the rapid expansion of optical surgery in the management of a large variety of pathologies in man. Most of the well-established laser applications rely on thermal processes: cutting of tissues, vaporization of pathological masses, coagulative necrosis of tumors are easily achieved with the laser operating as an "optical knife" or with endoscopic techniques. The photomechanical interaction of powerful laser pulses with ocular structures has proved to be an unique technique for treating complex pathologies of the anterior segment of the eye: Q-switched Nd-Yag lasers are currently used to perform iridectomy, capsulotomy, etc. Photodynamic therapy of cancer employing the photosensitizing action of a derivative of hemophorin (HpD) represents an important potential application of laser light. Reports on the clinical uses of this procedure are increasing in number, and great effort is devoted to the improvement of this technique by using new photosensitizers capable of greater absorption in the spectral region where light penetration in tissue is maximum. Applications based on photoablative interactions are still in the experimental stage, in particular in ophtalmology, dermatology, and angioplasty.

A brief selection of laser applications described by some of the most active and authorative clinicians in this field is now presented.

Surgical applications

Lasers in gynecology. (G. Bandieramonte, Istituto Nazionale per lo Studio e la Cura dei Tumori, Milano, Italy). Intra-abdominal and lower genital lesions can be primarily treated by CO_2 laser surgery with microsurgical and freehand-technique. In pelvic surgery, the CO_2 laser is used to vaporize or reduce tumor mass in advanced cases and to perform fallopian tubae microsurgery. Laparascopic laser surgery can be used with the rigid laparascope, which allows direct vision. Tubal sectioning for sterilization, treatment of endometriotic nodes, and adhesiolysis can be performed with CO_2 laser or KPT/532 laser laparoscopic application under direct vision of the lesions. But, since the laser has to be perfectly located in the axis of the tube, extreme prudence should be used and some limitations of handling are recognized. Substantial improvement will be obtained as soon as fiberoptic bundles for CO_2 laser surgery are available for routine use. Tumor volume reduction and debulking has been performed successfully[17] in conjunction with radiation therapy for palliative care of cancer patients with presacral and lower prelumbar extention of pelvic tumors. Although complete tumor vaporization is often technically impossible, reductive CO_2 laser surgery can render the tumor more responsive to combined therapy (i.e. radiation and/or chemoterapy). It is therefore, preferable to perform exploratory surgery for pelvic recurrence of gynecological tumors with one lasing instrument available and set in the operating room. The same technique can be applied for intraperitoneal benign or malignant tumors, such as ovarian tumors, or pelvic fibrotic and inflammatory adhesions, with either microsurgical and freehand applications.

Bloodless enucleation of myomas[18] can be performed; the incision is easy and large vessels must be ligated. Total bleeding has been found to be significantly less than with conventional surgery; there is nevertheless very little advantage in using CO_2 laser for uterine surgery. Excessive endometrial bleeding has been successfully treated with Nd:YAG photocoagulation through hysteroscope[19].

In gynecology the microsurgical equipment and procedure have been the main improvement. Operations for which microsurgery can offer substantial improvement have increased the operative precision and the possibility of conserving the function of healthy tissue. Modern cervical and fallopian tubae microsurgery are the most important examples of this technical advance. The linking of CO_2 laser instruments with microscopes, colposcopes, and micromanipulation of the beam, under constant visual control, have opened a new era of treatment modalities and, after completing the experimental phase, offers an up-to-date and important addition to the gynecologyc armamentarium. The precision of the treatment is the fundamental property of CO_2 laser action and it may be measured in millimeters both in depth and surface area, controlled by the power setting and the speed at which the laser beam is micromanipulated over the tissue.

Fallopian tubae microsurgery gains the maximum advantage from the precision of CO_2 laser surgery. Actually no other conventional method offers the possibility of treating only one layer of the walls of these extremely delicate organs without minimal adjacent tissue damage. Moreover, the sealing of blood and lymph vessels at the incision borders prevents disseminations in cases of chronic salpingitis[20].

Cervical laser mircrosurgery is the most common application in gynecological surgery. Today, the standardization of methods and indications for laser vaporization or cone resection of the cervix which have reduced postoperative sequelae of treated cervical preneoplastic lesions and result in rapid healing, have and should replace diathermy, cryosurgery, and also knife procedure[21-27].

Based on statistical consideration, it has been determined[28] that over 70% of the cases with preneoplastic cervical (CIN-CIS) and high vaginal lesions (VAIN) can be successfully treated on the outpatient basis. As general anesthetics are usually not necessary, postoperative complications are minimal or absent, time in hospital is reduced, thereby reducing the cost to the patient.

Vaginal microsurgery has approached excellent results especially in inaccessible anatomic regions, such as the fornices or the superior third of the vaginal vault. Diathermy, cryosurgery, and surgical excision have not satisfying results because of the risk of such frequent postoperative complications as retraction, fibrosis, stenosis, and fistulae. With the absence of these complications, laser surgery can be considered the treatment of choice. Small areas of neoplastic lesions can be easily removed with the CO_2 laser and bloodless operative mode preserves microscopic visualization throughout the procedure, precluding the need for clamping and suturing[29,30]. Laser surgery on the vulva with the CO_2 scalpel, when colposcopically directed, has obtained the best results in the therapy of condylomata[31]. The vaporization procedure is considered here the most important, since the depth of penetration of the laser distruction can be easily controlled. Furthermore, excision of early neoplasia and dystrophy is considered effective.

The industry for surgical applications of the laser is still expanding, and more technologically advanced equipment is continuously being offered. Future substantial improvement can be expected; for example, fiberoptic conducting CO_2 laser beam or new delivery systems for endocervical, endometrial, or endopelvic applications are being developed, therefore the

benefits of more precise instruments for the care of patients will be
available.

Detailed knowledge, continued practice, rigid safety precautions, and
judicious clinical considerations are essential for competent, safe and
simple use of lasers in surgery, as with the classical cold knife.

Lasers in laryngeal microsurgery. (G. Motta, Clinica Otorinolaringo-
iatrica, Università di Napoli, Italy). The CO_2 laser is used in the treatment of several laryngeal diseases and offers considerable advantages over
traditional techniques. New applications of this surgical tool are currently
under study for other laryngeal and ENT pathologies.

Over the last six years our team has used the CO_2 laser in direct
microlaryngoscopy to treat 621 patients affected by laryngeal pathological
processes, in order to evaluate the advantages and disadvantages of laser
surgery in comparison with traditional techniques and to define its limitations and indications. The results are as follows:

- vocal nodes: the laser does not yield any advantages in comparison
  to traditional surgery and, since there are a few advantages, it is
  not advisable to use this technique in such pathologies;
- peduncolate laryngeal polyps: the laser is a good alternative to
  traditional surgery, even though it does not provide any great
  advantages;
- Reinke's oedema: the laser is more precise than traditional surgery;
- dyscheratosis: the laser is very precise and give better functional
  results; it is necessary in these cases to make a previous hystologi-
  cal examination to exclude any carcinomatous degeneration;
- cancer of the glottic region, also extended to the ventricles, to
  the anterior commissure, to the upper vocal cord or the hypoglottic
  region: the laser permits a radical operation with little surgical
  trauma avoiding tracheotemy with a short hospital day;
- laryngeal paralysis: the laser makes it possible to perform less
  difficult and less traumatizing operations than traditional surgery,
  and to obtain very satisfactory results as well;
- laryngeal stenosis by chronic oedemas (generally due to sopraglottic
  laryngectomies): with the laser it is possible to remove the oede-
  matous mucosa very easily; the results are always satisfactory;
- laryngeal stenosis by diaphragms or synechias: the removal of
  synechias is very simple; however, in these cases it is necessary
  to make very frequent postoperative controls to detach the fibrine
  and prevent recurrences;
- concentric stenosis: after the creation of a tunnel, a tube T-shaped
  by Montgomery is inserted for 6-12 months; in these cases frequent
  laryngeal check-up is required to avoid recurrences. In these regards, the laser provides a good alternative to traditional surgery
  permitting less complicated surgery.

In conclusion, our observations support data in the literature and
extend earlier reports, documenting the effectiveness of the CO_2 laser in
the treatment of several laryngeal processes. The success of laser treat-

ment however will depend on a preventive and critical analysis of its indications and possibilities, and, of course, on a precise knowledge of the techniques to be employed.

Lasers in dermatology. (G. J. Hruza, J. A. Parrish, Department of Dermatology, Harvard Medical School and Massachusetts General Hospital, Boston, USA). Over the past few years, lasers have become a useful therapeutic modality for the dermatologist. The argon laser is being successfully used in the treatment of port wine stains (PWS) and other superficial vascular lesions[32]. However, due to its long pulse duration, 0.1 sec to continuous, this laser causes relatively nonspecific thermal destruction of the epidermis and upper dermis[33], which in 2-11% of patients leads to scarring[34]. A modification of the standard treatment technique being developed in Australia may improve on these results[35]. Other disadvantages include incomplete lightning of lesions[32,34] and need for anesthesia[36].

A laser that would selectively destroy the abnormal blood vessels while leaving the remaining epidermis and dermis intact was needed. A tunable dye laser with a pulse duration of 400 μs , which is less than the 1 msec thermal relaxation time of 50 μm diameter blood vessels, cas developed[37]. As the major chromophore is oxyhemoglobin, the laser is tuned to 577 nm which is a major oxyhemoglobin absorption peak[38]. PWS treated with this dye laser fade to similar degree as PWS treated with the argon laser[36]. However, there is no need for anesthesia or postoperative wound care and scarring has not been noted[36]. Also, the lack of scarring or fibrosis make treatment of trunk lesions and of children with PWS more feasible. Histologically, the abnormal blood vessels disappear and are replaced with a substantially normal dermis[36]. A multicenter trial with the tunable dye laser for the treatment of PWS and other vascular lesions is currently underway in the United States.

Benign pigmented epidermal lesions, such as solar lentigines, have been treated with several destructive modalities including cryosurgery[39], electrodessication[39], and argon laser irradiation[40]. All of these modalities frequently result in incomplete pigment removal, scarring of hypopigmentation[40,41]. Several laser systems are under investigation to improve on these results. Low-fluence, 4 J/cm^2, CO_2 laser irradiation for 0.1 sec of lentigenes resulted in epidermal necrosis and mild dermal changes. Seventy-six percent of the lesions totally cleared or lightened substantially. Less than 2% of the lesions developed atrophy and no instances of scarring were noted[42].

The foregoing therapics rely on nonspecific destruction of the epidermis. The specific targeting of melanosomes may allow for selective destruction of pigmented lesions without significant dermal damage[43]. The theory of selective photothermolysis[44] predicts that laser pulses shorter than the thermal relaxation time of melanosomes, approximately 1 μs, should affect only the melanosomes if an appropriate wavelength is chosen[45]. Therefore, the effect of various nanosecond domain laser pulses were studied in black and brown guinea pigs. The Q-switched ruby (694 nm; 40 nsec)[46-48] the Q-switched Nd:YAG (1064, 532, 355 nm; 12 nsec)[49], and the tunable dye lasers (694, 560, 530, 488, 435 nm; 750 nsec)[50], were used to irradiate guinea pig skin at various energy fluences. Immediately after irradiation, photodisruption of melanosomes was noted by electron microscopy[46-51]. This was followed by temporary cutaneous hypopigmentation and follicular depigmentation at the highest energy fluences[44,49,50]. Damage thresholds increased as the wavelength increased in line with the thermal mechanism implied by selective photothermolysis and correlated with the absorption spectrum of DOPA-melanin[47,50]. The potential for the treatment of epidermal pigmented lesions in humans is being explored in ongoing human studies. In another approach, the argon laser was modified to rotate a 120 μm beam in a 2 mm diameter circle at up to 70 Hz, thus providing an exposure time of 300 μs. Seventy-six percent of 25 treated lentigines, ephelides and cafe-au-lait spots showed greater than 50% lightening without evidence of scarring[52].

The continuous wave CO_2 laser is used by the dermatologic surgeon for the vaporization or excision of various cutaneous lesions such as verrucae, keloids, tattoos, epidermal nevi, PWS and vrious skin tumors[53]. Due to the large 200-750 μm, zone of thermal damage left behind by this laser[54], poor graft take[55,56], weaker early scars[57], increased bacterial colonization of the wound[56] and possibly larger than necessary scars have been noted[54,57,59]. A laser that leaves less thermal damage behind would be desirable. Tissue ablation has been studied with various lasers in the guinea pig model. The ArF (193 nm. 14 nsec) and the KrF (248 nm; 14 nsec) excimer lasers have been s'own to ablate skin precisely and to leave behind minimal residual damage[60,61]. However, neither laser is hemostatic[60,61] and as they are both ultraviolet light lasers, their potential mutagenicity[62,63] and carcinogenicity are unclear at present. Simple thermal models suggest that precise ablation of tissue, with minimal residual thermal damage, can be produced by pulsed radiation at wavelengths absorbed strongly by water[54]. A 2 μs pulse duration CO_2 Transverse Electric Atmospheric (TEA) laser can hemostatically ablate skin, leaving behind a 50-100 μm zone of thermal damage[54,64]. Even better results have been obtained with the 2.94 μm, 1 μs pulse duration Er:YAG laser. The 2.94 μm wavelength coincides with a major water absorption peak allowing easy skin ablation. This laser cuts relatively hemostatically while leaving only a 40-50 μm zone of thermal damage behind[64]. Healing and grafting studies are underway with these lasers in order to assess their potential utility in the treatment of various skin lesions in humans.

Laser welding of hairless mice skin has been accomplished with the continuous wave Nd:YAG, argon and CO_2 lasers[65]. The dermis is thermally altered, somehow causing the wound edges to "stick" together while early healing takes place obviating the need for skin sutures[66]. If laser welding proves effective, it might have advantages over suturing, since it is sterile and nontactile, does not require introduction of foreign material into the wound, and provides, in the animal mode, subjectively better cosmetic results[65,66].

Recently, reports of successful laser treatment of psoriatic plaques have appeared[67,68]. The CO_2 laser was used to superficially vaporize several psoriatic plaques on three patients. The healed areas were similar to normal skin and have remained free of psoriasis during 3-5 years of follow-up[67].

In another report[68], systemic hematoporphyrin derivative (HPD) in combination with 630 nm dye laser light, usually used in the treatment of extensive cutaneous malignancies by dermatologists[69], cleared two psoriatic plaques without scarring. These reports are intriguing, but they should be interpreted with caution in view of the well known tendensy of psoriasis to Koebnerize with trauma[70]. Also, the risk of permanent scarring must be considered[71]. The prolonged phototoxicity of systemic HPD could be avoided with the development of an effective topical form of HPD[72].

The future of lasers in dermatology is very bright. As existing techniques are further refined and the foregoing new techniques are perfected, the usefulness of lasers to the dermatologist will be greatly enhanced.

Lasers in haemophilia. (A. Musajo Somma, Cattedra di Chirurgia Plastica Ricostruttiva, Università di Bari, Italy). Haemophilias are hereditary bleeding disorders resulting in the deficiency of specific plasma factors coagulant activity. Even minor wounds bleed profusely unless specific transfused factor is promptly used. However blood products bring the risk of severe liver diseases and acquired immune deficiency diseases.

The advent of Aids has added another risk to the list that haemophiliacs take when treating with plasma products. Nowadays, laser surgery allows photocoagulation of bleeding areas avoiding, in many cases, relevant blood products transfusion and inherent risks. Since 1979 we used laser surgery in haemophiliacs according to the following scheme:

(1) assessment of type of haemophilia A (factor VIII), B (factor IX), von Willebrand disease;
(2) severity of haemophilia (circulating factor less than 1%);
(3) the presence of an antibody to relevant clotting factors;
(4) skin and mucosal traumas of disorders;
(5) oral bleeding and dental extractions.

Every patient was investigated preoperatively about general health and haemostatic function. On the day of laser surgery the patient was given antifibrinolytic intravenously and local or general anaesthesia administered. CO_2 laser was used in a focused mode at 30 W output power to remove soft tissues, pseudotumours or various skin disorders[73]. Incision and simultaneous coagulation were successfully obtained. Because the argon laser wavelengths are well absorbed by hemoglobin, we selected mainly oral bleeding lesions to be treated. We used a spot of 1 mm and pulsed power of 3 W lasting 0.2 s each pulse; the argon laser radiation was flexibly transported through fiberoptics even to the bleeding gengival socket of to the nasal mucosa. We treated overall 85 haemophiliacs in 215 laser surgery sessions, mainly on a day-hospital basis (95% of cases). Among them only 14 patients received infusion of cryoprecipitates related to surgical opportunity. Optimizing laser systems available in a multidisciplinary university centre was a necessity inherent to the cost of the technology; it was possible to save to public finance as much as 200,000 US dollars using laser surgery and not giving to the haemophiliacs expensive blood product (especially in those patients with inhibitors), over a period of 8 years.

Clinical results in this selected pool of haemophiliacs were excellent, with only two minor complications (regional hematomas due to local anesthesia). In every case laser surgery achieved adequate performance. In our experience cost/benefit ratio prompted us to suggest new adequate therapeutic scheme without transfusion risks (i.e. hepatitis, Aids, etc.).

Endoscopic photocoagulative applications

Lasers in gastroenterology. (S. G. Bown, Department of Surgery, The Rayne Institute, Faculty of Clinical Sciences, University College Hospital, London, U.K.). Gastroenterologists first looked to lasers as a means of controlling gastrointestinal haemorrhage endoscopically. For appropriate lesions, the results have been very successful. There are cheaper non-laser devices which may be equally effective although their efficacy has not yet been proven. The major future of lasers in gastroenterology is in tumor therapy. Current clinical practice can provide extremely effective palliation for malignant obstruction of the oesophagus, stomach and rectum when non other therapeutic options are open, and with a minimum of general disturbance to patients with a short life expectancy. Current research is aimed to ablating earlier gastrointestinal tumours in their entirety and also looking at the local treatment of hepatic and pancreatic neoplasms.

Control of haemorrhage. Most episodes of acute upper and lower gastro-intestinal (GI) haemorrhage cease spontaneously. Many of whose that do not occur in elderly patients who have other serious medical conditions. The most promising non-surgical approach to their management is the endoscopic applications of thermal haemostatic devices. These includes laser and non-laser techniques, but to date, the laser techniques have received the most scientific study.

The most important non-neoplastic lesions at risk of further bleeding are peptic ulcers, varices and vascular anomalies (angiodysplasia and telangiectasia).

i) Peptic ulcers. In U.K., this represents the most important group. About 50% of emergency admission for upper GI bleeding have peptic ulcers, and of these 25-30% will not stop bleeding spontaneously. The high risk cases can be identified at emergency endoscopy as long as the ulcer crater is cleared of overlying clot to identify the precise bleeding point, and are those in which an artery can be identified[74]. Lasers were the first modality to be assessed in these high risk cases in a series of controlled clinical trials and the results have recently been reviewed[75]. The most difficult part of the endoscopy is identifying the precise blending point, but once this has been confirmed as a vessel (bleeding or non-bleeding) the technique is to ring the vessel with 6-8 laser shots (80 W, 0.5 s with the Nd-YAG laser) as close to the point where the vessel breaks through the crater as possible to seal the feeding artery. In our trial with the Nd-YAG laser, 4 of 35 ulcers with visible vessels treated with the laser rebled, compared with 21 of 38 controls ($p<0.005$) with significant reduction in the need for emergency surgery. Comparable trials with non-laser techniques have yet to be reported, although experimental[76] and uncontrolled clinical results[77] with monopolar, bipolar and liquid diathermy electrodes and the heater probe are good.

ii) <u>Varices and vascular anomalies</u>. Effective arrest of variceal haemorrhage with no contact laser techniques seems inherently unlikely as haemostasis depends on thermal contraction of the vessel wall and the sorrounding tissue and varices are large and thin walled. There are some studies that suggest that the Nd-YAG laser may give temporary haemostasis from varices both experimentally[78] and clinically[79], but the vessels are not permanently occluded and sclerotherapy seems a much better option. Laser results with angiodysplasia and in hereditary telangiectasia are much better[80]. Most of the lesions that bleed in these patients occur in the stomach, duodenum or colon and are readily accessible endoscopically, and in the great majority, blood transfusion requirements can be drastically reduced or eliminated.

<u>Upper gastrointestinal tumours</u>. Lasers had been used endoscopically to control haemorrhage from non-neoplastic lesions for several years before their potential for relief of dysphagia in advanced obstructing cancers of the upper GI tract was proposed[81], but the latter is now becoming the most important application. The high power of the Nd-YAG laser, delivered under direct endoscopic vision to a small target area is used to bore a passage through obstructing tumours of the oesophagus and stomach. It has many advantages over alternative therapies. It can be performed under sedation, and if the patient's general condition permits, as a day case. The treatment is enterly local so does not carry the systemic side effects of chemotherapy nor the prolonged morbidity of surgery or radiotherapy. General disturbance to the patient is comparable to that of the best established technique for endoscopic palliation, namely insertion of a prosthetic tube, but the quality of palliation seems to be better. The main limitation of the technique is that it is only suitable for exophytic neoplasms. It is unwise to treat submucosal lesions and quite inappropriate for stenosis due to pressure from extrinsic masses.

<u>Lower gastrointestinal tumours</u>. The majority of obstructing lower gastrointestinal tract can be removed surgically, even if this is only a palliative procedure. However, this may be difficult with advanced rectal tumours. Endoscopic Nd-YAG laser palliation provides symtomatic relief better than other techniques, and can be repeated as often as required, with the minimum of disturbance to the patient. As with the upper GI tumors, the aim is to ablate the most friable areas and those causing the worst obstruction. Improvement is possible in most such cases, and in some can be dramatic[82]. In addition to malignant lesions, lasers may also have an important role to play in the precise destruction of benign colonic tumours. These include recurrent polyps in the rectal stump of patients with familial polyposis who have had a previous colectomy with ileorectal anastomosis and villous adenomas.

Photodynamic applications

<u>Endoscopic therapy</u>. (P. Spinelli, M. Del Fante, Istituto Nazionale

per lo Studio e la Cura dei Tumori, Milano, Italy). The main purpose of cancer therapy is to treat malignant tissue with the least damage to normal sorrounding structures. Photodynamic therapy (PDT) seems to be able to fulfill these simple but fundamental premises. PDT consists of activation of a photosensitizing drug by light irradiation. The drug is injected intravenously, and after 24 to 72 hours it is concentrated in malignant tissue at a variable rate of 4 to 10 times more than in normal tissue[83]. The light used to irradiate the tumor is generally a red light at a wavelength of 630 nm produced by filtered arc lamps or laser sources[84,85]. Lasers are used mainly for endoscopic purposes because of the possibility of obtaining sufficient power on a small surface, such as the proximal end of a flexible optic fiber. In addition to the photodynamic process a thermal mechanism related to light absorption and consequent temperature rise also seems to be involved in malignant necrosis by PDT[86,87].

From 7/82 to 3/87 thirty patients have been submitted to endoscopic PDT. Treated lesions were located in the tracheo-bronchial tree, in the gastro-intestinal tract, or in the urinary bladder. All patients were treated after that a surgical resection of the lesion was escluded on the basis of a multidisciplinary discussion and after that an informed consent about the modality of treatment was given by the patients. Hematoporphyrin derivative (HpD) was intravenously injected at a dosage of 3 mg/Kg b.w. in all patients. Forty-eight hours after HpD administration the lesions were exposed to 630 nm light from argon-dye laser system delivered into the operative channel of a conventional endoscope by means of a 400 micron fiber.

Tumour response was considered if a reduction more than 50% of the primary volume was obtained. Results were divided into: CR = Complete Response of total disappearance of the lesion at endoscopy and at histological examination of biopsy specimen; PR = Partial Response or total disappearance at endoscopy with positive biopsy.

Advanced gastro-intestinal cancers in 9 patients were treated with 13 PDT sessions. Eight PR and 1 CR were observed. The latter received a course of radiotherapy after DPT. In this group the median follow-up is 6 months (range 2-20 months). No difference in the survival was observed relating to the amount of energy delivered on tumour surface. Early GI tumours were treated in 7 patients during 12 PDT sessions. Six CR an 1 PR were obtained. The median follow-up of this group is 8.5 months (range 1-24 months). An estimated energy dose of 40 J/cm^2 to 170 J/cm^2 was delivered in all cases where a CR was observed. The patient who responded partially received a dose of 120 J/cm^2. Early tracheo-bronchial tumours were treated in 8 patients. All cases responded well and total disappearance was obtained. Median follow-up is 9.5 months (1 to 20 months) and the estimated energy delivered from 90 to 600 J/cm^2. Advanced tracheo-bronchial tumours were treated in 5 patients with 6 PDT sessions; 90 to 490 J/cm^2 were delivered to the lesions. Only PR were obtained. A median of 5 months (range 1 to 20 months) of follow-up was observed. The only case of urinary bladder tumour treated was lighted with energy dose of 40 J/cm^2. No response was observed.

No major complications were noted after treatments. Patients treated for esophageal carcinoma experienced pain in the treated area, which lasted

3-4 days. A minor hemorrhage was observed in few cases after treatments of advanced cancer, especially in the lower GI tract.

PDT appears to be a new and very promising possibility in the treatment of cancer. The facts that the photosensitizer is retained in the malignant more than in normal tissue and that the light radiation is endoscopically aimed on the malignancy account for the selectivity of the photodynamic effect. The selection of patients is very important for PDT, because for palliation of advanced cancers deeply infiltrating the walls of hollow viscera, thermal laser action is preferred to obtain a controlled necrosis. In such a way, the incidence of the most serious complications such as perforation and hemorrhage can be limited. Cancers in large hollow viscera that can be filled with refractive medium, as for example the urinary bladder, offer the best possibility of treatment. A quite homogeneous distribution of the light power can be obtained in this way. On the contrary, in a narrow channel with a stenotic tortuous segment, as in the esophagus or colon-rectum, light distribution is a problem.

Phototechnology

A new progress in laser opthalmology: the diode laser photocoagulator. The new generation of high-power semiconductor diode lasers[88,89] have been successfully used for producing photocoagulation of the retina[90]. The emission wavelength of the most powerful GaAlAs lasers ranges from 780 to 950 nm, and the power output available commercially exceeds 1 W in continuous way operation.

A very compact diode laser photocoagulator employing a standard slit-lamp configuration is under development at IEQ-CNR in Florence in collaboration with HS Raffaele in Milan. Diode lasers are expected to replace argon ion lasers in a near future[90]. The advantages are potential low cost, high efficiency, extreme compactness, good reliability, and reduced maintainance. The improvement of the output beam quality and the operation at shorter wavelength will definitively make these lasers the ideal sources for ophtalmology.

REFERENCES

1. T. B. Fitzpatrick, ed., "Sunlight and Man", University of Tokyo Press, Tokyo (1974)
2. A. G. Giese, "Living with our Sun's ultraviolet rays", Plenum Press, New York
3. I. A. Magnus, "Dermatological photobiology", Blackwell, Oxford (1976)
4. J. A. Parrish, R. R. Anderson, F. Urbach, D. Pitts, eds., "UV-A", Plenum Press, New York (1978)
5. B. L. Diffey, "Ultraviolet Radiation in Medicine", A. Hilger Ltd, Bristol (1982)
6. J. D. Regan, J. A. Parrish, "The Science of Photomedicine", Plenum Press, New York (1982)
7. W. L. Morison, "Phototherapy and Photochemotherapy of Skin Disease", Praeger, Westport (1983)

8. J. Jagger, "Solar-UV actions on living cells", Praeger, Westport (1985)
9. R. J. Wurtman, M. J. Baum, J. T. Potts, Jr., eds., "The medical and biological effects of light", The New York Academy of Sciences, New York (1985)
10. F. Urbach, R. W. Gange, eds., "The biological effects of UVA radiation", Praeger, Westport (1986)
11. J. A. S. Carruth, A. L. McKenzie, "Medical Lasers", A. Hilger Ltd., Bristol (1986)
12. J. L. Boulnois, Photophysical processes in recent medical laser developments: a review, Lasers Med. Science, 1:47 (1986)
13. F. F. Rubaltelli, G. Jori, "Neonatal Jaundice", Plenum Press, New York, (1984)
14. R. Pratesi, C. A. Sacchi, eds. "Lasers in Photomedicine and Photo-biology", Springer, Heidelberg (1980)
15. F. Hillenkamp, R. Pratesi, C. A. Sacchi, eds. "Lasers in Biology and Medicine", Plenum Press, New York (1980)
16. S. Martellucci, A. N. Chester, eds. "Laser photobiology and photo-medicine", Plenum Press, New York (1985)
17. H. F. Schellhas, Laser surgery in gynecology , Surg. Clin. of North Am., 58, 151:166, 1978
18. D. S. MacLaughlin, Metroplasty and miomectomy with CO_2 laser for maximizing preservation of normal tissue and minimizing blood loss , J. Reprod. Med., 30, 1:9, 1985
19. M. H. Goldrath, T. A. Fuller, S. Segal, Laser photovaporization of endometrium for the treatment of menorrhagia , Am. J. Obstet. Gynecol., 140, 14-19, 1981
20. J. J. H. Bellina, A. C. Fick, J. D. Jackson, Application of the CO_2 Laser to infertility surgery, Surg. Clin. N. Am., 64, 899:904 (1984)
21. M. C. Anderson, Treatment of cervical intraepithelial neoplasia with the carbon dioxide laser. Report of 543 patients , Obstet. Gynecol. 59, 720:725 (1981)
22. E. Rylander, A. Isberg, I. Joelsson, "Laser vaporization of cervical intraepithelial neoplasia. A five-year follow-up", Acta Obst. Gyn. Scand. Suppl., 125, 33-36 (1984)
23. P. D. Indman, B. C. Arndt, Laser treatment of cervical intraepithelial neoplasia in an office setting, Am. J. Obstet. Gynecol., 152, 674: 676 (1985)
24. R. Totani, T. Karasawa, Y. Syzuoki, et. al., "Nd:YAG laser conization of uterine cervix utilizing ceramic scalpel", Proc. XIV Int. Cancer Cong., Budapest, August 22-26 (1986), Abstr. 3187
25. M. S. Baggish, A comparison between laser excisional conization and laser vaporization for the treatment of cervical intraepithelial neoplasia, Am. J. Obstet. Gynecol., 155, 39:44,(1986)
26. J. H. Bellina, G. Bandieramonte, "Principles and practice of gynecologic laser surgery", Chapter 4: Applications in gynecology with emphasis on the cer ix, pp. 111-164, Plenum Publ. Corp., New York,(1984)
27. G. D. Sadoul, T. M. Beuret, Management of 633 cervical intraepithelial neoplasia by CO_2 laser: persistent disease and recurrences, Laser Surg. Med., 6, 110:118 (1986)
28. A. J. Jordan, M. J. Mylotte, The treatment of cervical intraepithelial neoplasia by CO_2 laser vaporization, in: "Gynecologic laser surgery", Proceedings Int. Congr. on Gynec. Laser Surg. and related works,

New Orleans, 1979, J. H. Bellina et al., eds., Plenum Press, New York (1980)

29. C. V. Capen, B. J. Masterson et al., Laser therapy of vaginal intra-epithelial neoplasia, Am. J. Obstet. Gynecol., 142, 973:976 (1982)

30. D. E. Townsend, R. U. Levine et. al., Treatment of vaginal carcinoma in situ with the carbon dioxide laser, Am. J. Obstet. Gynecol., 143, 565:568 (1982)

31. M. S. Baggish, Carbon dioxide laser treatment for condyloma acuminata venereal infection, Obstet. and Gynecol., 55, 711:715 (1980)

32. D. B. Apfelberg and E. McBurney, Use of the argon laser in dermatologic surgery, in: "Lasers in Cutaneous Medicine and Surgery", Ratz J. L. eds., Chicago, Year Book Medical Publishers, Inc., pp. 31:63 (1986)

33. O. T. Tan, M. Corney et. al., Histologic responses of port-wine stains treated by argon, carbon dioxide, and tunable dye lasers, Arch. Dermatol., 122, 1016:1022 (1986)

34. V. L. R. Jonquet and J. A. S. Carruth, Review of the treatment of port-wine stains with the argon laser, Laser Surg. Med., 4, 191:199 (1984)

35. A. Scheibner and W. H. McCarthy, Argon laser treatment of superficial blood vessel malformations on the trunk and extremities in adults and the face in children, Lasers Surg. Med., 6, 244 (1986)(abstracts)

36. J. G. Morelli, O. T. Tan et. al., Tunable dye laser (577 nm) treatment of port-wine stains, Lasers Surg. Med., 6, 94:99 (1986)

37. J. M. Garden, O. T. Tan et al., Effect of dye laser pulse duration on selective cutaneous vascular injury, J. Invest. Dermatol., 87, 653:657 (1986)

38. J. Greenwald, S. Rosen et al., Comparative histological studies of the tunable dye (at 577 nm) laser and argon laser : The specific vascular effects of the dye laser, J. Invest. Dermatol., 77, 305:310 (1981)

39. A. L. Lorincz, Disturbances of melanin pigmentation, in: "Dermatology", S. L. Moschella and H. J. Hurley, eds., Philadelphia, W.B. Saunders Company, 1273:1305 (1985)

40. D. B. Apfelberg , M. R. Maser et al., The argon laser for cutaneous lesions, J. Am. Med. Assoc., 245, 2073:2075 (1981)

41. H. M. Crumay, Electrosurgery, ultraviolet light therapy, cryosurgery, and hyperbaric oxygen therapy, in: "Dermatology", S. L. Moschella H. J. Hurley, eds., Philadelphia, W. B. Saunders Company, 1980:2001 (1985)

42. J. S. Dover, B. Smoller et al., Low-fluence CO_2 laser irradiation of lentigines, Laser Surg. Med. (abstract, in press)

43. G. F. Murphy, R. S. Shepard et al., Organelle-specific injury to me-lanin-containing cells in human skin by pulsed laser irradiation, Lab. Invest., 49, 680:685 (1983)

44. R. R. Anderson and J. A. Parrish, Selective photothermolysis: precise microsurgery by selective absorption of pulsed radiation, Science 220, 524:527 (1983)

45. J. A. Parrish, R. R. Anderson et al., Selective thermal effects with pulsed irradiation from lasers: from organ to organelle, J. Invest. Dermatol., 80, 75:80s (1983)

46. J. S. Dover, R. J. Margolis et al., Morphologic and Histologic findings in pigmented guinea pig skin irradiated with Q-switched ruby laser pulses, J. Invest. Dermatol. (abstract, in press)

47. J. S. Dover, L. L. Polla et al., Pulse width dependence of pigment cell

damage at 694 nm in guinea pig skin, SPIE Proceedings, 712, 200:205 (1987)

48. L. L. Polla, R. J. Margolis et al., Melanosomes are a primary target of Q-switched ruby laser irradiation in guinea pig skin, J. Invest. Dermatol., (in press)

49. G. J. Hruza, R. J. Margolis et al., Selective photothermolysis of pigmented cells by Q-switched Nd:YAG laser pulses at 1064, 532 and 355 nm wavelengths, J. Invest. Dermat., (abstract in press)

50. R. J. Margolis, J. S. Dover et al., Visible action spectrum for melanin specific selective photothermolysis, J. Invest. Dermatol. (abstract, in press)

51. S. Watanabe, T. J. Flotte et al., The effect of pulse duration on selective pigmented cell injury by dye lasers, J. Invest. Dermatol., (abstract, in press)

52. S. V. Tang, D. Bourgelais, et al., Modification of the argon laser to specifically treat epidermal pigmented lesions, Clin. Res., 34, 419a (1986), abstract.

53. P. L. Bailin and J. L. Ratz, Use of the carbon dioxide laser in dermatologic surgery, in: "Lasers in Cutaneous Medicine and Surgery", J. L. Ratz, eds., Chicago, Year Book Medical Publ., Inc., 73:104 (1986)

54. J. T. Walsh, T. F. Deutsch et al., Precise cutting of tissue using short-pulse CO_2 lasers, Lasers Surg. Med., 6, 167 (1986) (abstr.)

55. F. J. Lejume and G. V. Van Horf, Impairment of skin graft take after CO_2 laser surgery in melanoma patients, Br. J. Surg., 67, 318:320 (1980)

56. T. L. Fry, R. W. Gerbe et al., Effects of laser, scalpel and electro-surgical excision on wound contracture and graft "take", Plast Reconstr. Surg., 65, 729:731 (1980)

57. B. R. Buell and D. E. Schuller, Comparison of tensile strength in CO_2 laser and scalpel skin incisions, Arch. Otolaryngol., 109, 465:467 (1983)

58. J. E. Madden, R. F. Edlich et al., Studies in the management of the contaminated wound. IV. Resistance to infection of surgical wounds made by knife, electrosurgery and laser, Am. J. Surg., 119, 222: 224 (1970)

59. T. C. Montgomery, J. B. Sharp et al., Comparative gross and histological study of the effects of scalpel, electric knife and carbon dioxide laser on skin and uterine incision in dogs, Lasers Surg. Med.,3, 9:22 (1982)

60. R. J. Lane, R. Linsker et al., Ultraviolet-laser ablation of skin, Arch. Dermatol., 121, 609:6017 (1985)

61. J. Morelli, O. T. Tan et al., Effect of wavelength (193 and 248 nm) and repetition rate"in vivo"guinea pig skin, Lasers Surg. Med., 6, 252 (1986)

62. H. Green, J. Boll et al., Cytotoxicity and mutagenicity of low intensity 248 and 193 nm excimer laser radiation in mammalian cells, Cancer Res., 47, 410:413 (1987)

63. J. A. Parrish, Ultraviolet-laser ablation, Arch. Dermatol., 121, 599: 600 (1985) (editorial)

64. J. T. Walsh, G. J. Hruza et al., Comparison of tissue ablation using TEA CO_2 and Er:YAG lasers, Conference on lasers and electro-optics

27 April - 1 May, 1987, Baltimore, Maryland. Digest of Technical
Papers (in press)

65. R. P. Abergel, R. Lyons et al., Use of lasers for closure of cutaneous
    wounds: experience with Nd:YAG, argon and CO_2 lasers, J. Dermatol.
    Surg. Oncol., 12, 1181:1185 (1986)

66. R. P. Abergel, R. F. Lyons et al., Skin closure by Nd:YAG laser welding,
    Am. Acad. Dermatol., 14, 810:814 (1986)

67. Z. BeKassy and B. Astedt, Carbon dioxide vaporization of plaque
    psoriasis, Br. J. Dermatol., 114, 489:492 (1986)

68. M. W. Bemo, M. Rettenmaier et al., Response of psoriasis to red laser
    light (630 nm) following systemic injection of hematoporphyrin
    derivative, Lasers Surg. Med., 4, 73:77 (1984)

69. T. J. Dougherty, Photoradiation therapy for cutaneous and subcutaneous
    malignancies, J. Invest. Dermatol., 77, 122:124 (1981)

70. D. L. Ramsay and H. J. Hurley, Papulosquamous eruption and exfoliative
    dermatitis, in: "Dermatology", Moschella and Hurley eds., Phila-
    delphia, W. B. Saunders Company, pp.499:556 (1985)

71. S. M. Olbricht, R. S. Hen et al., Complications of cutaneous laser
    surgery: a survey, Arch. Dermatol., 123, 345:349 (1986)

72. M. W. Berns and J. L. McCullough, Porphyrin sensitized phototherapy,
    Arch. Dermatol., 122, 871:874 (1986) (editorial)

73. A. Musajo Somma, N. Ciavarella, S. Scaraggi, T. Ripa, Laser Surgery
    in Haemophilia, Haemostasis, 10 (Suppl.), 229 (1981)

74. D. W. Storey, S. G. Bown, C. P. Swain et al., Endoscopic prediction of
    recurrent bleeding in peptic ulcers, N. Eng. J. Med.,305, 915:6
    (1981)

75. S. G. Bown, Controlled studies of laser therapy for haemorrhage from
    peptic ulcers, Acta Endoscopica, 15, 1:12 (1985)

76. C. P. Swain, T. N. Mills, E. Shemesh et al., Which electrode? A com-
    parison of four endoscopic methods of electrocoagulation in
    experimental bleeding ulcers, Gut., 25, 1424:31 (1984)

77. J. H. Johnston, J. Q. Sones, B. W. Long et al., Comparison of heater
    probe and YAG laser in endoscopic treatment of major bleeding from
    peptic ulcers, Gastrointest. Endosc., 31, 175:80 (1985)

78. D. M. Jensen, M. L. Silpa, J. I. Tapia et al., Comparison of different
    methods for endoscopic haemostasis of bleeding canine oesophageal
    varices, Gastroenter., 84, 14-55:61 (1983)

79. D. Fleischer, Endoscopic Nd YAG laser therapy for active oesophageal
    variceal bleeding: a randomised controlled study, Gastroint. Endosc.,
    31, 4:9 (1985)

80. S. G. Bown, C. P. Swain, D. W. Storey et al., Endoscopic laser treat-
    ment of vascular anomalies of the upper GI tract, Gut. (in press)

81. D. Fleischer, F. Kessler, O. Hayes, Endoscopic Nd:YAG laser therapy
    for carcinoma of the oesophagus. A new palliative approach, Am. J.
    Surgery, 143, 280 (1982)

82. E. M. H. Mathus-Vliegen, G. N. J. Tytgat, Nd:YAG laser photocoagulation
    in gastroenterology - its role in palliation of colorectal cancer,
    Lasers in Medical Science, in press

83. G. Sabben, A. Bosshard, R. Lambert, Time dependence of 3 H hemato-
    porphyrin derivative distribution in the digestive tract of the
    rat, in: "Porphyrins in tumor phototherapy", A. Andreoni, R.
    Cubeddu eds., Plenum Publ. Co., N.Y., London, 423:426 (1984)

84. T. J. Dougherty, J. E. Kaufman, A. Goldfarb, K. R. Weishaupt, D. G. Boyle, A. Mittelman, Photoradiation therapy for the treatment of malignant tumors, Cancer Res., 38, 2628:2635 (1975)

85. Y. Hayata, H. Kato, C. Konaka, J. Ono, N. Tazikawa, Hematoporphyrin derivative and laser photoradiation in the treatment of lung cancer, Chest 81, 269:277 (1981)

86. K. R. Weishaupt, C. J. Gomer, T. J. Dougherty, Identification of singlet oxygen as the cytotoxic agent in photo-inactivation of a murine tumor, Cancer Res., 36, 2326:2329 (1976)

87. M. W. Berns, J. Coffey, A. G. Wile, Laser photoradiation therapy of cancer: possible role of hyperthermia, Lasers Surg. Med., 4, 87:92, (1984)

88. R. Pratesi, Diode Lasers in Medicine, J. Quant. Electr., 20:1433 (1984)

89. R. Pratesi, Semiconductor (diode) lasers: basic principles and potential applications in the biomedical field, Photobiochem. Photobiophys., Suppl. 57 (1987)

90. R. Brancato, R. Pratesi, Applications of diode lasers to ophthalmology, Lasers Ophthalm., 2 :34 (1987)

VASOVASOSTOMY BY MEANS OF A ND:YAG LASER

P. Gilbert and W. Thon

Department of Urology, Military Hospital of Ulm

Oberer Eselsberg 40, D-7900 Ulm, West Germany

For many years vasectomy has offered a simple and safe procedure for male contraception. Since the operation represents only a very small risk to the patient, it has been accepted by many couples all over the world as an appropriate means for birth control. In the United States, for example, about one-half million vasectomies are performed each year[1]. Considering the high divorce rate in this century, one cannot be surprised that a good many of the men concerned want a new family and therefore require vasectomy reversal. The successful achievement of an anastomosis of the vas deferens is not only dependent on the operative technique and the skills of the surgeon, but also on such important factors as the age of the patient[2], the time interval between vasectomy and vasovasostomy[2,3], the existence of sperm granulomas at the distal end of the vas deference[3], and the detection of spermatozoon antibodies in the seminal fluid[2].

Several authors have reported on various operative techniques concerning vasovasostomy with potency rates ranging from 70% to 90%[4,5,6,7] Microsurgical procedures, especially the technique described by Silber[8], provide the best results. The operation time, is long however if an accurate and potent anastomosis of the vas is to be obtained[5].

In search of less time-consuming techniques, the use of lasers has been explored by some surgeons. In 1984, Rosenberg reported on vasovasostomies in 20 dogs[9], ten of which had been performed by CO_2 laser welding and ten by suture techniques. All animals of the laser-treated group had good postoperative sperm counts and showed no sign of sperm leaks or granulomas. The author also reported on the use of the laser technique in a 27-year-old patient 5 years after vasectomy. The apposed vasal ends were welded with the CO_2 laser at "very low power". Unfortunately, the post operative results was not described.

More recently, Jarow and co-workers presented a study of laser-assisted anastomoses of human vas deferens in vitro and the rat vas deferens in vivo[10]. They also used a carbon dioxide laser. The conclusion of their scientific work was that laser assisted vasal anastomosis is a fast and simple technique and a successful as conventional suture anastomosis despite a higher incidence of sperm granulomas.

Due to the wavelength of the CO_2 laser beam, its energy is strongly absorbed by water, thus limiting its effect to the tissue surface. The biophysical properties of the Nd:YAG laser are quite different. Its wavelength of 1.06 µm allows this laser beam to penetrate the surface and generate its thermal effects on deeper tissue layers, which are coagulated according to the amount and duration of energy. Impressive experiments on this subject have been carried out by Keiditsch and coworkers,using liver tissue and rat and rabbit urinary bladders for Nd:YAG laser coagulation[11]. In 1980, Kewal K. Jain published a paper in the Journal of Microsurgery which described his experimental work on sutureless microvascular anastomosis of rat carotid arteries with the help of a Nd:YAG laser[12]. The average diameter of the blood vessels concerned was about 1mm. The potency rates compared favourably with those achieved by suture techniques.

Based on these experimental experiences and relying on their good results, we dared to transfer the method of laser-assisted vasal anastomosis to our first patient in August, 1986. A 22-year-old man, who had been sterilized two years earlier for ideological reasons, came to our department of Urology requiring vasectomy reversal. He had become engaged and wanted to have a family. We used a Nd:YAG laser model Medilas 2 from the firm of MBB, Munich (Fig. 1 and 2).

The operation was carried out under general anaesthesia. Beginning with a scrotal incision, both vasal ends were identified and dissected (Fig. 3). They then were transsected at a distance of approximately 5 mm from the ligation site (Fig. 4). Stay sutures were placed to prevent their withdrawal. At the distal end of the vas a white fluid emerged, proving potency of the lumen. Reapproximation of both vasal ends was achieved by placing an intraluminal 6-0 prolene suture which was held under tension (Fig. 5a, 5b). Welding of the apposed tissue edges followed using a continuous laser mode which, by footpedal-control, emitted pulses of 10 W in power and 0.5 sec in duration (Fig. 6a, 6b). At the anastomatic site the vas deferens showed dark grey spots representing coagulation marks produced by by the impact of the laser beam (Fig. 7). After withdrawal of the prolene suture the anastomosis was secured by two transmural 5-0 PDS sutures placed on diametrically opposed sites of the vasal circumference. Finally, the welded vas deferens was replaced into the scrotum and the incision closed by 3-0 chromic catgut sutures. The time necessary for the achievement of the anastomosis was about 8 minutes for both vasa, and the entire operation time was less than 1 hour. Postoperative wound healing was uncomplicated. The patient was dismissed three days after the operation. Two months later, a sperm count was performed which proved normal.

Meanwhile, two other patients underwent laser-assisted vasovasostomy (See Table I). They had both divorced their wives and had married again. The first patient, aged 28, had been vasectomized 1½ years earlier; the second, a 35 year-old man, 5 years prior to vasectomy reversal. Operative

286

Fig. 1. Nd:YAG laser (Medilas 2)

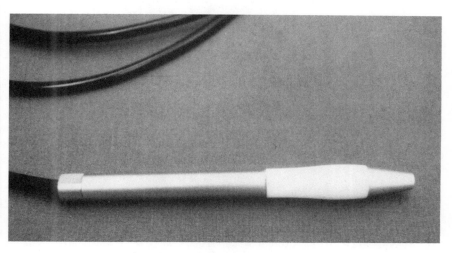

Fig. 2. Handpiece of the laser

conditions for both patients were the same as for the first patient. No
complications occurred, neither during nor after surgery. The 28 year–old
man had a good postoperative sperm count whereas the 35 year–old man had
cryptozoospermatism; i.e. only a few spermatozoons in the ejaculate. In
this case, potency of the vasal anastomosis was proven but it remains to
be seen whether fertility can be restored, since the time interval between
vasectomy and vasovasostomy had been much longer for this patient than for
the other two patients. We hope that future sperm analyses of this patient
will show an improvement of semen quality.

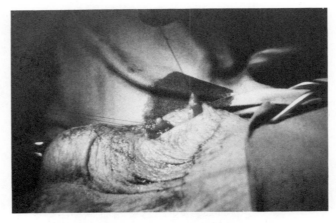

Fig. 3. Dissection of the vasal ends

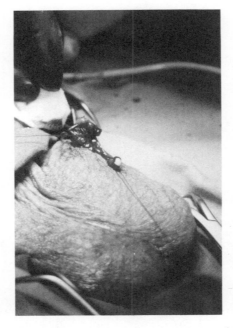

Fig. 4. Status after resection of the distal and proximal end of the vas.
Excretion of white seminal fluid at the distal end of the vas

DISCUSSION

The three clinical examples of laser-assisted vasovasostomy clearly demonstrate the feasibility of this simple surgical procedure in male patients. Unlike conventional microsurgical techniques, this method does not demand any special skills or training from the surgeon. All that has to be done is to guide the laser beam to the anastomotic site where its thermal effect causes fusion of the collagen in the smooth muscle

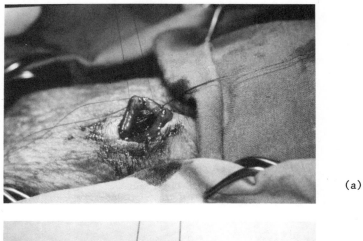

(a)

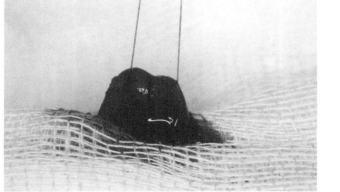

(b)

Fig. 5a. Reapproximation of the vasal ends
5b. Detail from Fig. 5a

of the vas deferens, thus welding the apposed vasal ends. The anastomosis
is achieved very quickly: this gain in time is the main advantage compared
to microsurgical suture techniques. In all three patients we operated on,
the operation time did not exceed one hour. The anastomosis itself can be
completed within a few minutes. It was demonstrated that laser pulses of
10 W in power and 0.5 sec in duration are sufficient to provide stability
of the anastomosis. Higher energies and pulses of longer duration introduce
the danger of entirely coagulating the smooth muscle wall and the mucosa of the vas, thus leading to subsequent stenosis of the lumen.

The question of whether sperm leaks and sperm granulomas are more
or less pronounced in the laser technique as compared to the suture technique cannot be definitely answered yet. A question of even grater importance concerns the potency rate. As the precision of the laser beam does not
depend on special manual skills, a higher consistency of positive results,
i.e. of patent anastomoses, should be expected when compared to the results obtained by suture techniques. A definite assessment of laser-assisted vasovasostomy will be possible when many more patients have been
treated by this method. In any case, these preliminary promising results
warrant further clinical work using the Nd:YAG laser for vas deferens re-
anastomosis.

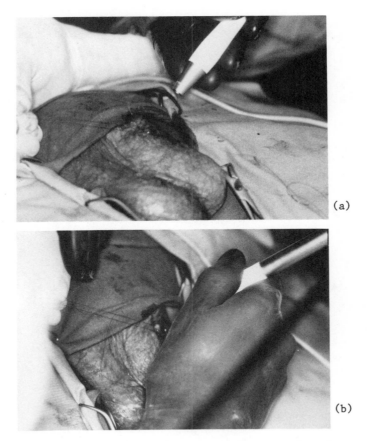

(a)

(b)

Fig. 6a, 6b. Laser-assisted vasovasostomy (LAVV)

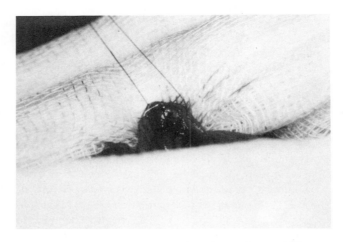

Fig. 7. The anastomotic site after LAVV

Table 1. Laser-assisted vasovasostomy (LAVV) in 3 patients

No. of patient	1	2	3
age (years)	22	28	35
time interval between vasectomie and LAVV (years)	2	$1\frac{1}{2}$	5
postoperative sperm count	normal	normal	crytozoospermatism

REFERENCES

1. L. Liskin, J. Pile and W. Quillin, Vasectomy-safe and simple, Population Reports, Series D, 4:63 (1983)
2. K. Bandhauer, E. Senn, Die Vaso-vasostomie-Faktoren, welche die Ergebnisse beeinflussen, Akt. Urol., 18, 7-10 (1972)
3. A. M. Belker, J. W. Konnak, J. D. Sharlip, and A. J. Thomas Jr., Intraoperative observations during vasovasostomy in 334 patients, J. Urol., Vol. 129, No. 3, 524-527 (1983)
4. J. F. Redman, Clinical Experience with vasovasostomy utilizing absorbable intravasal stent, "Urology", Vol. XX, No. 1, 59-61 (1982)
5. K. W. Kaye, R. Gonzalez and E. E. Fraley, Microsurgical vasovasostomy = an aoutpatient procedure under local anaesthesia, J. Urol., Vol. 129, No. 5, 992-994 (1983)
6. L. V. Wagenknecht, Verschlussazoospermie - operative andrologische Massnahmen, Z. Allg. Med., 58, 451-456 (1982)
7. E. Owen, H. Kapila, Vasectomy reversal, (Med) Aust.,Vol. 140, No. 7, 398-400 (1984)
8. S. J. Silber, Microsurgery for vasectomy reversal and vaso-epididymostomy, "Urology", Vol. XXIII, No. 5, 505-524 (1984)
9. S. K. Rosemberg, First laser vasectomy reversal, Clin. Laser Monthly, 2:60 (1984)
10. J. Jarow, B. C. Cooley, F. F. Marshall, Laser-assisted vasal anastomosis in the rat and man, J. Urol., Vol. 36, 1132-1135 (1986)
11. F. Keiditsch, A. Hofstetter, J. Zimmermann, J. Stern, F. Frank, and J. Babaryka, Histological investigation to substantiate the therapy of bladder tumors with the Neodymium-YAG-laser, Laser,1:19-23 (1985)
12. K. K. Jain, Sutureless microvascular anastomosis using a neodymium-YAG laser, Microsurgery, 1:436 (1980)

ADVANCES IN LASER FUSION

C. Yamanaka

Institute of Laser Engineering

Osaka University

INTRODUCTION

After the invention of the laser in 1960, laser fusion research began in several countries. In early days, laser-plasma interactions were principally studied. Absorption physics was a target for keen interest. In 1972 the laser implosion concept was publicly disclosed. Since then, inertial confinement fusion (ICF) research has progressed remarkably to compete with magnetic confinement fusion.

The concept of inertial confinement fusion is to implode a fuel pellet by the irradiation of laser beams, as shown in Fig. 1.

The pellet surface is heated to produce a plasma, whose ablation causes the pressure which compresses the fuel. The core of the fuel pellet experiences a pressure of 10^9 bar and the fuel temperature and the density reach 4 kev and $10^{26}/cm^3$, respectively. Under these conditions we can satisfy the Lawson inequality for fusion reaction.

 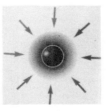 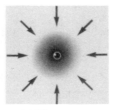

Heating
Lasers irradiate the fuel pellet and heat the surface of the pellet, then plasma ablation starts.

Compression
The fuel is compressed by rocket-like blowoff of the surface material.

Ignition
The full core reaches 1000 times liquid density and ignites at 10^8K.

Burn
Thermonuclear burn spreads rapidly through the compressed fuel yielding many times the driver input energy.

Fig. 1. The scenario for implosion of an inertial confinement fusion target.

293

CONDITIONS FOR BREAK-EVEN

The Lawson inequality is used as a measure to maintain the fusion reaction at a temperature of 10 Kev,

$$n_\tau > 10^{14} \text{ s/cm3}$$

where n is the plasma density and τ is the reaction time. If we designate the radius of the fuel pellet by R and the sound velocity in the fuel by v_s, the reaction time τ is given by R/v_s. Then the Lawson condition becomes $\rho R > 0.2 \text{g/cm}^2$ for DT fuel.

The necessary laser energy E_L to heat up a solid fuel pellet to a temperature of 10 keV is given by

$$E_L = \frac{4}{3} \pi \frac{(\rho R)^3}{\rho^2} \times 4.6 \times 10^8 \times \frac{1}{\varepsilon} \text{ (J)}, \tag{1}$$

where ε is the coupling efficiency between laser and pellet and ρR is a constant. The necessary laser energy E_L is inversely proportional to ρ^2. If one increases the fuel density by 1000 times, the laser energy decreases by 10^{-6}.

The burning rate η is estimated by the burning velocity and the disintegration speed of the fuel. For the solid pellet target, the burning rate is

$$\eta = \frac{\rho R}{6 + \rho R} . \tag{2}$$

The fusion output energy E_{out} is

$$E_{out} = \frac{4}{3} \frac{(\rho R)^3}{\rho^2} \eta \times 4.2 \times 10^{11} \text{ (J)}. \tag{3}$$

The scientific break-even condition is given by the pellet gain $Q = E_{out}/E_L = 1$, where $R = 0.2 \text{g/cm}^2$ at $\varepsilon = 10\%$. The necessary laser energy E_L is easily estimated, using a pellet fuel density 0.2g/cm^3 and $\rho R = 0.2 \text{g/cm}^2$, to be 10^9 J. This is beyond the present state or laser technology. However, if one can compress the DT fuel pellet 1000 times, $\rho = 200 \text{g/cm}^3$, the necessary laser energy E_L decreases to only 1kJ. This estimate depends on too simple assumptions to give the correct answer. Actually, the scientific break-even condition will require a 100kJ laser.

The final goal of laser fusion is to attain self-sustained operation of the reactor. This requires higher pellet gain. One D-T reaction can produce an energy of 17 MeV, which is divided 14 MeV (80%) to the neutron and 3 MeV (29%) to the a particle. The pure fusion reactor uses the neutron energy as a heat source. The hybrid fusion fission scheme uses the high energy neutron as an agent for fission reaction. The self-sustained condition for the reactor is given by

$$E_{out}/E_{in} = \eta_L Q \eta_T (0.8M + 0.2) \tag{4}$$

294

where M is the fission blanket gain for neutrons, η_L is the laser efficiency, and η_T is the thermal conversion efficiency. As shown in Fig. 2, the fission blanket gain M depends on the fission material surrounding the reactor vessel. In Fig. 2 a circulating energy of 100%, 24%, 10% indicates the inner reactor consumption for self-sustaining operation. The engineering break-even condition is given by $E_{in}/E_{out}=1$.

For a pure fusion reactor M=1, self-sustained operation is attained for $\eta_L Q=2.5$ ($\eta_L=5\%$, Q=50). For a fusion-fission reactor we can use M=16, $\eta_L Q=0.2$ ($\eta_L=0.2\%$, Q=100). For a practical reactor, one should have the condition $\eta_L Q$ 10, where a quarter of the produced energy is consumed in the reactor.

LASER PLASMA INTERACTIONS

Laser light penetrates the plasma to the cutoff density region and is absorbed. The rest of the light is specularly reflected, as shown in Fig. 3.

In the case of weak laser intensity, the plasma temperature in the absorption region is so low that coulomb colisions between electron and ion are dominant. The laser light is absorbed by inverse bremsstrahlung. This is so-called classical absorption, where the energy distribution of electrons is maxwellian.

If the direction of the density gradient is along the x coordinate, then the variation of laser intensity I(x) is given by

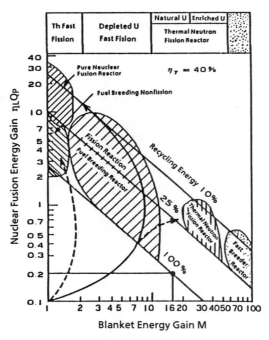

Fig. 2. Blanket gain (pellet gain Q_p x laser efficiency η_p) vs. nuclear fusion gain.

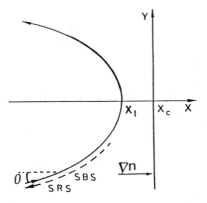

Fig. 3. Ray trace of the laser.

$$\frac{dI}{dx} = \mp \frac{v(x)}{kxC^2} \frac{n(x)}{n_c} I, \tag{5}$$

where $v(x)$ is the collision frequency between electron and ion. $-(+)$ represents the incident (reflected) light, and $k_x C^2/\omega$ is the group velocity of light.

From the dispersion relation

$$k_x = (\cos^2\theta - n(x)/n_c)^{1/2} k_0,$$

where θ is the incident angle in Fig. 3 and k_0 is the wave number in vacuum.

The density distribution is given by the characteristic length

$$L = \left| \frac{1}{n_c} \frac{dn}{dx} \right|^{-1},$$

and the absorption coefficient η_{ab}^c is given by (Eq. 5)

$$\eta_{ab}^c = 1 - \exp(-\frac{32}{15} k_0 L \frac{v_c}{\omega} \cos^5\theta) \tag{6}$$

where the temperature T is constant[1], and v_c is the electron ion collision frequency at the cutoff density, which is proportional to n_c and $T^{-3/2}$. Since $n_c \propto \lambda_L^{-2}$, shorter wavelength lasers have higher absorption coefficients. At lower temperatures with lower laser intensities, the absorption is high. And a low density gradient ($\cong 1/L$) with a longer laser pulse shows the highest absorption.

Using the energy balance

$$\eta_{ab}^c I_L = f n_c T v_e, \tag{7}$$

we can determine the plasma temperature. Then the laser intensity dependent absorption coefficient is derived, where f is a so-called thermal flux limiting factor. For normal incidence Eq. (6) tends to

$$\eta_{ab}^{c} \cong \frac{32}{15} k_0 L \frac{v_c}{\omega} \; ;$$

using Eq. (7),

$$\eta_{ab}^{c} \cong 0.1 \left(\frac{ZLf}{I_L}\right)^{1/2} \lambda_L^{-2} \; ,$$

where Z is ionic charge, L and λ_L is given in μm, and I_L is given in units of 10^{14} w/cm^2. If we set L=30 m, f=0.03, Z=3, at $I_L\lambda_L^4 > 10^{14}$w(μm)^4/cm^2, classical absorption is not applicable. However, from the requirement of ablation pressure, $I_L > 5 \times 10^{14}$w/cm^2 is expected. Thus the laser wavelength should be $\lambda_L < 0.5\mu$m.

The absorption characteristics have been well investigated theoretically and experimentally[2,3,4]. Fig. 4 shows these data, where the increase of absorption at higher laser intensity is due to excitation of electron plasma waves (so-called resonance absorption); this effect was investigated by Ginzburg[5]. However, this kind of absorption introduces hot electrons, which is undesirable because it induces preheating in the fuel.

ABLATION AND IMPLOSION

The internal energy of solid DT fuel corresponds nearly to the Fermi energy of a degenerate electron gas: $W_0 = 1.2 \times 10^5$J/g. The internal energy increases to a W_0 by preheating, then the fuel is compressed adiabatically. By the equation of state of an ideal gas, the internal energy w is

$$w = aw_0(\rho_m/\rho_s)^{2/3} \; ,$$

where ρ_s and ρ_m are the solid density and the maximum compressed density, respectively. If preheating is suppressed, a will be 2∿4 and the internal energy is $(2.4 \sim 4.8) \times 10^7$J/g in the case of 1000 times compression. This

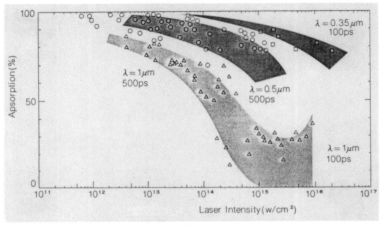

Fig. 4. Laser intensity and wavelength dependences of absorption.

energy is $(2.4\sim4.8)\times10^7$J/g in the case of 1000 times compression. This corresponds to a kinetic velocity of $v=2\sim3\times10^7$cm/sec. Then the pellet shell is accelerated to a velocity of 2×10^7cm/sec and this energy is converted to the internal energy of DT fuel. We can expect 1000 times compression in this way.

In Fig. 5 the structure of laser ablation plasmas is depicted. At the implosion, the fuel pellet shell is accelerated by the reaction of the ablating plasma produced by laser irradiation. The ablating pressure Pa is given by the rocket equation

$$\frac{dMv}{dt} = - \frac{dM}{dt} (u-v) = SPa, \tag{8}$$

where $M = SPa\Delta x(t)$ is the total mass accelerated inwards, S is the area of pellet, ρ_a is the average density, $\Delta x(t)$ is the thickness, and u and v are the velocity of ablating plasma and pusher respectively (u is the velocity on the pusher system). From this relation, the ablation pressure is proportional to the mass ablation rate $-dM/dt$ and ablating velocity $u - v$. The energies consumption rate due to ablating plasma is

$$\left|\frac{dM}{dt}\right| (u - v)^2 \approx \left|\frac{dM}{dt}\right| u^2 \qquad (u>v).$$

If the laser absorption energy is a constant, the smaller the ablating velocity u, the larger the mass ablation rate, yielding a larger ablation pressure Pa. Actually u and $|dM/dt|$ depend upon the laser intensity, wavelength, pulse duration and target materials. The laser-plasma interaction governs the implosion process.

The kinetic energy of ablating plasma and accelerated pusher is equal to the laser absorbed energy

$$E_a = \int_0^t dt \eta_{ab} I_L S$$

$$\eta_{ab} I_L S = \frac{d}{dt} (\frac{1}{2} Mv^2) + \frac{1}{2} (- \frac{dM}{dt})(u-v)^2$$

with Eq. (8),

$$E_a = \frac{1}{2} \Delta M u^2$$

where ΔM is the ablated total mass. Then

$$P_a = 2(1- \frac{v}{u}) \eta_{ab} I_L \Big/ u. \tag{9}$$

In the case of $u \gg v$, a smaller ablation velocity, we get a larger pressure. The hydrodynamic efficiency (pusher kinetic energy/absorbed laser energy) is

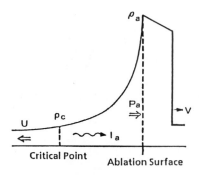

Critical Point   Ablation Surface

Fig. 5. Ablation structure driven by a laser; pressure Pa is generated
at the ablation surface whose density is ρa by the absorbed laser
intensity of KIa which is converted at the critical point (cutoff
density).

$$\frac{\eta}{H} = \frac{\frac{1}{2}(M_0-\Delta M)v^2}{\frac{1}{2}\Delta M u^2} = (\frac{M_0}{\Delta M} - 1)\left|\ln(1-\frac{\Delta M}{M_0})\right|. \tag{10}$$

When $M/M_0 \ll 1$, $\eta_H \cong \Delta M/M_0$, and we see that a larger ablating mass gives
a larger hydrodynamic efficiency. Fig. 6 shows the comparison between
theoretical estimate and experimental results at ILE Osaka[6,7].

The final velocity of the accelerated pusher is $v=2\sim 3\times 10^7$cm/sec. At
Eq. (8), Pa=const, $M \sim M_0$(const) and $S=4\pi R^2(t)$; one integrates to obtain

$$v^2 = \frac{2}{3}\frac{Pa}{\rho_a \Delta R_0}R_0,$$

where ΔR_0 is the initial thickness and R_0 is the initial radius of a pellet.
The pressure Pa is written

$$Pa \cong \frac{3}{2}\rho_a v^2 \frac{\Delta R}{R_0}. \tag{11}$$

The acceleration time of a pellet is so finite, so the pressure is
determined by the final velocity. As an example, Pa=45Mbar is necessary for
$v=3\times 10^7$cm/sec, $R_0/\Delta R_0=30$, $\rho_a=1$g/cm^3. According to Eq. (11), as the aspect
ratio of a target $R_0/\Delta R$ is larger, the necessary pressure is reduced.
However, implosion stability imposes a limitation on the aspect ratio $R_0/\Delta R_0$. From Eq. (10) we can estimate the lowest threshold energy for laser
absorption at the pellet surface. The detailed threshold value is determined
by scaling between u and $\eta_{ab}I_L$, or Pa and $\eta_{ab}I_L$.

The plasma ablation velocity yielding the ablation pressure depends
upon the expanding plasma structure, which is determined by the laser
absorption mechanism, energy transport in the plasma, and some micro
processes. The relationship between laser wavelength and ablation pressure
is one of the most important determinants for selection of a suitable laser.

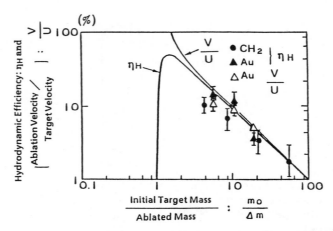

Fig. 6. Ablative acceleration: comparison of experiment and the Rocket model (solid line). Solid and open data points correspond to η_H and V/U.

We can find a simple scaling law for this relation in the case of classical absorption with a steady state ablation structure.

According to Bernoulli's law, the total pressure Pa is

$$Pa = \rho_{cJ} u_{cJ}^2 + P_{cJ} = 2\rho_{cj} c_S^2 \tag{12}$$

where ρ_{cJ} is the plasma structure between the ablation front and CJ point is in steady state. This is valid[8] when the laser pulse is long enough compared with L_s/C_s, where L_s is the distance between the ablation front and the CJ point. The heat transport flux is equal to the energy flux of ablating plasma.

The heat flux is

$$q = \left| k \frac{dT}{dx} \right| \cong \frac{2}{7} \frac{kT_{cJ}}{L} ,$$

The energy flux at the CJ point is

$$Q = \frac{1}{2} \rho_{cJ} c_S^3 + \frac{1}{y - 1} P_{cJ} = \frac{y + 1}{2(y-1)} P_{cJ},$$

where k is the thermal conductivity $= 8\sqrt{2\pi} \, l_e v_e n_e$, y is the adiabatic constant, l_e is the electron mean free path, v_e is the electron mean thermal velocity, and

$$L \cong 2\sqrt{m_i/Zm_e} \, l_e \qquad at \quad y = \frac{5}{3}.$$

Using these results, if we put T_{cJ} = 1keV, $n_{cJ} = 10^{21}$ cm^{-3}, m_i/Zm_e =3600, Z=4.5, the steady condition is attained by 0.2 nsec.

In the classical absorption case, the CJ point nearly corresponds to the cutoff density, and the region downstream from the CJ point (the low density region) becomes isothermal due to the laser heating and the long mean free path of electrons. The plasma expansion is well described by a self-similar solution of isothermal expansion,

$$\rho(x,t) = \rho_{cJ} \exp \left| -(x-x_{cJ})/tc_s \right|$$

$$v(x,t) = c_s \left| 1+(x-x_{cJ})/tc_s \right|$$

Then the total energy of the isothermal expanding plasma is

$$W(t) = \int_x^\infty dx \left| \frac{1}{2} \rho v^2 + \frac{3}{2} c_s^2 \right| = 4 \rho_{cJ} c_s^3 t.$$

The balance between the laser absorption intensity and dw/dt gives

$$\eta_{ab} I_L = 4 \rho_{cJ} c_s^3 .$$

The ablation pressure is

$$Pa = \frac{\eta_{ab} I_L}{2 c_s} = \left(\frac{\rho_{cJ}}{2}\right)^{1/3} (\eta_{ab} I_L)^{2/3}. \tag{13}$$

This shows that Pa is proportional to (laser absorption intensity)$^{2/3}$ and, considering that ρ_{cJ} is approximately equal to the laser cutoff density $\rho_{cr} \sim \lambda^{-2}$, Pa is proportional to $\lambda^{-2/3}$:

$$Pa = 12 (\eta_{ab} I_L / \lambda_L)^{2/3} \text{ Mbar,}$$

where $I_{ab} = \eta_{ab} I_L$ is expressed in 10^{14}w/cm^2 and λ_L is in μm. In Fig. 7 these results are shown with experimental data for $\overset{\cdot}{=} 1$: solid line, $\lambda_L = 0.5 \mu$m: broken line, $\lambda_L = 0.33 \mu$: chain line. From these values we can estimate the necessary laser absorption for various implosion exponents.

LASER DRIVERS

There are several choices for the energy driver in ICF. However, the laser is most promising driver. Table I shows the present status of fusion lasers. Fig. 8. is a picture of the GEKKO XII glass laser system at ILE Osaka University, which can deliver 30kJ and 55 TW. As previously shown, shortening the laser wavelength is one of the key issues. Nonlinear frequency conversion techniques are now higly developed.

The output power of lasers is specified depending upon the research program: the 10kJ class laser is used for fundamental research of implosion physics; the 100kJ laser is expected for break-even experiments. The reactor laser belongs to the MJ class. In table 2, specifications for the reactor laser are given. The glass laser is expected to have the efficiency 1∿2%.

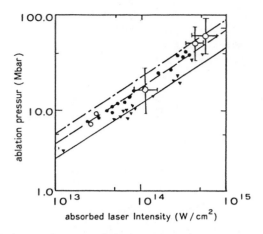

Fig. 7. Absorbed intensity scaling for ablation pressure. The solid, the broken and the chain lines correspond to the theoretical prediction at λ_L=1.06μm, λ_L=0.53μm, and λ_L=0.35μm. Experimental results are shown as follows: (∇) and (O) , 1.06μm and 0.53μm at Rutherford Lab, Osaka; (O), 1.96μm at NRL; (O), 0.35μm at Rochester Univ.

The CO_2 laser will be used for special target implosion, such as the MICF target described later. The KrF laser and new Gernett lasers are under development to attain high repetition rate and high efficiency. The free electron laser is also a newcomer.

TARGET

In laser implosion we can expect compression of the DT fuel up to $\rho R > 0.3 g/cm^2$ and T>4keV, where a particles can produce self heating of the fuel to ignition.

Today we have two different types of target for research, direct implosion targets and indirect implosion targets. The former produces the ablation plasma at the outer surface oa a pellet which compresses the fuel. The uniformity requirement on the laser irradiation is very stringent. However the LHART target (Large High Aspect Ratio) attained[10] neutron yields up to 10^{13}. The CD shell target[11] was also compressed to $40g/cm^2$. The latter target is a double shell structure, a hole in whose outer shell introduces the laser light into a cavity. The Xrays and plasmas produced between the double shells can implode the inner fuel shell. This is known as the Cannonball target and was invented by ILE Osaka.

A picture of the Cannonball is shown in Fig. 9. Another indirect implosion target is the MICF[12] target (Magnetically insulated inertial confinement fusion target). This is a similar target to the Cannonball. The CO_2 laser irradiation produces hot electrons which induce a strong magnetic field. The magnetic lines merge to form a closed wall confining the energy. The particles are confined inertially. This is a very interesting type of target.

Table 1. Laser systems for inertial confinement fusion

Country	Laboratory	Name of the system	Beam number	Output power (TW)	Output energy (kJ)	Pulse width (ns)	
U S A	LLNL	Nova	10	50	50	0.1~3.0	10^{13} neutrons 20g/cc
	LLE U.Rochester	Omega-X	24	15	5	0.03~0.1	10^{10} neutrons 6g/cc
	NRL	Pharos-II	2	1.3	1.3	0.1~1.0	
	KMS	Chroma-I	3	0.6	1	0.1	
U R S S	Kurchatov	Mishen	4	–	1	1.0	
		Delfin	216	33	10	0.2~3.0	
		Aurora	20	–	50 500	0.03~10	under construction plan
JAPAN	ILE Osaka	Gekko-IV	4	2	1	0.1~1.0	10^{8} neutrons 2g/cc
		Gekko-MII	2	7	3	0.1~1.0	2×10^{8} neutrons 4g/cc
		Gekko-XII	12	55	30	0.1~1.0	10^{13} neutrons 40g/cc
U K	Rutherford	Vulcan	6	3.6	1.2	0.1~1.0	
FRANCE	Limeil	Phebus	2	20	20	0.1~1.0	
		Octal	8	2	1		
	Ecole Poly. Tef.	Greco	1	0.25	0.25	0.1~2.5	

Fig. 8. GEKKO XII glass laser system: twelve beam 50TW, 30kJ.

Table 2. Parameters of Laser Development Program

Lasers	Wavelength (m)	Energy (MJ)	Pulse Width (ns)	Efficiency (%)	Remarks
Nd Glass	1.05 0.35	1	>10 ns	1 2	Design for efficient extraction using convensional technology
CO_2	10.6	2	20 ns	10 12	High extraction efficiency with multi-line, multi-pass amplification
KrF	0.25	1	>10 ns	5	Simple use Compression technique
Novel Solid State	< 1	1	> 10 ns	10 12	Search for new materials for high repetition operation and new excitation light sources for efficient operation
Free Electron	-	-		-	Too early to evaluate

Let us summarize the necessary conditions for implosion:

(1) The coupling between the fuel and the laser should be better than 4 to 5%;

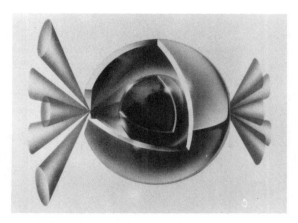

Fig. 9. The Cannonball target was invented at ILE Osaka University. The Cannonball target can display good absorption, high compression efficiency and excellent uniformity.

(2) Preheating before the implosion should be as small as possible; the initial temperature should be less than 10eV.
(3) Ablative pressure should be larger than a few 10Mbar.
(4) The uniformity of the implosion is important; variation in implosion velocity should be less than $\pm$ 3%.

Various kind of targets are used to meet these conditions. (1), (2), and (3) are nearly satisfied by using shorter wavelength lasers. (4) is fulfilled by using Cannonball type targets.

COMPUTER SIMULATION

As the implosion process is a strong nonlinear phenomenon, computer simulation is a very important tool to classify the processes. The purposes of this approach are:

(1) For pellet design, simulation can develop scaling laws for implosion using information on absorption, temperature, density and fusion reaction.
(2) It can treat the equation of state of implosion plasmas and the transport of Xrays which originally lay in the realm of astrophysics; laboratory plasmas now reach these regimes.
(3) The experimental data can be compared with simulation data to understand the behavior of imploded dense plasmas.
(4) The fusion reactor can be designed using this type of simulation.

Several simulation codes have been made to carry out these calculations; a Super-Computer is essential for this purpose.

INERTIAL CONFINEMENT FUSION REACTOR

We can design a hybrid reactor by using lasers at the present state of the art, that is $\eta_L Q=0.2$. When high gain pellets attain $\eta_L Q=2.5$, the zero

305

power pure fusion reactor will achieved. One of the most critical design problem is the first wall which is exposed to the fusion products and the Xrays. The inertial confinement fusion reactor is very insensitive to the plasma impurities which are critical to the magnetic confinement fusion reactor. In ICF one can introduce a liquid metal blanket inside the reactor chamber. The life of the reactor may be long enough to resemble the fission reactor. This is a great advantage for the ICF reactor system.

As an example, Fig. 10 shows a conceptual design for the laser fusion reactor "Senri I". The specification of "Senri I" reactor is given in Table 3.

CONCLUSIONS

The laser can deliver very high energetic flux to a small volume. If you use a 1TW laser at 1 m wavelength, the intensity at the focus is 10^{20}w/cm^2, the electric field is 3×10^{14}v/m, and the magnetic field is 10^{10} oersted, and we can expect extrordinary fusion effects. Laser fusion is one of the most exciting research areas for tne future of the humankind. This guest may give us not only a near-infinite energy source, but also very wide impact across the frontiers of Science and Technology.

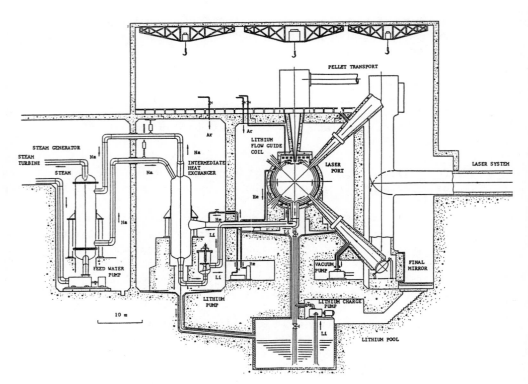

Fig. 10. Inertial Confinement Fusion Reactor SENRI I; the remarkable characteristic of this reactor is the Liquid Lithium Blanket to collect the fusion products.

Table 3. Parameters of laser fusion reactor SENRI I

Lasers	Gas laser	
	Beam number	8
	Output energy	1∿5 MJ
	Repetition	1 Hz
	Peak power	200 TW
	Pulse width	100 ns
	Efficiency	10%
Output power	1,240 MW (Thermal Output Power)	
	426 MW (Net Electric Power)	
Cooling system	Li – Na – H_2O	
Tritium cycle	Critical concentration of Tritium	10 ppm
	Recovery from Lithium	
	Initial inventory	∿ 1.3 Kg
	Doubling time	∿ 35 days
	Consumption rate	51.6 Ci/s
	(Burning rate Y=0.33)	
Breeding rate of Tritium	1.6	

REFERENCES

1. W. L. Kruer,"Progress in Lasers and Laser Fusion", Plenum Publishing, N.Y. p. 5 (1975)
2. E. Fabre et al., IAEA Tech, Comm. Meeting, Takarazuka, Japan, Rev. Laser Eng. 8, 28 (1980)
3. W. C. Mead et al., Phys. Rev. Lett., 47:18 (1981)
4. C. Yamanaka et al., Proceed. of 8th Inter. Conf. on Plasma Physics and Controlled Nuclear Fusion Research, IAEA (1980) IAEA-CN-38.
5. V. L. Ginzburg, "The Propagation of Electromagnetic Waves in Plasma", Pergamon Press, N.Y. p.260 (1964)
6. R. Decoste et al., Phys. Rev. Lett., 42:1673 (1979)
7. H. Nishimura et al., Phys. Rev., A, 23, 2011 (1981)
8. C. E. Max et al., Phys. Fluids, 23:1620 (1980)
9. H. Takabe et al., J. Phys. Soc. Jpn., 45, 2001 (1978)
10. C. Yamanaka et al., Phys. Rev. Lett., 56, 1575 (1986)
11. C. Yamanaka et al., 11th Inter. Conf. on Plasma Physics and Controlled Nuclear Fusion Research, IAEA (1986), IAEA-CN47/B-I-4
12. ibid, IAEA-CN47/B-I-5
13. C. Yamanaka, 18th ECLI Meeting, 1987,Prage

NON LINEAR LASER PLASMA INTERACTIONS WITH APPLICATIONS

TO ELEMENTARY PARTICLE ACCELERATION

J. L. Bobin

Université Pierre et Marie Curie

Paris, France

INTRODUCTION

Since the days of Ampère, who was the first to accelerate objects using electromagnetic interaction[1], many methods based on this same force have been invented to impart large energies to microscopic bodies. By this means, the fundamental laws of our Universe are investigated in detail: the larger the energy, the more basic and the more accurate are the results. The evolution of accelerators is best seen on the Livingston chart[2] as recently updated[3]. It shows that the optimal performances of accelerators has followed an exponential growth (Fig. 1).

Nevertheless, elementary particle physicists are demanding still higher energies. Gigantic machines are being built. In Europe, at C.E.R.N., the L.E.P. (Large Electron Positron Collider) is under construction. This circular ring whose circumference is 27 Km will operate in 1989 at the 2x80 G.e.V. level. In the United States, at S.L.A.C. the S.L.C. (Stanford Linear Collider) is in the testing stage with energy 2x50 G.e.V. while the S.S.C. (Superconducting Super Collider) project is to be approved. However, the location of this 96 Km circular proton-antiproton collider (2x20 T.e.V.) is not yet decided.

In the next generation of accelerators, energies up to a few tens of T.e.V. are needed. In that range, a number of interesting physical problems are worthy of experimental investigation[4,5]. It is generally agreed that, in the future, electron-positron colliders should be built. Indeed, these particles are, thus far, elementary: therefore, their entire acceleration energy is available for collision processes between elementary objects. On the contrary, protons and antiprotons are built up of three quarks each: consequently only about 1/6th of the acceleration energy is available for a fundamental process. Electrons in linear colliders are the best choice, in order to avoid losses through synchrotron radiation[6]. At present, no more than about 20 M.e.V./m can be reached using conventional R.F. techniques applied to electron acceleration, and this results in tremendously large

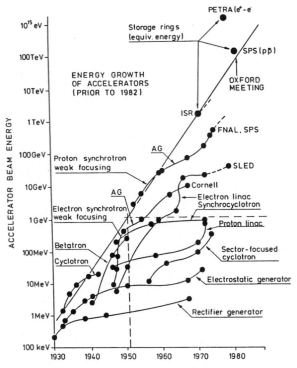

Fig. 1. The Livingston chart after Voss[3].

accelerators. Much higher accelerating gradients are thus mandatory.

Another requirement is also important. Cross sections between elementary objects such as quarks and leptons vary as the reciprocal energy squared: see Fig. 2[7]. In order to observe a reasonable number of events per unit time, it is advisable to increase the luminosity accordingly:

$$\mathcal{L} = N^2 \, hf/A \,. \tag{1}$$

In this definition, N is the number of accelerated particles in a bunch, f the repetition of bunches passing through an experimental region, A the beam cross section, and h a reduction factor due to the pinching when encountering bunches of particles with the opposite charge. The number of expected events is the product of the cross section times the luminosity. The largest luminosity ever, 10^{32} cm^{-2} s^{-1}, was achieved at C.E.R.N. with proton-antiproton intersecting storage rings (I.S.R.).

The challenge accelerator builders are faced with can be stated as follows:

accelerate along straight lines

- electrons
- up to energies over 1 T.e.V.
- in a machine the size of which compares with that of present

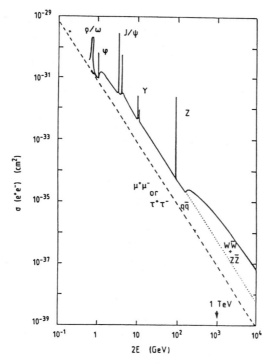

Fig. 2. Cross sections for e^+ e^- events after Amaldi[7].

day devices
- with a luminosity 10^{34} cm^{-2} s^{-1}.

To reach this goal, new concepts have been devised in the past few years, some of which make use of plasmas and laser plasma interaction.

The reason for using plasmas is the following: in every accelerating device, particle travel "in vacuo" actually occurs within a rarefied gas, inside pipes whose walls are made of either conducting or dielectric material. Larger gradients imply higher electric fields, which may cause electrical breakdown in the residual gas or along the walls. Thus, the necessary fine tuning of the cavities is destroyed. This is the main limitation on increasing electric field gradients. Inside a plasma, there is no such limitation since matter is already ionised. Furthermore, plasmas are able to propagate large electric fields in electron plasma waves. Laser plasma interactions can then be used, among other methods, to drive high amplitude plasma waves.

Since breakdown thresholds in high frequency electromagnetic systems increase with frequency, optical frequencies are especially attractive. When focusing a laser beam with a lens or a mirror, high intensities, I, are readily obtained. The corresponding electric fields are given, independently of wavelength, by

$$E = (2/\varepsilon_0)^{1/2} (I/c)^{1/2} \qquad (2)$$

311

with the result: 1.8 Gigavolt/m for 10^{12} Wcm^{-2} (a 1967 achievement) and 1.8 Teravolt/m for 10^{18} Wcm^{-2} (today's performances with the most advanced high power lasers). Here again, laser-plasma interaction provides a way to take advantage of the laser's capability for producing high fields.

When a high intensity laser beam is focused onto a solid surface, matter is strongly heated. The temperature is so large that a plasma is formed and set into motion. It was observed long ago in such experiments that energetic charged particles (ions or electrons) are emitted. Their velocities are inconsistent with the measured temperatures in the plasma plume. This phenomenon was never satisfactorily explained. Many mechanisms were proposed for particle acceleration in the laser driven plasma: gasdynamical rarefaction, the influence of self-generated magnetic fields, electrostatic fields associated with charge separation, wave particle interaction... None of these could be identified as a single cause of the observed results. Quite recently it was shown by Tajima and Dawson[8], with the help of particle simulation, that a particular process, the resonant beating of two laser waves whose frequency difference is equal to the plasma frequency, is able to drive a high amplitude longitudinal plasma wave in which electrons can be accelerated to very high energies. This was the starting point for many theoretical, numerical and experimental investigations which show that laser-plasma interaction is indeed a promising way to reach the large accelerating gradients required in future devices.

CHARGED PARTICLE ACCELERATION IN A PLASMA WAVE

Relativistic equations of motions

Consider a particle with electric charge q moving under the influence of an electromagnetic field. The momentum energy 4-vector

$$(p, W) = (m_0, \gamma v, m_0 \gamma c^2) \tag{3}$$

and the 4-potential $(A, \emptyset)$, are related through the usual equations of motion

$$dp/dt = q(E+v\times B) = q\left[-\nabla\emptyset -\partial A/\partial t-(v\nabla)A+\nabla(vA)\right] =-q\left[-\nabla\emptyset-dA/dt+\nabla(vA)\right]$$

$$dW/dt = q(E+v\times B)v = -q\left[-\nabla\emptyset-\partial A/\partial t\right]v. \tag{4}$$

Whenever the scalar potential $\emptyset$ has a purely longitudinal gradient and the vector potential A is purely transverse, one obtains the following 4 equations for the 4-vector components:

$$d(p_x+qA_x)/dt = d(p_y+qA_y)/dt = 0,$$

$$dp_z/dt = m_0\gamma dv_z/dt+m_0 v_z d\gamma/dt = q\left[-\partial\emptyset/\partial z +\partial(A_x v_x+A_y v_y)/\partial z\right]$$

$$dW/dt = m_0 c^2 d\gamma/dt = -q\left[v_z\partial\emptyset/\partial z + v_x\partial A_x/\partial t+v_y\partial A_y/\partial t\right] . \tag{5}$$

The equations for the x and y components are readily integrated to give

$$v_x = -qA_x/m_0\gamma \quad ; \quad v_y = -qA_y/m_0\gamma \qquad (6)$$

and consequently

$$A_x v_x + A_y v_y = -q|A|^2/m_0\gamma \qquad (7)$$

Finally, the relevant equations are

$$d\gamma/dt = -(q/m_0c^2)v_z \, \partial\phi/\partial z + (q^2/2m_0^2c^2\gamma)\partial|A|^2/\partial t$$

$$dv_z/dt = -(q/m_0\gamma^3)\partial\phi/\partial z-(q/m_0\gamma)^2\partial|A|^2/\partial z-(q/m_0\gamma)^2(v_z/2c^2)\partial|A|^2/\partial t. \qquad (8)$$

These equation will serve as a starting point throughout this paper. They will be applied to individual electrons, isolated or in bunches, and to the electron fluid in a plasma as well.

Electrons in a longitudinal plasma wave

In a purely electrostatic plane wave propagating in the z direction through a cold plasma, there is no vector potential. The electric field is

$$E = -\partial\phi/\partial z - E_0\sin(k_D z-\omega_p t+\phi_0) \qquad \omega_p =(n_0e^2/\epsilon_0 m_0)^{1/2} \qquad (9)$$

where ω_p is the plasma frequency for the electron density n_0. When the phase velocity is very close to c, such a wave can be effective in accelerating relativistic electrons. Indeed, the equations of motion for a single electron are

$$d\gamma/dt = -(e/m_0c^2)v_z \, E_0\sin(k_D z \, \omega_p t+\phi_0)$$

$$dv_z/dt = -(e/m_0\gamma^3) \, E_0\sin(k_D z-\omega_p t+\phi_0). \qquad (10)$$

Making use of the similarity variable

$$\xi-\xi_0 = k_D z-\omega_p t+\phi_0, \qquad (11)$$

these equations can be expressed as an autonomous first order ordinary differential system

$$d\gamma/dt = -(e/m_0c^2)v_z \, E_0\sin(\xi-\xi_0)$$

$$dv_z/dt = -(e/m_0\gamma^3) \, E_0\sin(\xi-\xi_0)$$

$$d\xi/dt = k_D v_z -\omega_p \qquad (12)$$

where $_0$ is the initial relative phase. The corresponding phase variation in space is displayed in Fig. 3. The figure shows that electrons can be accelerated to high energies either if:

i) they have trapped trajectories and initial low energy, or
ii) they have a sharply defined initial phase in passing trajectories

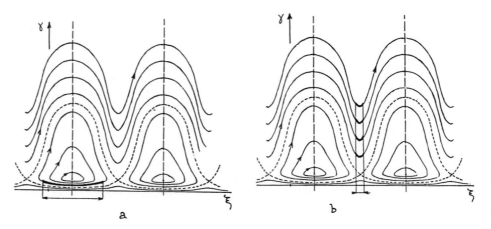

Fig. 3. γ,ξ phase plot. (a) acceleration of trapped electrons, (b) acceleration of passing electrons.

associated with initial velocities greater than the phase velocity of the wave

$$u_R = \omega_p/k_D. \tag{13}$$

Then one can imagine an accelerating device with several stages. In the first stage, an electron beam with a uniform longitudinal density and velocity u_R is bunched within a quarter period and half wavelength in space (see Fig. 4b). After energy filtering, high energy bunches are injected with proper phasing into successive stages (Fig. 4c). Note that an initially spatially uniform monoenergetic electron beam with an energy well above trapping is slightly bunched with a global energy loss: this is the free electron laser regime (Fig. 4d).

Acceleration energy and length

Since is the Lorentz factor corresponding to v_z, the differential system[3,4] actually consists of only 2 independent autonomous equations, viz

$$d\gamma/dt = -(e/m_0 c^2)(1- 1/\gamma^2)^{1/2} E_0 \sin\xi \tag{14}$$

$$d\xi/dt = K_D(1-1/\gamma^2)^{1/2} -\omega_p.$$

The first ordinary differential equation in $d\gamma/d\xi$ resulting from Eq. (14) is readily integrated to give

$$ck_D\gamma-\omega_p(\gamma^2-1)^{1/2}-ck_D\gamma_0+\omega_p(\gamma_0^2-1)^{1/2} = (eE_0/m_0c)(\cos\xi-\cos\xi_0). \tag{15}$$

The maximum value of the second factor in the right side is 2. It corresponds to the largest increase in γ along a passing trajectory in phase space. Introducing

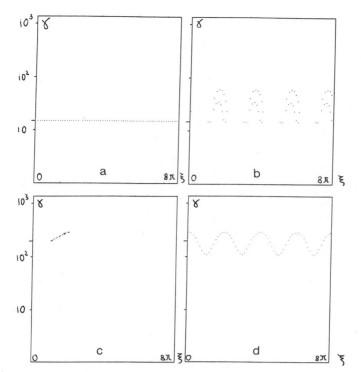

Fig. 4. Electron behaviour in γ,ξ phase space. (a) initial state of a
completely trapped electron beam; (b) bunching of the trapped
electrons of fig. a); (c) acceleration of a bunch of passing
electrons; (d) behaviour of a high energy electron beam: slight
bunching and global energy loss, free electron laser regime.

$$\beta_R = (1-1/\gamma_R^2)^{1/2} = \omega_p/ck_D, \tag{16}$$

and assuming that both γ_0 and γ are much greater than 1, yields

$$\Delta\gamma = \gamma - \gamma_0 \cong (2eE_0/m_0c\omega_p)\beta_R/(1-\beta_R). \tag{17}$$

Finally, expanding Eq. (16) one finds the maximum energy that an electron
may acquire in the process

$$W_A = m_0c^2\Delta\gamma \cong (m_0c\omega_p)m_0c^2\gamma_R^2 = 4eE_0\gamma_R^2c/\omega_p. \tag{18}$$

Now, there exists an absolute maximum for the plasma wave amplitude:
density perturbation equal to the background density, the so called wave-
breaking condition. An equivalent statement is: all electrons in the plasma
are trapped in the wave. In a reference frame moving with the phase velocity
(henceforth called the "moving frame"), most of the plasma electrons are
travelling with a velocity $-\omega_p/ck_D$. They will be trapped provided the
potential energy in the wave is at least equal to their energy $\gamma_R m_0c^2$.
Integrating the Poisson equation over half a wavelength and noting that
the particle number in a wavelength is a relativistic invariant, as well as

the longitudinal field, one finds the condition:

$$2eE_0\gamma_R/k_D = m_0\gamma_R c^2, \quad \text{i.e.}, \quad E_0 = m_0 c\omega_p/e \tag{19}$$

Substituting in Eq. (18) the upper limit for the electron energy turns out to be:

$$W_{Amax} = 2m_0 c^2 \gamma_R^2 \ . \tag{20}$$

In the moving frame, electrons are accelerated over at most half a wavelength. Since there is no phase locking, electrons in the wave are decelerated whenever they propagate beyond that distance. The corresponding acceleration length 1_A in the laboratory frame is given by

$$W_A = eE_0 1_A, \quad \text{hence} \quad 1_A = 2\gamma_R^2 c/\omega_p \ . \tag{21}$$

The length of the accelerating plasma column should not exceed 1_A and the extension of the bunches along the z axis has to be less than a small fraction (e.g. 1/10th) of the wavelength $2\pi/k_D$. From Eq. (18) and Eq. (21), the accelerating gradient is in general

$$W_A/1_A = 2eE_0 = cm_0\omega_p(n/n_0), \tag{22}$$

where n is the amplitude of the density perturbation. The upper limit is

$$W_A/1_A)_{max} = m_0 c\omega_p. \tag{23}$$

In all cases, $(W_A/1_A)$ is proportional to the square root of the background plasma density.

The "surfatron"

The existence of an acceleration length has some unpleasant consequences: in order to reach the energies particle physicists are aiming at, one has to use high amplitude waves in high density plasmas. Now, electron acceleration in a longitudinal wave bears some analogy with the case of a surfer on ocean waves. To stay in phase as long as possible, the skilled surfer gives his board transverse momentum. The same trick can be used with electrons, thanks to the presence of a transverse magnetic field[9]. Assume as a constant B_y field. The electrons then move in both the x and z directions according to

$$d\gamma/dt = \gamma^3(v_x dv_x/dt + v_z dv_z/dt) = -e(v_z/m_0 c^2)E_z,$$

$$d\gamma v_x/dt = v_x d\gamma/dt + \gamma dv_x/dt = \gamma^3|(1-v_z^2/c^2)dv_x + (v_x v_z/c^2)dv_z/dt =$$

$$= (e/m_0)v_z B_y,$$

$$d\gamma v_z/dt + \gamma dv_z/dt = \gamma^3 \left| (v_x v_z/c^2) dv_x/dt + (1 - v_x^2/c^2) dv_z/dt \right| =$$

$$= -(e/m_0)(E_z + v_x B_y). \tag{24}$$

Using the form of the Eq. (9) for the travelling longitudinal electric field, one obtains the non autonomous first order differential system

$$dv_x/dt = (e/m_0)(1-v_x^2/c^2-v_z^2)^{1/2}\{(v_x/c^2)E_0\sin|k_D(z-u_Rt)+\phi_0|+B_y\}v_z$$

$$dv_z/dt = (e/m_0)(1-v_x^2/c^2-v_z^2/c^2)^{1/2}\{(1-v_z/c^2)E_0\sin k_D(z-u_Rt)+\phi_0 +B_y v_x\}$$

$$dz/dt = v_z \tag{25}$$

whose solutions have been investigated with the help of a computer[9]. Phase locking occurs provided that suitable initial conditions are fulfilled: obviously, one should have

$$dv_z/dt = 0 \qquad v_z = u_R \qquad z = u_R t + z_0. \tag{26}$$

Now, from the second equation of (24) we get

$$d\gamma v_x/dt - eB_y v_z/m_0 = \Omega u_R, \tag{27}$$

where Ω is the electron cyclotron frequency. Eq. (27) is integrated trivially and solved for v_x:

$$v_x = (\Omega u_R t/\gamma_R)(1+\Omega^2 u_R^2 t^2/c^2)^{-1/2} \rightarrow c/\gamma_R - (c^2 - u_R^2)^{1/2} \quad (t \rightarrow \infty). \tag{28}$$

The accelerated particle trajectory is at an angle θ ($\sin = 1/\gamma_R$) with respect to the direction of propagation of the wave (Fig. 5). The energy increment can be expressed in 2 ways:

$$d\gamma/dt = \Omega\beta_R\gamma_R = -(eE_0/m_0)(u_R/c^2)\sin(k_D z_0 + \phi_0). \tag{29}$$

The initial location and phase should be choosen so that $\sin(k_D z_0 + \phi_0)$ is negative. Furthermore, there is an upper limit for the magnetic field

$$B_y = E_0\sin(k_D z_0 + \phi_0)/(c^2\gamma_R) \quad E_0/(c^2\gamma_R), \tag{30}$$

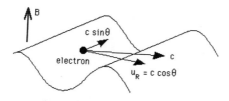

Fig. 5. Particle and wave propagation in the surfatron.

quite a small value! It turns out that a comparatively low magnetic field can be set up over large distances, making the surfatron concept rather attractive. Furthermore, can be shown that acceleration is improved when the magnetic field is at an angle with respect to the normal to the x0z plane[10].

LASER DRIVEN PLASMA WAVES

Electrons in the field of 2 electromagnetic waves

In a purely transverse field, $\emptyset$ is zero and the vector potential A is a relativistic invariant. Assume for convenience that A results from 2 circularly polarized waves in the laboratory frame:

$$A_1 = (A_1 \cos(k_1 z - \omega_1 t + \phi_1), A_1 \sin(k_1 z - \omega_1 t + \phi_1), 0)$$

$$\omega_1 > \omega_2$$

$$A_2 = (\pm A_2 \cos(k_2 z \mp \omega_2 t + \phi_2), A_2 \sin(k_2 z \mp \omega_2 t + \phi_2), 0) \tag{31}$$

upper sign: waves propagate in the same direction; lower sign: waves propagate in opposite directions. The vector potential enters the equations of motion through its modulus squared only:

$$|A|^2 = A_1^2 + A_2^2 - 2A_1 A_2 \cos \left[(k_1 \mp K_2) - (\omega_2 - \omega_1) t + \Delta\phi \right]. \tag{32}$$

The expression for $|A|^2$ in a reference frame moving with the relativistic velocity u_R, is obtained by the Lorentz transform:

$$z = \gamma_R (\xi + u_R \tau), \qquad t = \gamma_R (u_R \xi / c^2 + \tau), \tag{33}$$

Substituting in Eq. (32), it turns out that $|A|^2$ has no τ dependence provided that

$$u_R = (\omega_1 - \omega_2) / (k_1 \mp k_2). \tag{34}$$

Waves A_1 and A_2 propagate in opposite direction in the moving frame, with the same frequency and wave number (in the case of the free electron laser, $\omega_2 = 0$, this is known as the Bambini-Renieri reference frame[11]). In such a frame, (primed quantities) the electron velocity is given by

$$dv_z'/dt = 2(e/m_0\gamma)^2 \partial A_1 A_2 \cos \left[(k_1 \mp K_2) \xi / \gamma_R \right] / \partial\xi = (e/m_0\gamma)^3 \partial\Psi / \partial\xi \tag{35}$$

where Ψ is the ponderomotive potential. The corresponding force results from the $v_i x B_j$ terms of the Lorentz force.

Now, assume that many electrons are present, the case of a plasma or dense electron beam. Then, the electrostatic potential $\emptyset$ which is associated with density perturbations must be accounted for. Due to the factor v_z/c^2, the term in $\partial|A|^2/\partial\tau$ is negligible. In the moving frame:

318

$$dv_z'/d\tau \cong (e/m_0\gamma^3)\partial(\emptyset+\Psi)/\partial\xi = -(e/m_0\gamma^3)E_S - 2|(k_1 \mp k_2)/\gamma_R|(e/m_0\gamma)^2 A_1 A_2 \sin$$

$$|(k_1-k_2)\xi/\gamma_R|. \tag{36}$$

Using fluid dynamical and Poisson equations yields a second order differential equation for the density perturbation in the electron oscillation ($k_D = k_1 - k_2$; $\omega_p = \omega_1 - \omega_2$)

$$d^2n'/d\xi^2 + k_D'^2 n' = -2n_0'(e/m_0 c\gamma_R)^2 k_D'^2 \; A_1 A_2 \cos(k_D'\xi) \tag{37}$$

It describes the behaviour of an electron plasma wave driven by the panderomotive potential associated with two electromagnetic waves. This problem can be dealt with in two ways. First it may be assumed that all oscillations involved have a slowly varying amplitude and phase. A set of coupled first order differential equations is then derived. Alternatively, (Eq. 37), is a typical case of forced oscillations to be investigated as such.

Stokes and antiStokes stimulated Raman scattering

Resonant coupling between two electromagnetic waves and an electron plasma wave is an important nonlinear effect in laser-plasma interaction. Whenever one of the E.M. modes has a large initial intensity (e.g. a laser beam), while the other mode and the plasma oscillation are very weak, we have the case known as stimulated Raman scattering. We will restrict ourselves to the case of all waves propagating in the same direction in the laboratory frame (forward scattering). In this frame, the vector potentials obey the usual propagation equation in the Lorentz gauge; for example, the complex potential A_1 obeys

$$\partial^2 A_1/\partial t^2 - c^2 \partial^2 A_1/\partial z^2 = j_1/\varepsilon_0 \tag{38}$$

where the current density j_1 in lowest order is

$$j_1 = -(e^2 n_0/m_0)A_1 - i(e^2/m_0)nA_2. \tag{39}$$

The first term on the right hand side is the polarisation current due to A_1; the second term comes from the resonant coupling of A_2 to the plasma wave. Setting

$$A_1 = A_1(z,t)\exp i(k_1 z - \omega_1 t), \tag{40}$$

neglecting the second derivatives of the slowly varying amplitude $A_1(z,t)$, and denoting by v_1 the group velocity, one gets a first order partial differential equation

$$2\omega_1(\partial A_1/\partial t + v_1 \partial A_1/\partial z) = -(e^2/\varepsilon_0 m_0)nA_2. \tag{41}$$

By the same token,

$$2\omega_2(\partial A_2/\partial t + v_2 \partial A_2/\partial z) = (e^2/\varepsilon_0 m_0)A_1 n'*$$ (42)

Eqs.(41) and (42) are readily transformed for quantities in the moving frame. The second order equation (37) is also reduced to first order by giving n a form analogous to Eq. (40). The final system of ordinary differential equations for the problem at hand is thus:

$$(v_1 - u_R)dA_1/d\xi = -(e^2/2\varepsilon_0 m_0 \omega_1)n'A_2$$

$$(v_2 - u_R)dA_2/d\xi = (e^2/2\varepsilon_0 m_0 \omega'_2)A_1 n'*$$

$$dn'/d\xi = n'_0 (e/m_0 c\gamma_R)^2 k'_D A_1 A_2*.$$ (43)

Such equations have been thoroughly investigated in various situations. In all cases, the amplitudes obey the Manley Rowe relations[12], which are best expressed quantum mechanically as

1 photon ω_1 ↔ 1 photon ω_2 + 1 plasmon ω_p.

First, assume that A_1 requests a large fixed potential (pump). Initially small values of A_2 and n' will then increase exponentially. This is an example of a convective instability with growth rate

$$\Gamma_c = (\omega_p e/\gamma_R^{1/2} m_0 c^2)A_1.$$ (44)

Several situations are of interest when A_1 is variable. Assume now that $A_1(0)$ is still a high intensity pump with zero $A_2(0)$ and a small n'(0). The solution of Eq. (43) is then expressed in terms of Jacobian elliptic functions[13]. The corresponding time history is displayed in Fig. 6a. It exhibits a non linear period which depends on n'(0). When this quantity is zero, the nonlinear period is infinite and a soliton is obtained (Fig. 6b). Since at the begining, a good approximation to the solution is an exponential growth, the physical process is known as Stokes Raman scattering by analogy with the behaviour of the solution of the Airy equation. The saturation of the plasma wave is due to pump depletion. When, on the contrary, $A_1(0)$ is zero, A_2 is large and n'(0) is still small, the regime is sinusoidal with comparatively low amplitude variations (fig. 7). In this case a photon ω_2 combines with a plasmon ω_p to form a photon ω_1: antiStokes Raman scattering. Since plasma waves can be damped either collisionally or via the Landau mechanism, Raman scattering will occur provided the pump intensity exceeds a threshold value which has been evaluated for both homogeneous and inhomogeneous plasmas[14].

In Stokes and antiStokes scattering, the wave generated can grow an intensity above threshold valuue, so that it also takes part in a further Raman process. Thus, satellite waves with frequency shifts which are multiples of ω_p can appear on both sides of a given high intensity laser line (Fig. 8): this Raman cascade can lead to turbulence and plasma heating[15]. This should be considered also as a possible saturation mechanism[16].

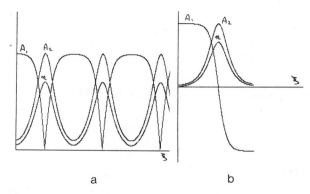

a   b

Fig. 6. Time history of amplitudes in Stokes Raman scattering. (a) periodic regime; (b) solitons.

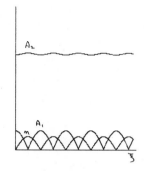

Fig. 7. AntiStokes Raman scattering.

Resonant beat waves

We now consider the case with large initial amplitudes for both A_1 and A_2. This situation was investigated first by Montgomery[17] who set up equations similar to Eq. (42), and later in more detail by Rosenbluth and Liu[18], who took the forced oscillator approach. Assume that both laser frequencies are much greater than the plasma frequency (underdense plasma). Then, the phase velocity u_R of the plasma wave is equal to the group velocity of the laser waves:

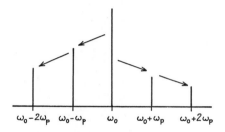

$\omega_0-2\omega_p$ $\omega_0-\omega_p$ ω_0 $\omega_0+\omega_p$ $\omega_0+2\omega_p$

Fig. 8. Radiation spectrum induced by a Raman cascade.

$$u_R = c(1-1/\gamma_R^2)^{1/2} = \delta\omega/\delta k = c(1-\omega_p^2/\omega_0^2)^{1/2}, \qquad (45)$$

where ω_0 is the average laser frequency and use was made of the plasma dispersion relation for E.M. waves. Consequently

$$\gamma_R = \omega_0/\omega_p. \qquad (46)$$

Since the acceleration energy varies as γ_R^2, this result shows that high laser frequencies are most desirable.

In order to study beat wave dynamics, the same procedure as in previous section can be used. The result is shown in Fig. 9. Besides pump depletion, the actual amplitude of the longitudinal field is expected to saturate through various possible processes: wavebreaking, relativistic oscillatory motion of the electrons in the wave, cascading, collisional or Landau damping ... Choose the longitudinal electric field E as the significant variable. Using Poisson's equation, a forced oscillator equation is readily derived for E from Eq. (37). Accounting for relativistic electron oscilla-tions requires us to reintroduce and expand the γ^{-3} factor in the equation for v_z. This results in a cubic nonlinear term added to the left hand side of the forced oscillator equation:

$$\delta_r |E|^2 E = -(3/2) \ (\varepsilon_0/m_0 n_0 c^2)|E|^2 E. \qquad (47)$$

Furthermore, modeling the damping by a phenomenological coefficient Γ times the first derivative of E eventually yields a normalized Duffing's equation, which then reads in the moving frame:

$$d^2E/d\xi^2 + \Gamma \ dE/d\xi \ + \left| 1-(3/2)(\varepsilon_0/m_0 n_0 c^2)E^2 \right| E = -2(e/m_0)k_D' A_1 A_2 \cos\xi. \quad (48)$$

In reference (18), damping was overlooked, and a solution was obtained in which the amplitude varies slowly with a nonlinear period much longer than the linear oscillation period. The solution relevant to growth from random noise is almost a singular integral in phase space (Fig. 10). The maximum field is then

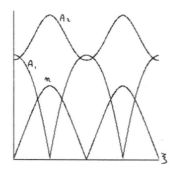

Fig. 9. Beat wave saturation by pump depletion.

322

$$E_{max} = (cm_0\omega_p/e)|(16/3)(e^2 A_1 A_2/m_0^2 c^2)|^{1/3} = (cm_0\omega_p/e)(16\Lambda/3)^{1/3}, \quad (49)$$

where Λ is a quantity known as the interaction parameter.

As shown by Karttunen and Salomaa[16], Raman cascade also brings a cubic nonlinearly into the equation, however with the opposite sign. The corresponding correction involves the laser frequency ω_0 and electric field amplitude E_0:

$$\delta_c E^2 = (5/4)(\omega_0/\omega_p)^{3/2}(\varepsilon_0/n_0 e E_0)E^2 \quad (50)$$

The correction is inversely proportional to the square root of the laser intensity. Since ω_0/ω_p is equal to the Lorentz factor γ_R, this quantity has to be large for efficient acceleration. Thus δ_c might be of the same order as δ_r. In any case, the coefficient $\delta_r - \delta_c$ is a small quantity, as is readily seen by putting numbers into Eq. (47) and Eq. (50). The prediction of a chaotic state was obtained in reference (19), by using a Duffing's equation with a positive cubic term $\alpha|E|^2 E(\delta_r < \delta_c)$. This is consistent with the physical picture of turbulence resulting from successive cascading through stimulated Raman scattering.

Compensation of relativistic detuning

It was shown in reference (20) that a properly choosen initial detuning towards higher frequencies can compensate the progressive detuning towards lower frequencies induced by the negative cubic term. Cancellation occurs when the field amplitude passes through its first maximum, which is then enhanced by a factor up to 1.6. Now, in experiments to be reported in section "Preliminary experiments", the electron density increases with time. In numerical simulations[21], a linear approximation was used to account for this effect. Subsequently, a moderate increment was found to provide further enhancement of the longitudinal field amplitude. Since the beat wave growth is a convective process, such a model might also apply to spatially inhomogeneous plasma.

In Duffing's equation replace the constant eigenfrequency squared with a function of the similarity coordinate ξ:

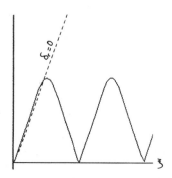

Fig. 10. Beat wave saturation by relativistic detuning.

$$d^2E/d\xi^2 + \Gamma dE/d\xi + \left| f(\xi) - \alpha E^2 \right| E = -F\cos\xi . \tag{51}$$

$f(\xi)$ is chosen to mimic the spatial or temporal variations of the electron density, and F is the amplitude of the external force. A term like $f(\xi)E$ is commonplace in oscillator equations. It is usually dealt with through the so called W.K.B. approximation. Here the situation is complicated by the presence of the cubic term. Accordingly, there will be no attempt to investigate the solutions analytically. Setting

$$Y = dE/d\xi , \tag{52}$$

the modified Duffing's equation is rewritten

$$dY/d\xi = -\Gamma Y - f(\xi) - \alpha E^2 \left| E - F\cos\xi \right| . \tag{53}$$

Eqs. (52) and (53) form a non autonomous differential system of the first order, which was solved computationally by means of a centered finite difference scheme[22]. In all runs, we took A=1 and α=.001. As in ref. (21), assume there is no damping and choose a linear $f(\xi)$:

$$f(\xi) = 1 + K\xi \tag{54}$$

e.g., electron density increasing linearly with time.

The results are shown in Fig. 11. When K is varied, a sharp transition is observed for a critical value K_c slightly above .0077, as is better seen on the graph of Fig. 12a, which plots the first maximum of the E field amplitude versus K. Other plots in Fig. 12 correspond to increasing damping coefficients. The transition is very steep at low damping, and much smoother when the damping is comparatively high. This is typical of a fold catastrophe of the kind exhibited by the ordinary Duffing's equation[23].

The influence of damping was also investigated in longer runs. The chosen value for K, i.e. .001, corresponds to a small enhancement of the first maximum. Without damping, the average value of the amplitude steadily increases, as in Fig. 13a. A strong damping is seen to smooth out the oscil-

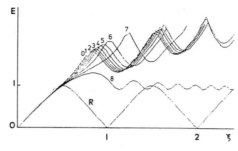

Fig. 11. Evolution of the plasma amplitude for a linearly increasing density without damping. The curve labelled R is the reference, with neither damping nor relativistic detuning. In curves labelled 0 to 8, K increases from .0070 to .0078.

324

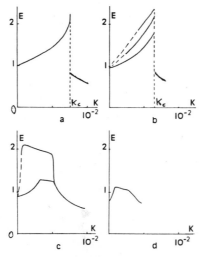

Fig. 12. (a) $\Gamma = .001$: first maximum of the amplitude vs K; (b) $\Gamma = .01$: first, second and third maxima; (c) $\Gamma = .025$: first and last maxima; (d) $\Gamma = .05$: maximum vs K.

lations of the amplitude, Fig. 13b. Similar results were also obtained independently in a numerical calculation closer to experimental conditions[24]. When $f(\xi)$ is different from linear (e.g. a parabolic variation) the results are quite similar to those obtained with a linear ξ dependence.

Detuning due to the nonlinearities can be compensated in a different way. Indeed one can act on the driving frequency[25]. Since short laser pulses are needed, one or both of the beating wave might be produced after some compression technique thus exhibiting a time varying frequency: chirped pulse. The dynamics of the electric field are now described by another modification of the Duffing's equation viz.

$$d^2E/d\xi^2 + \Gamma dE/d\xi + (1-\alpha E^2)E = - Fcos|g(\xi)\xi| , \qquad (55)$$

where $g(\xi)$ is a function chosen to model the time dependence of the driving frequency. A solution with linear g is displayed in Fig. 14. Again, a catastrophic transition is observed. However, the enhancement of the amplitude is limited to less than a factor of 2.

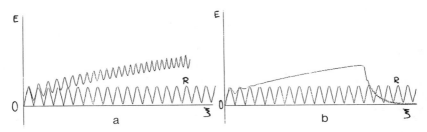

Fig. 13. Evolution of the plasma wave amplitude for a linearly increasing density (K=.0001). (a) no damping; (b) $\Gamma = 0.25$.

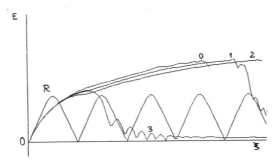

Fig. 14. Linearly decreasing driving frequency (pulse chirping) with damping $\Gamma=.05$. R: reference, 0 to 3: $K=7.10^{-4}$ to 10^{-3}.

Thus, the compensation of relativistic detuning can be very effective. This is also apparent if we use a different approach[26,27]: the first order equations for the amplitudes of Eq. (43) can be rewritten in canonical form as

$$dA_1/d\xi = -g_1 A_2 A_3,$$

$$dA_2/d\xi = g_2 A_1 A_3{}^*$$

$$dA_3/d\xi = g_3 A_1 A_2{}^* - i\,\Delta\omega - 3i\lambda\omega_p |A_3^2| A_3. \tag{56}$$

Then, properly adjusting the mismatch $\Delta\omega$, spatial solitary waves (i.e., waves which are time independent in the moving frame) can be obtained in which saturation is still the result of pump depletion.

Scaling laws for acceleration

The maximum longitudinal electric field in the plasma wave is given by the second equation (19). It is obtained when the saturation level matches the wavebreaking condition, independently of the physical mechanism involved. In the case of saturation by relativistic detuning, the corresponding interaction parameter is

$$\Lambda = (E_1/\omega_1)(E_2/\omega_2) = 3/16, \tag{57}$$

indicating an $|\lambda^2|$ scaling familiar to those who study plasma interactions. Now, using Eqs.(20), (21) and (46), the acceleration energy and length are

$$W_A = 2(\omega_0/\omega_p)^2 m_0 c^2, \qquad 1_A = 2(\omega_0/\omega_p)^2 c/\omega_p \tag{58}$$

In practical units (energies in M.e.V., lengths in cm, wavelengths in μm, electron densities in cm^{-3}) these quantities read, provided Eq. (59) is satisfied (i.e. $|\lambda_0^2 = 1.3\ 10^{17}\ W\mu^2\ cm^{-2})$,

$$W_A = 10^{21}\lambda_0^{-2} n_0^{-1}, \qquad 1_A = 1.2\ 10^{27}\lambda_0^{-2} n_0^{-3/2}. \tag{59}$$

These formulas were used to derive Table I[28], in which a cell is defined

Table I. Maximum energies and acceleration lengths

Laser wavelength (μm)	10(CO_2)	1 (Nd)	0,25 (KrF)	
Intensity (Wcm^{-2})	1.3×10^{15}	1.3×10^{17}	2×10^{18}	
n_e				Gradient
10^{15} cm^{-3}	10 GeV	1 TeV	16 TeV	
	4 m	400 m	6.4 Km	2.5 GeV/m
10^{16} cm^{-3}	1 GeV	100 GeV	1.6 TeV	
	13 cm	13 m	210 m	6 GeV/m
10^{17} cm^{-3}	100 MeV	10 GeV	160 GeV	
	4 mm	30 cm	6.4 m	25 GeV/m
10^{18} cm^{-3}	10 MeV	1 GeV	16 GeV	
	0.13 mm	1.3 cm	21 cm	60 GeV/m
10^{19} cm^{-3}		100 MeV	1.6 GeV	
		0.4 mm	6.4 mm	250 GeV/m

by the values of the electron density and the type of high power laser which is employed.

Observe the upper right hand corner of Table I: it seems impractical, even in the remote future, to accelerate electrons up to 16 TeV via the interaction of an excimer laser (KrF, λ_o= 0.25μm) with a low density plasma over several kilometers with such a high intensity. However, figures in the lower left corner represent exactly the same physics thanks to the scaling laws: a 10 MeV acceleration energy over a fraction of a millimiter, via the interaction of a high intensity Neodymium glass (λ_o = 1 μm) or CO_2 laser (λ_o = 10 μm) with a small size dense plasma, is within the reach of all laboratories involved in laser-plasma interactions.

The limited acceleration length is due to the phase slippage of the accelerated particle. It was shown in the section "The surfatron" that in the presence of a transverse magnetic field, the Surfatron effect provides phase locking so that the acceleration pay proceed indefinitely. There is no need then for the maximum possible laser intensity. The requirement on laser power is somewhat relaxed in such conditions. The situation can be improved even further by using laser beams propagating at an angle instead of colinearly[29]. Another improvement was proposed by Tajima[30], who noted that phase locking can result from obliquely propagating laser beams inside a plasma fiber: since the plasma refractive index is smaller than 1, a plasma with an intermediate electron density, embedded inside a larger density blanket, provides total reflection at the interface just as in the case of an optical fiber.

Preliminary experiments

High energy electrons of a few MeV energy were observed long ago as a by-product of laser-plasma interaction experiments. In most cases, a visible or infrared nearly single line high power laser was aimed at a solid surface. It is difficult to identify a given process in such a brute force action: many interactions take place at the same time and compete in the energy transfer from the field to the dense, hot and usually inhomogeneous resulting plasma. However, it has been possible to interpret some data on electron energy spectra in terms of laser-induced acceleration[31], by means of comparison with numerical simulations.

Specific beat wave experiments have been designed using CO_2 lasers, which can easily emit 2 lines (wavelengths 9.6 and 10.6μm) whose frequency difference matches the plasma frequency at a density of 10^{17} cm^{-3}. The objectives of the work were:

i) to observe the driven plasma wave and evaluate its amplitude;
ii) to accelerate electrons coming from a source outside the plasma.

Early attempts at U.C.L.A. on the first point were made using 2 counter-propagating laser beams along the axis of a θ-pinch[32]. More recently, 2 laser beams propagating in the same direction were used to transversally irradiate a high pressure He arc. Thomson scattering of a ruby laser line is the canonical diagnostic. From the scattering spectrum, a longitudinal electric field

$$0.3 < E_D < 1 \text{ GeV/m}$$

was inferred[33].

A different approach for verifying electron acceleration by a resonant beat wave was implemented in Canada by F. Martin and co-workers[34]. In this experiment, a two frequency high power CO_2 laser is used in three different ways:

i) it produces the plasma by breaking down a supersonic gas jet;
ii) it drives the beat wave;
iii) it creates an electron emitting auxiliary plasma by impact onto a nearby aluminum target.

A small group of these electrons in a narrow energy band around 0.64 MeV is selected by a focusing dipole magnet, then sent through the plasma in the same direction as the beating laser beams. Finally, the electrons enter an analysing spectrometer (Fig. 15). Preliminary measurements have evidenced electron energies up to 2 MeV, indicating acceleration by about 1 GeV/m over 1.5 mm. This experiment also shows that a laser created plasma might be a very convenient medium for electron acceleration. Gas breakdown through multiphoton ionization has created fairly uniform plasmas[35]. Other possibilities are foils exploding after laser impact or long (~1cm) plasma plumes obtained by focusing a laser beam onto a metal surface with a cylindrical lens[36].

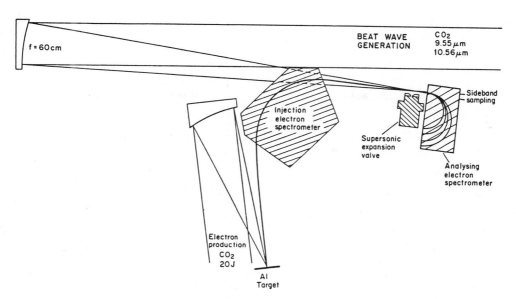

Fig. 15. Experimental setup for electron acceleration, after Ref. (34).

LASER-WIGGLER BEAT WAVES

Wave coupling and cut-off

In the free electron laser (F.E.L.) operating either in the Compton[37] or the Raman regime[38], laser-wiggler beating in presence of a relativistic electron beam has been successfully used. Inside a plasma, the same kind of wave coupling as in the F.E.L. is expected, together with a longitudinal wave generation similar to the one obtained by beating between two laser waves.

The wave couplings we are interested in deal with two high frequency modes plus a low frequency one. In the usual beat wave scheme, the low frequency oscillation is the plasma wave. Whenever a wiggler with wave-length λ_2 takes part in the process, it is obviously the low (actually zero) frequency mode. The E.M. wave and the plasma wave have the same high frequency ω_1; the only way they can match is in presence of a relativistic electron beam.

Consider a motionless neutral plasma with density n_0 in which a beam of electrons with density n_{Ob}, plasma frequency ω_{pb}, and velocity u_b (close to c) propagates in the same direction as a laser wave. The longitudinal wave dispersion relation is

$$\varepsilon(\omega_p, k_D) = 1 - (\omega_p/\omega_1)^2 - \omega_{pb}^2/\gamma_b^2 \ (ku_b - \omega_1)^2. \tag{60}$$

Due to the existence of two branches in the Brillouin plot (Fig. 16) two kinds of couplings are possible between the laser (wavenumber k_1), the wiggler (wavenumber $k_2 = 2\pi/\lambda_2$) and the plasma wave (wavenumber $k_D = k_1 + k_2$). Then:

329

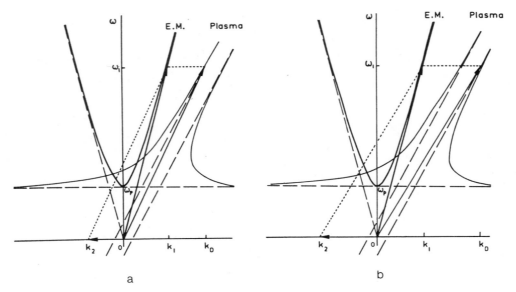

Fig. 16. Matching of electromagnetic and plasma waves: (a) inverse Cerenkov
regime; (b) Cerenkov regime.

i) either the phase velocity is larger than the beam velocity
(inverse Cerenkov regime)

$$u_b = u_R - \omega_{pb}/\gamma_b k_D (1-\omega_p^2/\omega_1^2)^{1/2} = \omega_1/(k_1+k_2)-\omega_{pb}/\gamma_b k_D (1-\omega_p^2/\omega_1^2)^{1/2}$$

(61)

or

ii) the beam velocity is larger (Cerenkov regime)

$$u_b = u_R + \omega_{pb}/\gamma_b k_D (1-\omega_p^2/\omega_1^2)^{1/2}.$$

(62)

In both cases the Lorentz factor of the beat is given by

$$\gamma_R^2 = (1-u_R^2/c^2)^{-1} = (k_1+k_2)^2/((2k_1+k_2)k_2-\omega_p^2/c^2)$$

(63)

which, using the plasma dispersion relation for E.M. waves, implies a
divergence for

$$\omega_p^2 - 2c\omega_1 k_2 + c^2 k_2^2 = 0.$$

(64)

In other words, there exists a critical density associated with a
cut-off frequency

$$n_{0c} = \varepsilon_0 m_0 (2\omega_1 - ck_2)ck_2/e^2 \qquad \omega_1 \geq \omega_{1off} = (\omega_p^2 + c^2 k_2^2)/2ck_2.$$

(65)

The Lorentz factor depends upon the plasma frequency (i.e., the density)
and upon the laser frequency, as depicted in figures 17a and 17b respec-

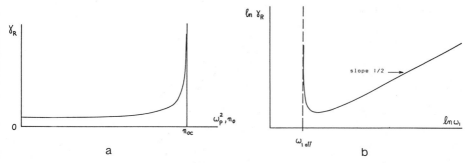

Fig. 17. Lorentz factor for the moving frame: (a) versus plasma density; (b) versus laser frequency.

tively. At a given plasma density and wiggler wavelength, the minimum value of the Lorentz factor and corresponding laser frequency are

$$\gamma^2_{R\ min} = 1 + \omega^2_p /c^2 k^2_2 \qquad \omega_{1\ min} = \omega_p (1+\omega^2_p /c^2 k^2_2). \qquad (66)$$

In ref. (18), a second order differential equation was derived in the moving frame to describe the growth and saturation of plasma waves resulting from the beating of two laser beams. The same procedure applied to the laser wiggler beat leads to

$$d^2 n'/dz'^2 + k'^2_D n' = -2|n'_0/\gamma^2_R + (\omega_1/\omega_p)^2 n'_{ob}/\gamma^2_R|k'_D \Lambda \cos k'_D z'. \qquad (67)$$

The first term on the right hand side is the almost non-resonantly driven oscillation of the motionless plasma. The second term is the resonantly driven oscillation in the beam. In Λ, the vector potentials are

$$A_1 = E_1/\omega_1 \quad \text{laser (E field } E_1 \text{), } A_2 = B_2/k_2 \text{ undulator (B field } B_2 \text{).}$$
$$(68)$$

As k_2 is small, A_2 can easily be made large; thus, when relativistic electrons pass through it, an undulator is equivalent to a very intense electromagnetic wave, and the laser intensity required to accelerate electrons up to a given energy decreases accordingly.

Inverse Cerenkov regime: electron acceleration

Assuming that the saturation is due to relativistic detuning, the maximum electric field amplitude in the plasma wave is obtained after some algebra[39] as

$$E_D = (cm_0 \omega_p /e)|(\omega_p/\omega_1)+(\omega_1/\omega_p \gamma_R)|(16\Lambda/3)^{1/3}, \qquad (69)$$

with the wave breaking condition

$$(16\Lambda/3)^{1/3} = \gamma_R \omega^2_1/(\gamma_R \omega^2_p + \omega^2_1). \qquad (70)$$

The acceleration energy and length are

$$W_A = 2m_0c^2|(\omega_p/\omega_1)^2\gamma_R^2+\gamma_R|(16\Lambda/3)^{1/3}, \qquad 1_A=2c\gamma_R^2/\omega_1.\qquad(71)$$

The gradient is then

$$W_A/1_A = m_0c\omega_1|(\omega_p/\omega_1)^2+1/\gamma_R|(16\Lambda/3)^{1/3}.\qquad(72)$$

Given k_2, this is a function of the laser frequency, as shown in Fig. 18.

The gradient grows proportionally to the (1/6)th power of the laser frequency when the latter tends to infinity. Actually, the acceleration length decreases to asymptotic limit $1/K_2$, which is obviously too small. Unless some trick is used (e.g., the "surfatron"), one cannot fully exploit the possibilities of high gradients offered by the laser-wiggler beat scheme. However, setting the acceleration length equal to a reasonable number of the undulator wavelengths (e.g. 10), one still obtains interesting results, which are presented in table 2. Energy evaluations are made for a value 1/20 of the interaction parameter Λ, which ensures that a large number of electrons will be trapped in the longitudinal wave. The magnetic field amplitude in the wiggler is 0.6 Tesla, a quite standard value: accordingly, the laser intensities are indeed moderate. Conditions in the table are close to the cut-off value of the laser frequency.

The laser-wiggler beat wave in presence of an electron beam is also able to accelerate charged particles directly, thanks to the ponderomotive potential. Indeed, on the right hand side of the second order equation of the second order equation of motion, the forcing term can also be considered as an equivalent longitudinal electric field, whose amplitude is

$$E_{eq} = 2(e/m_0\gamma_R)k_DA_1A_2 = 2(\omega_p m_0c/e\gamma_R)\quad,\qquad(73)$$

and whose periodicity is the same as that of the driven plasma wave, if any. If no plasma wave is to be generated, the corresponding accelerating mechanism is the so called inverse free electron laser[40] (I.F.E.L., a rather misleading designation). The field E_{eq} is proportional to Λ, whereas E_D varies as $\Lambda^{1/3}$. The two are equal for a critical value

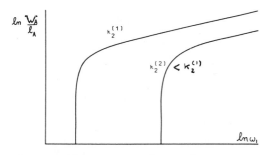

Fig. 18. Accelerating gradient versus laser frequency at a given wiggler wavelength.

Table 2. Acceleration by laser-wiggler beat wave in a plasma

(Wiggler: $\lambda_2 = 10cm$, $1_A = 1m$, $B_2 = .6T$)

LASER TYPE	CO_2	Nd	KrF
Wavelength m	10	1	0.25
Intensity Wcm^{-2}	5×10^{10}	5×10^{12}	8×10^{13}
Plasma density cm^{-3}	2×10^{15}	2×10^{16}	8×10^{16}
γ_R	540	1.7×10^3	3.4×10^3
W_A GeV	0.6	2.3	7

$$\frac{\Lambda}{c} = (2/3)^{1/2} (\omega_p {}_R / \omega_1 + \omega_1 / \omega_p)^{3/2} . \tag{74}$$

It turns out that Λ is exceedingly large . For $\Lambda \leq \Lambda_c$, E_D is larger than E_{eq}. The presence of a plasma thus greatly enhances the accelerating power of the I.F.E.L. Since the $\Lambda^{1/3}$ variation results from the same saturation mechanism via relativistic detuning, the 2 laser beat wave scheme also rates better than the inverse free electron laser for electron acceleration, at least for available and predictable laser intensities.

An advantage of the laser wiggler over the 2 laser beat wave is a laser intensity less by a factor 10^4 in order to obtain a given interaction parameter. However, in the conditions of table 2 the gradients would be higher for two laser beating with the same interaction parameter. Another difference lies in the Lorentz factor; it depends upon two quantities: the laser frequency and the plasma density. In the case of a laser wiggler beat wave there is a supplementary degree of freedom associated with the undulator wavelength. The electron energy can thus be tuned by varying the laser frequency.

The Cerenkov regime

In this case, wave coupling takes place on the middle branch of the longitudinal dispersion relation (Fig. 16b). Then the Lorentz factor γ_b of the beam varies with respect to laser frequency ω_1 in much the same way as γ_R (Fig. 17b). For large ω_1 an approximation is

$$\gamma_b \cong \omega_{pb} / 2ck_2 + (\omega_1 / 2ck_2)^{1/2} . \tag{75}$$

Two consistent sets of parameters are listed in table 3 (n'_{ob} is the electron density of the beam in the moving frame).

Radiation amplification may result from Raman scattering just as in an ordinary F.E.L. Now, however, the longitudinal waves represented by the middle branch of the dispersion relation have a negative energy. These waves fulfill the condition for the onset of another mechanism which is known among

Table 3. Two cases of radiation amplification in the Cerenkov regime

n'_{ob} (cm^{-3})	λ_2 (cm)	γ_R	ω_1 (s^{-1})	$\hbar\omega_1$ (e.V.)
10^{13}	10	10^3	4.10^{16}	20
10^{14}	1	10^4	4.10^{19}	2.10^4

plasma physicists as an explosive instability[41]. However, wave vector couplings in the Raman F.E.L. and in the explosive regime are fundamentally different, as shown in Fig. 19.

The three coupled waves obey equations similar to Eq. (43), but in which $dA_1/d\xi$, $dA_1/d\xi$ and $dn'/d\xi$ are positive. This allows the simultaneous growth of all three amplitudes. Such a system of equations can be solved analytically or on a computer for given initial conditions, e.g., large $A_2(0)$, small $A_1(0)$ and $n'(0)=0$. A divergence is found to occur at a time proportional to the reciprocal of the initial amplitude $A_1(0)$. Saturation may be due to finite length effects which act as damping[42], or to relativistic cubic detuning subsequent to the growth of the plasma wave amplitude. This is basically the regime which occurs in R.F. Klystrons. Also note that the explosive instability is similar to the behaviour of recently found solutions of the F.E.L. dynamical equations[43,44], in which laser light is amplified while electron bunching increases. A random noise can start the process (known as Self Amplification of Spontaneous Emission: S.A.S.E.).

CONCLUSIONS

Lasers for particle acceleration

Particle acceleration entails two kinds of requirements for the laser: those imposed by the needs of high energy physics, and others which result from the beat wave mechanism. Besides high energy, the accelerators should provide a sufficient number of expected events thanks to a high luminosity. Remember, this parameter is proportional to f (see Eq. 1). Therefore, a convenient repetition rate is 1 kHz, a condition the laser should certainly fulfill.

A high energy particle accelerator is obviously expensive. Routine operation is also costly. It is important that the machine be as efficient as possible. To this end, the overall laser efficiency has to be over 10%.

The beat wave machanisms outlined in Section "Resonant beat waves" lead to further requirements on wavelenght, power, and pulse duration. Equations (49) and (46) show that:

  i) the electric field E_0 in the plasma wave which determines the acceleration gradient increases as the square root of the plasma density, and
  ii) the Lorentz factor γ_R associated with the phase velocity is

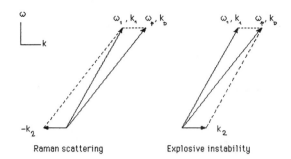

Fig. 19. In Raman back scattering, a pseudo photon (wiggler) combines
with a quantum of longitudinal oscillation (plasmon) in order
to create a photon (laser); in an explosive instability, a
plasmon decays into a photon and a pseudo photon.

proportional to ω_o/ω_p.

One wants high values for both E_D and γ_R, which implies a dense plasma and
consequently a short laser wavelength.

High laser power is also needed. In beat wave generation the saturation
amplitude, and hence the acceleration energy W_A, turns out to scale as
$(|\lambda^2)^{1/3}$. This constraint is somewhat relaxed in the "surfatron" and laser
wiggler beat wave schemes. However, considering the values in Tables 1 and
2, it appears difficult to match the focal volume of an optical system to
the acceleration lenght. Now, light self-focusing was evidenced in numerical
simulations[45]. The mechanism, which arises from either ponderomotive or
relativistic effects, provides high intensity over long distances. In both
cases the laser power (not the intensity!) has to exceed a wavelength
dependent threshold.

Finally, in Section "Compensation of relativistic detuning", it was
shown that the pulse duration should not be larger than a few nonlinear
periods. In practice, this condition requires 1-50 ns pulses.

The above requirements: high repetition rates over long periods of
time (days or months), 10% efficiency, short wavelength, high power, and
picosecond pulses, appear rather contradictory. No existing laser meets all
of them, as can be seen in Table 4.

The KrF laser offers some appealing features. The main issue is how to
efficiently extract the pump energy with picosecond pulses, a thus far unsolved problem. Angle multiplexing as used in Inertial Fusion is conceivable.
However, in recombining the beam one should be very careful about coherence,
which is essential in driving the plasma wave. Table 5 gives (after J. J.
Ewing[46]) the main properties of both a KrF and a CO_2 laser designed for
proof-of-principle experiments for a 10 TeV electron accelerator.

Preliminary experiments in the U.S. (U.C.L.A.) and Canada (I.N.R.S.)
have been successfully carried out using CO_2 lasers. Further attempts with

Table 4. State-of-the-art lasers and accelerator requirements

Laser	λ (μm)	Small λ	Picosecond Pulses	Energy 10^2–10^4 J	Rep. Rate 1 KHz	Efficiency $\geq$ 10 %
CO_2	10	NO	Possible	Proven	Possible	Proven
HF	2	NO	NO	Possible	Possible	Proven
Nd	1	YES	Proven	Proven	NO	NO
KrF	0.25	YES	Questionable	Possible	Possible	Possible

Table 5. Candidate lasers for a 10 TeV accelerator

Assume: laser/beat wave conversion efficiency is 25%

	CO_2	KrF
$\lambda_2 - \lambda_1$	1 μ m	37 Å*
plasma electron density	10^{17} cm^{-3}	4×10^{18} cm^{-3}
γ_R	10	68
1_A	0.6 cm	3.8 cm
Power for self-focusing	0.47 TW	21 TW
Pulse duration	3 psec	3 psec **
Energy/pulse	1.4 J	63 J.
Total length	660 m	104 m

(*) feasible by Raman shift in H_2; (**) DREAM!

CO_2 and Nd glass lasers are planned in Europe and in Japan.

Acceleration of elementary particles by laser-plasma interactions has been demonstrated on a very small scale: 1 GeV/m over 1mm only. It is being seriously considered by high energy physicists as a very promising way to reach energies beyond a few TeV. However, the subject is still in its infancy. The accelerators of the next generation will be designed and built by extrapolating known and reliable techniques. This gives us about 20 years, beginning now:

  i) to investigate all the physics relevant to laser-driven accelera-
     tion of particles in plasmas, and
 ii) to design laser sources suited to the job.

If one looks back at the progress in the physics and technology of high-power lasers designed for Inertial Fusion, one sees an increase in power by 6 orders of magnitude over the past 20 years. This is indeed a remarkable achievement. There is no doubt that, provided the demand and the motivation exist, a similar evolution will occur in the future: by A.D. 2007, laser properties could be close enough to the requirements of accelerator physics, to meet the needs of a future generation of machines.

REFERENCES

1. A. M. Ampère , "Théorie mathématique des phénomènes électrodynamiques uniquement déduite de l'expérience", Paris (1827)
2. M. Stanley Livingston, "High Energy Accelerators", Interscience (1954)
3. G. A. Voss, in: "The Challenge of Ultra-high Energies", ECFA-RAL Oxford (1982)
4. E. Eichen, I. Hinchlife, K. Lane. C. Quigg, Rev. Mod. Phys.,56:579 (1979)
5. U. Amaldi, Workshop on Advanced Accelerator Concepts, Madison, August 1986, to appear as A.I.P. Proceedings.
6. B. Richter, in: "Laser Acceleration of Particles", A.I.P. Proceedings n. 130 (1985)
7. U. Amaldi, Nucl. Inst. Met. A243:312 (1986)
8. T. Tjima and J. M. Dawson, Phys. Rev. Lett.,43:267 (1979)
9. T. Katsouleas and J. M. Dawson, Phys. Rev. Lett.,51:392 (1983)
10. R. Sugihara, S. Takeuchi, K. Sakai, M. Matsumoto, Phys. Rev. Lett., 52:1500 (1984)
11. A. Bambini, A. Renieri, Lett. Nuovo Cimento,21:239 (1978)
12. J. M. Manley, H. E. Rowe, Proc. I.R.E. 47:2115 (1959)
13. R. Bingham, C. N. Lashmore-Davies, Nucl. Fusion,16:67 (1976)
14. C. S. Liu, M. N. Rosebbluth, R. White, Phys. Fluids,16:1211 (1974)
15. B. I. Cohen, A. N. Kaufman, K. M. Watson, Phys. Rev. Lett.,29:581 (1972)
16. S. J. Karttune and R. R. E. Salomaa, Phys. Rev. Lett.,56:604 (1986)
17. D. C. Montgomery, Physica,31:693 (1965)
18. M. N. Rosenbluth e C. S. Liu, Phys. Rev. Lett.,29:701 (1972)
19. J. T. Mendonça, J. Plasma Phys.,34:115 (1985)
20. C. M. Tang, P. Sprangle and R. N. Sudan, Appl. Phys. Lett.,45:375 (1984)
    Phys. Fluids,28:1974 (1985)
21. J. P. Matte, F. Martin and P. Brodeur, INRS preprint (1986)
22. J. L. Bobin, Workshop on Advanced Accelerator Concepts, Madison August 1986, to appear as A.I.P. Proceedings
23. Jordan and Smith,"Nonlinar differential Equations", Oxford (1977)
24. F. Martin and J. P. Matte, private communication
25. C. Bordé, private communication
26. K. Mima et al., Phys. Rev. Lett., 57:421 (1986)
27. C. J. McKinstrie, D. F. Dubois, Phys. Rev. Lett., 57:2022 (1986)
28. J. L. Bobin, Ann. Phys. Fr., 11:593 (1986)
29. F. F. Chen, in: "The Phisics of Ionized Gases", S.P.I.G. 1984, World Scientific (1985)
30. T. Tajima, Laser and Particle Beams, 3:351 (1985)
31. C. Joshi, T. Tajima, J. M. Dawson, H. A. Baldis, N. A. Ebrahim, Phys.

Rev. Lett., 45:267 (1981)

32. B. Amini, F. F. Chen, Phys. Rev. Lett., 53:1441 (1984)

33. C. E. Clayton, C. Joshi, C. Darrow and D. Umstadter, Phys. Rev. Lett., 54:558 (1984)

34. F. Martin, P. Brodeur, J. P. Matte, H. Pépin and N. Ebrahim, S.P.I.E. Proc., 664:20 (1986)

35. A. E. Dangor et al., C.L.I.C. note 29, C.E.R.N. (1986)

36. F. Amiranoff, C. Labaune, private communication

37. J. M. J. Mady, J. Appl. Phys., 42:1906 (1971)

38. V. Granastein et al., Appl. Phys. Lett., 30:384 (1977)

39. J. L. Bobin, Optics Comm., 55:413 (1985)

40. C. Pellegrini, in: "The Challenge of Ultra-high Energies", ECFA-RAL Oxford (1982)

41. see e.g. J. Welland, H. Wilhelmsson, "Coherent Nonlinear Interaction of Waves in Plasmas", Pergamon (1977)

42. D. Pesme et al. Phys. Rev. Lett., 31:203 (1973)

43. R. Bonifacio, F. Casagrande, Optics Comm., 50:251 (1984)

44. J. B. Murphy, C. Pellegrini, Nucl. Instr. Methods, A 237:159 (1985)

45. C. Joshi, W. Mori, T. Katsouleas, J. M. Dawson, J. M. Kindel and D. W. Forslund, Nature, 311:525 (1984)

46. J. J. Ewing, Workshop on Advanced Accelerator Concepts, Madison August 1986, to appear as A.I.P. Proceedings.

LASERS IN NUCLEAR PHYSICS

E. W. Otten

Institut für Physik, Universität Mainz

6500 Mainz, Fed. rep. of Germany

1. INTRODUCTION

The application of lasers in fundamental nuclear research has been devoted predominantly to the investigation of nuclear structure of unstable isotopes through the measurement of isotope shift and hyperfine splitting of the atomic spectrum of the nuclei in question. From the isotope shift one deduces the change of nuclear charge distribution – or more precisely the change of the mean-squared nuclear charge radius $\delta <r^2> -$, whereas the hyperfine structure yields the spin (I), the magnetic moment (μ_I) and the electric quadrupole moment (Q_s) of the nucleus. These four nuclear properties form the cornerstones for any quantitative theory of nuclear structure. In this respect, systematic measurements, throughout the nuclear chart of elements are particularly important. This implies experiments on extended series of isotopes which reach far out into the region of unstable and short-lived nuclei.

Modern mass separators, on-line with a powerful accelerator, enable the production and separation of typically 20 – 30 isotopes of almost any element. For many years the ISOLDE facility at CERN has been the leader in this field[1].   However, the low production cross-section for isotopes far off stability and their short lifetimes force the experiments to be done on extremely small quantities of atoms, many orders of magnitude below the optimum of forming a vapour density of some reasonable optical thickness. Therefore, only spectroscopic methods of the highest sensitivity can cope with these unfavourable conditions. In addition, high, Doppler-free resolution is required in many cases in order to resolve the tiny hyperfine structure. Moreover, the experimental techniques have to be adapted to the typical on-line conditions, requiring fast, efficient and safe handling of the radioactive sample.

Although a few isolated on-line optical experiments could be performed already in the "age of spectral lamps", for instance by the so-called RADOP method (radiactive detection of optical pumping)[2] (see references therein), the field has been advanced principally by the advent of tunable lasers, which permitted simultaneous solution of the problems of sensitivity and resolution by a number of elegant methods. Some of them have been particularly developed for this purpose, whereas others could be adapted to the conditions met with on-line facilities. The first series of on-line laser spectroscopic experiments was initiated in 1975 by a measurement on a series of unstable Na-isotopes[3]. In this experiment a collimated atomic beam was spin-polarized by optical pumping (OP) with a resonant laser beam at right angles, thus reducing the Doppler-width. The pumping signal was monitored downstream by analyzing the hyperfine states by means of a Stern-Gerlach magnet. The method was applied later on to all alkali elements, including the first atomic spectroscopy results on francium. These experiments covered altogether about 100 isotopes. Also, Doppler-limited laser fluorescence spectroscopy from atomic vapours in resonance cells could successfully be applied on-line to isotopes of Hg an Cd[4,5,6]. The activities in this field up to around 1980 have been summarized in several reviews[7,8,9]. These experiments yielded important discoveries, such as sudden transitions of nuclear shapes from spherical to deformed, and the coexistence of different nuclear shapes in one and the same nucleus at almost degenerate energies. This work will be briefly discussed together with more recent results in section 3 of this paper.

The center section will be devoted to the description of two more recent experimental methods in the field of on-line optical spectroscopy, namely collinear laser spectroscopy and multiphoton resonance ionization spectroscopy (RIS). These new techniques have enlarged the range of accessible elements very much, so that as of now about 400 unstable nuclei (out of the total of 1500 known ones) can be investigated spectroscopy of unstable isotopes is closely connected to the more general problem of trace analysis.

To conclude this introduction, let us briefly recall the basic concepts of isotope shift and hyperfine structure and their relation to nuclear parameters. For heavy nuclei, the isotope shift of a s-electron, which is the only one to penetrate the nucleus, is dominated by the isotopic change of the nuclear charge distribution, as can be seen from Fig. 1. The (almost) constant electronic density builds up a parabolic potential at the site of the nucleus. Folding this potential with the nuclear charge distribution yields from its first, constant therm, $V_e(0)$, the Coulomb interaction as it would be expected from a point-like nucleus. The second, parabolic term yields the finite size correction, being proportional to the mean-squared nuclear charge radius $\langle r^2 \rangle$. In a particular optical transition the shift between isotopes A and A' is then proportional to the product of the change of the mean-squared radius $\langle r^2 \rangle^{AA'}$ and the change of the electronic density $\Delta|\Psi(0)|^2$ in the electronic transition:

$$\delta\nu_{\text{Fn.r.}}^{AA'} = -\frac{2\pi}{3} Z e^2 \Delta|\Psi(0)|^2 \langle r^2 \rangle^{AA'} \tag{1}$$

Eq. (1) is a non-relativistic expression of the so-called field shift. In addition, one observes the well-known mass shift

$$\delta\nu_m = \frac{M (A - A')}{A A'} \qquad (2)$$

which is difficult to calculate in complex spectra. The techniques for separating the mass from the field shift, as well as relativistic corrections to the latter, can not be discussed within the framework of this paper; we refer to the topical literature[10].

The hyperfine splitting of an electronic state with angular momentum J depends on the nuclear parameters I, μ_I and Q_s through the well-known formula

$$W (F) = A K/2 + B \frac{3}{4} \frac{K (K+1) - I (I+1) J (J+1)}{2 (2I-1) (2J-1) I \cdot J} \qquad (3)$$

with K = F(F+1) - I(I+1) - J(J+1). $A = \mu_I H_e(0)/I \cdot J$ is the magnetic dipole and $B = e Q_s \phi_{jj}(0)$ is the electric quadrupole constant. $H_e(0)$ is the magnetic field, $\phi_{jj}(0)$ the electric field gradient produced by the electrons at the site of the nucleus. For the calculation and calibration of hyperfine fields see the topical literature[11,12].

2. ON-LINE LASER SPECTROSCOPIC TECHNIQUES

2.1 Cross-Section, Excitation Rates and Efficiency

This methodological part of this paper will be introduced by a few basic considerations characterizing the laser spectroscopic situation encountered in these experiments.

Let the laser radiation drive transitions between atomic levels (1) and (2) which may (or may not) couple to a third level (or a group of levels) by spontaneous emission or relaxation rates Γ_{23}, Γ_{31} (compare Fig. 2). An atom with resonance frequency ν' in the lab frame will offer

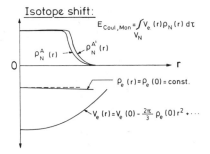

Fig. 1. Schematic drawing of a nuclear (ρ_n) and an electronic (ρ_e) charge distribution, the latter one for a s-wave function. In addition the potential (V_e) contributed by the electron is shown.

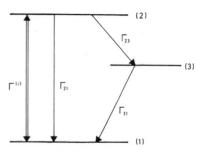

FIg. 2. Schematic level diagram for laser spectroscopy. The laser drives
transitions between levels (1) and (2) with the induced rate Γ^i.
Γ_{kl} are decay rates.

to a photon of frequency ν an absorption (or induced emission) cross-section, calculated in first-order perturbation theory to be

$$\sigma \, (\nu-\nu') \quad = \quad \frac{\lambda^2}{2\pi} \quad \frac{\Gamma_{21} \, / \, \Gamma_2}{1 + (\, 4\pi \, (\nu-\nu')/\Gamma_2)^2} \quad ; \tag{4}$$

λ is the wave length of the photon and $\Gamma_2 = 2\pi\Delta\nu_h$ the total decay rate of
level 2, which determines the homogeneous linewidth $\Delta\nu_h$ of the Lorentz
profile (4). At the resonance point $\nu = \nu'$, and under the assumption that
Γ_2 is comparable to Γ_{21}, the cross-section for an optical photon is of
the order

$$\sigma \, (\nu-\nu') \quad = \quad \Theta \, (10^{-9} \, cm^2). \tag{5}$$

This is an enormous dimension for an elementary system and explains
the effectivness of laser spectroscopy. Assuming for instance a laser
power of 10 mW/cm^2, corresponding to a photon beam intensity $j_{Ph}(\nu)$ of
the order of 10^{17}/cm^2 s, the induced transition rate $\Gamma^{(i)}$ already reaches 10^8/s, which is comparable to the spontaneous decay rate $_{21}$ of
allowed optical transitions. $\Gamma^{(i)} > \Gamma_{21}$ marks the regime of saturation
where the populations in the excited (2) and the ground level (1) states
are equalized in the absence of branching ($\Gamma_{23} = 0$). A single atom excited to saturation will hence emit about 10^8 fluorescence quanta per second, providing a strong signal. In an experiment on a single, trapped
Ba ion this fluorescence signal has even been observed by the naked eye[13].

In many cases, however, branching leads to OP into a third state (3)
which is metastable for the time of observation T, that is $T \cdot \Gamma_{31} \ll 1$.
This "silent" level can be one of the hfs levels of the ground state, for
instance. In these cases, the excitation emission cycle stops after a total number of fluorescence quanta equal to $(1 + \Gamma_{21} + \Gamma_{23})$ has been emitted per atom, on the average. Usually this number does not very much
exceed 1 Therefore, many attempts have been made to search for another
signature of optical resonance excitation which is more efficient and
less sensitive to background than the detection of fluorescence quanta.

RIS is one example, and collinear laser spectroscopy also has a number of variants with non-optical single detection (see below). Experience has taught us, however, that in the absence of OP and for long observation times, fluorescence detection is not easily beaten in sensitivity. Moreover, it is the easiest and most general detection method.

In case of Doppler broadening the cross-section is a convolution of a Lorentzian and a Gaussian (so-called Voigt profile):

$$\sigma_D(\nu) = \frac{(\lambda^2/2\pi)\ (\Gamma_{21}/\Gamma_2)}{\sqrt{\pi}\ \ \Delta\nu_D} \int_{-\infty}^{+\infty} \frac{\exp\left[-(\nu' - \nu_o)/(\Delta\nu_D)^2\right]}{1 + (4\pi\ (\nu - \nu')/\Gamma_2)^2}\ d\nu', \quad (6)$$

where ν_o is the resonance frequency for the atom at rest. For a thermal gas the $1/e$ half-width

$$\Delta\nu_D = (\nu_o/c)\ \sqrt{2kT/m} \quad (7)$$

is of the order of 1 GHz and exceeds the natural width by a factor of 100 to 1000. The cross section (6) is flattened by the same factor, as compared with the Doppler-free case ($\nu' = \nu_o$). That means that only a small fraction of the order $\Delta\nu_h/\Delta\nu_D$ of the atoms can participate at a given excitation frequency ν in the transition. Thus the Doppler-width spoil is not the only resolution, but also the sensitivity.

Let us examine the relation between spectroscopic resolution and sensitivity for the particularly simple and clear example of exciting a collimated, thermal atomic beam by a laser at right angles (see Fig. 3). This geometry obviously leads to the elimination of the atomic velocity in the direction of the light. It is also clear that resolution can only be gained at the cost of signal height, since the collimation cuts down the solid angle of the atomic beam observed. For a simple geometry as sketched in Fig. 3, the necessary integrations can be carried out analytically. For instance, in the absence of OP and for moderate light intensities below the saturation point, the rate of fluorescence photons emerging from the illuminated volume V is given by

$$R = j_{Ph}(\nu)\ n_V V \sigma_{Dc}(\nu), \quad (8)$$

where

$$n_V \sim n_o\ \frac{A}{4\pi r^2} \quad (9)$$

is the total density of atoms in V, whereas n_o is the atomic density in the oven. $\sigma_{Dc}(\nu)$ is the frequency-dependent cross-section (6), but now taken for the "collimated" Doppler width

$$\Delta\nu_{Dc} = \Delta\nu_D \sin\delta_o = \frac{\nu_o}{c} \sqrt{2kT/m}\ \sin\delta_o. \quad (10)$$

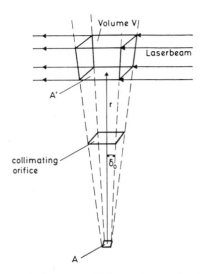

Fig. 3. Geometry for exciting a collimated atomic beam in the volume V. The beam emerges from an oven with orifice A.

The other symbols are explained in Fig. 3. The gain in resolution is given by the sine of the collimation angle δ_o, and the loss in density by the solid angle through which the orifice of the oven is seen in the observatione volume.

Let us assume now more realistically that in practice OP limits the number of fluorescence photons to about 1 for each atom passing V; furthermore, the laser intensity may be adjusted such that this limit is just reached at the resonance peak ν_o. Then the peak count rate is given by

$$R(\nu_o) \sim I \; \frac{A'}{\pi r^2} \, , \tag{11}$$

where I is now the total current of atoms leaving the oven and A' is the cross-section of the observation volume. Eq. (6) sets also an upper limit for any non-optical detection rate as in RIS, for instance, where each atom can give rise to one signal evento only and cannot be re-used many times as in fluorescence detection.

Assuming a typical collimation of the beam of 25 mrad in both directions, eq. (11) implies a geometric reduction factor of ~ 5000. Fluorescence photons may be detected with a solid angle of $\Delta\Omega/4\pi \sim 5$ % and a quantum efficiency of the photomultiplier of 10 %. From these numbers one calculates a total detection efficiency for an atom of only 10^{-6}. Nevertheless, this technique has proved to be very fruitful in off-line applications to long isotopic chains of Ca, Sr, Sn, Ba and Pb, including many unstable isotopes[14,15].

2.2 Resonance Ionization Spectroscopy (RIS)

Ionization of atoms and molecules by stepwise resonant excitation with two or more quanta is a well-established, quite generally applied method in trace analysis[16,17,18]. The method has earned its first acceptance in the field of unstable nuclei by an experiment at the Leningrad on-line mass-separator on a series of Eu isotopes $141 \leq A \leq 150$[19,20]. Fig. 4 shows the relevant part of the level scheme. Three photons from three different dye lasers lead from the atomic ground state via two relay states into the continuum. Ionization efficiency is optimized if all three transitions are satured. This requirement is easily met for the two first steps, but poses a problem for the third one, since the cross-section for photo-ionization into the plane continuum is of the order of 10^{-17} cm^2 only. Therefore, it is advisable to tune this last step to an autoionizing state; in complex spectra plenty of these states are available, just above the ionization limit. From equations (4) and (6) one finds that the excitation cross-section for such a state is the higher, the smaller is its autoionizing width (equivalent to $\Gamma_{23} \sim \Gamma_2$ in this case). The same purpose can be fulfilled by exciting a Rydberg state just below the ionization limit which can easily be field-ionized.

Hfs and IS are measured in the first step. The spectral width of the laser which serves this transition is therefore adjusted to its Doppler-width. Since the latter depends on the collimation of the atomic beam, one compromises between resolution and efficiency once again. The two other

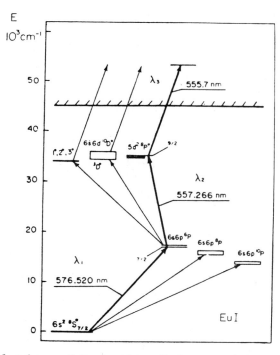

Fig. 4. Partial scheme of Eu levels and the transitions used for three-step laser photoionization[20].

lasers are run broadband (a few GHz wide) as to match steps 2 and 3 for all of the isotopes without tuning.

Fig. 5 shows some details of the apparatus used. The ion beam from the mass-separator is captured in an oven, from which it evaporates as atomic beam into the ionization region. The light intensity is amplified by a pair of mirrors providing multipassing of the laser through the atomic beam. The ions are then extracted, focussed onto a channeltron, and counted as a function of laser frequency. Fig. 6 shows a scan of [144]Eu, together with reference lines from stable isotopes and fringes from a calibrating interferometer. The group has extended these measurements recently to still lighter Eu isotopes and neighbouring elements, and proceeded to high resolution by using in the first step a narrow-band cw laser amplified in a pulse-excited dye[21].

The RIS technique is usually based on pulsed, synchronized lasers, preferably with high repetition rate. We quote five principal reasons for this choice:

1) The power necessary to saturate the intermediate and ionizing transition with broadband irradiation is only available from pulsed lasers.
2) The pulse width, of the order of 10 ns, is smaller than or at least comparable to the spontaneous decay times into any other atomic

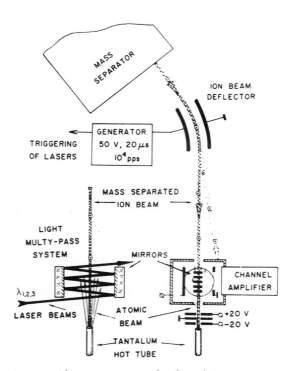

Fig. 5. Set-up for on-line resonance ionization spectroscopy at the Leningrad mass separator. The inset on the left shows the ionization region blown up[20].

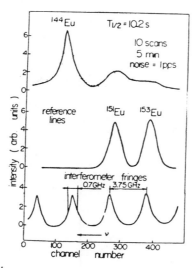

Fig. 6. Scan of the [144]Eu resonance by the RIS technique together with
reference signals from stable isotopes.
Bottom: Interferometer fringes for calibrating the laser frequency[20].

state. By running up the energy ladder very rapidly one therefore
prevents losing atoms on the way by OP.

3) The demand for a high pulse rate is self-explanatory in view of
   the high efficiency desidered. In the experiment discussed above
   a copper vapour laser was used to pump the dye lasers. Its repetition rate of the order of 10 kHz already comes close to the
   optimum at which each atom experiences at least one laser pulse
   during its transit time through the interaction region. An overall efficiency of $3 \cdot 10^{-4}$ was reached in the Eu experiment[20], which
   is a very good number in comparison to others.

4) A sharp (delayed) coincidence between the laser pulse and the ion
   signal allows us to cut down the background by the factor of the
   duty cycle of the laser, which was of the order of 10^{-4} in the
   Eu experiment; this resulted in a noise level of only 0.1 count/s[20].

5) Pulsed operation provides the very important advantage of combining a mass-separation with the RIS technique by a time-of-flight
   measurement. For this purpose, the acceleration voltages and the
   geometry of acceleration and drift regions of the ions are adjusted
   to each other such that a time focus is created at the site of the
   ion detector. In that focus the total time-of-flight is independent
   (to first or even higher order) of the spatial extension of the
   ionization region. Fig. 7 shows the time-of-flight spectrum obtained in an experiment on neutron-deficient Au isotopes at
   ISOLDE[22,18]. The peak of [186]Au is clearly separated from the
   stable [197]Au and also from the background of lighter molecular
   masses. The latter point was very important since the high laser
   power, necessary for the ionization step in Au, led also to non-
   resonant multiphoton ionization of rest gas molecules.

2.3 Collinear Laser Spectroscopy

2.3.1 The Standard Method with Fluorescence Detection

In the optical on-line experiments mentioned so far an ion beam has been converted by capture and reevaporation into thermal atoms, forming e.g. a vapour in a cell or a collimated beam. In the former case, chemical stability against reaction or adsorption at the wall is a restrictive condition; in the latter a large factor is lost by collimation if high resolution is required. The original idea behind collinear laser spectroscopy was simply to avoid these problems by directly using the mass-separated beam as the spectroscopic sample. Since at typical ion velocities of 10^7 cm/s the interaction time with the laser in a crossed geometry would shrink to about 10^{-8} s, a collinear superposition of laser and ion beam seemed favourable. In this way, interaction time and length are easily increased by a factor of 100, when observing fluorescence light from a pathlength of 10 - 20 cm, as can be achieved, e.g., by means of a cylindrical lens which images the beam onto the entrance slit of a light pipe, which in turn adapts the image to the geometry of the photocathode (see Fig. 8).

A surprising, crucial advantage of the collinear geometry has been found by Kaufman and published in his proposal[23] and independently by Wing at al.[24]: the spread of kinetic energy (δE) in the beam remains unchanged under electrostatic acceleration,

$$\delta E = \delta(mv^2/2) = mv\delta v = (mc^2/v^2)\Delta v_D \delta v_D = \text{const.} \qquad (12)$$

Therefore, the product of the average velocity v and the velocity spread δv, or, likewise, the product of the Doppler shift Δv_D and the Doppler-width δv_D, are constants of the motion. In other words: acceleration reduces the Doppler-width along the beam direction by a large factor

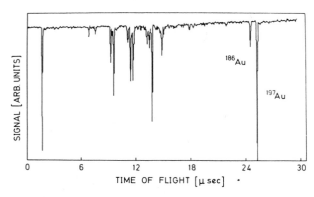

Fig. 7. Time-of-flight spectrum obtained in a resonance ionization mass spectrometry experiment (RIMS). ^{186}Au was collected in an atomic beam oven. Stable ^{197}Au was added to serve as a mass marker. The medium-mass ions are due to photoionization of rest gas and can be suppressed by setting an appropriate time window[18,22].

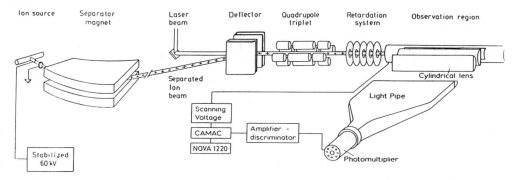

Fig. 8. Schematic of a standard collinear laser spectroscopy experiment
using fluorescence detection[25].

from its original value in the ion source. Assuming ideal starting conditions, i.e., thermal velocity distribution at the ion source, one observes
after acceleration a reduced Doppler-width of

$$\delta\nu_D(v) = 1/2 \ (kT/eU)^{1/2} \ \delta\nu_D(0). \tag{13}$$

The reduction factor is about 10^3 for a source temperature of 2000 K,
mass A = 100 and an acceleration voltage of U = 60 kV. Under these conditions the residual Doppler-width of the green barium resonance line at
535 nm, for example, would be about 1 MHz, far below the natural linewidth
of 19 MHz. In practice the linewidth ranges from 10 to 50 MHz, depending
on the the type and performance of the ion source, the stability of the
acceleration voltage, and the quality of the beam optics. The angular emittances of the laser and ion beams which deteriorate collinearity do not
contribute significantly to the linewidth, if they do not exceed the order
of 1 mrad.

The effect of velocity bunching increases not only the resolution but
also the sensitivity: since the total area under the resonance curve remains unchanged, the peak intensity rises by the same factor by which the
linewidth narrows.

Another very useful feature of the concept is the charge exchange
cell containing an alkali vapour, which neutralizes the beam in flight.
By this means, laser spectroscopy is significantly facilitated, since most
ions do not have resonance lines in the visible. It is presumed, of course,
that the charge exchange process does not disturb the velocity distribution of the beam. This in fact is guaranteed by the large charge-exchange
cross section of about $10^{-15} - 10^{-14}$ cm^2. Since this exceeds the kinetic
cross-section by 2 orders of magnitude, most exchange collisions are peri-
pherical, with little momentum transfer. As seen below, charge exchange
may also populate a metastable state of the atom, which can serve as the
lower spectroscopic state from which laser light is adsorbed. Collinear

laser spectroscopy owes its wide application nowaday to these four afore-
said qualities:

1) good adaptation to mass-separators
2) high resolution
3) high sensitivity due to velocity bunching
4) great versatility due to preparation of suitable spectroscopic
   states by charge exchange.

Fig. 8. shows in more detail the procedure of measurement for the
example of the collinear set-up at the ISOLDE[25], which was built with the
experience of pilot experiments on stable Na[26] and on unstable fission
isotopes of Rb and Cs[27]. The determination of the atomic transition frequency requires precise knowledge of the laser frequency, the acceleration voltage, and the atomic mass. Instead of measuring the two former
quantities independently, it is safer and easier to run alternatively
stable isotopes through the apparatus for calibration. For this purpose,
the separator magnet is switched periodically from one to the other mass,
and an appropriate correction voltage is fed to the post acceleration
stage just in front of the charge exchange cell. It generates a Doppler
shift such that the resonances of the different isotopes coincide in the
lab system, that is, for one and the same laser frequency. Scanning of
the resonances, finally, is achieved by adding another small voltage stepwise to the charge exchange cell. Thus it is sufficient to run the laser

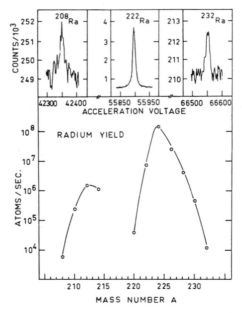

Fig. 9. Lower part: Yield curve of Ra isotopes separated at the ISOLDE
facility at CERN.
Upper part: Scanning signals from collinear laser spectroscopy
of some even Ra isotopes. The integration time is about 30s/chan-
nel for ^{208,232}Ra and a fraction of a second per channel for the
abundant ^{222}Ra [28].

at constant frequency in a stabilized but uncalibrated mode.

Fig. 9 shows a set of resonances for three even Ra isotopes, recorded in the manner described above[28]. In Ra the atomic resonance leads from a single, diamagnetic ground state $(7s^2 \ {}^1S_0)$ to an excited ${}^1P_1$ state, from which branching into a metastable ${}^1D_2$ state is only a few percent. Therefore, the atom emits many fluorescence quanta in the observation region before decaying into the principle). The total detection efficiency of an atom under these conditions has been found to be about 1%. At typical background levels of 10 kHz on the photomultiplier from stray flight, surrounding radioactivity, etc., measurements were feasible down to currents of about 10^4 atoms/s. The resonance of ^{208,232}Ra shown in Fig. 9 were taken under these ultimate conditions.

In complex spectra the sensitivity is cut down by the multiplicity of (metastable) fine structure and hfs levels. The example of the hfs

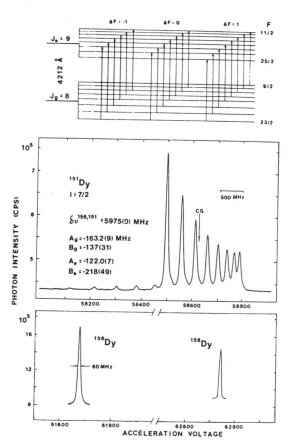

Fig. 10. Scan of the hfs of ^{151}Dy by collinear laser spectroscopy. The strong group of resonances belongs to $\Delta F = +1$ transitions, the weak one to $\Delta F = 0$ (compare level scheme on top). $\Delta F = -1$ transitions are too weak to be seen. Bottom: Reference signals from stable Dy isotopes[29].

spectrum of ^{151}Dy with $J_1 = 8$ and $J_2 = 9$ and $I = 7/2$ shows, nevertheless, that such cases can also be handled with an on-line mass-separator of sufficient intensity, although the detection efficiency per separated atom has now dropped to the range of 10^{-4} to 10^{-5} (see Fig. 10)[29].

The high resolution of the collinear method also pays in the case where several isomers are produced and separated simultaneously with the nuclear ground state. Fig. 11 shows the example of ^{122}In; 13 out or the 15 hfs components are resolved, arising from spins $I = 8, 5, 1$[30]. In this way, altogether 37 nuclear states could be investigated in the In chain $104 \leq A \leq 127$.

In a number of cases one can leave out the neutralization by charge exchange, considering the fact that ion sources (especially plasma sources) also produce ions in metastable states which display resonance lines in a convenient spectral range. One of the first collinear experiments was conducted this way on a metastable Xe^+ - beam[31]. The same technique was applied to a series of Eu isotopes at the Leningrad on-line mass separator[32].

Access to metastable states of noble elements is also provided by charge exchange with an alkali vapour. Consider for instance the case of noble gas Rn, which has a metastable state $(7s[3/2]_2)$ bound by $-3,9$ eV. This is very close to the ionization potential of a Cs atom in its ground state (see Fig. 12). Since charge exchange prefers resonant conditions, the transfer rate into the metastable state is muche stronger than into the ground state which is bound by 10.7 eV. This metastable state is connected by a group of lines in the red spectral range to levels of the

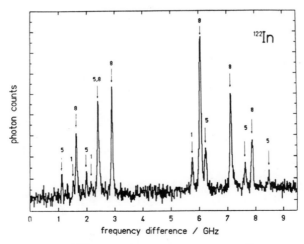

Fig. 11. Scan of hfs of ^{122}In in the transition $5p\ ^2P_{3/2} \rightarrow 6s\ ^2S_{1/2}$ ($\lambda = 451$ nm) obtained by collinear laser spectroscopy[30]. The hfs components of the nuclear ground state ($I = 1$) and the two isomers are marked by their spin values. Fluorescence is detected on the other doublet component down to the $5p\ ^2P_{1/2}$ state ($\lambda = 410$ nm). Stray light from the laser can thus be cut off by a colour filter.

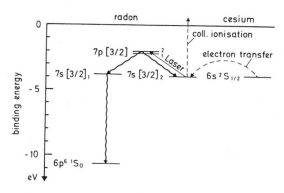

Fig. 12. Level diagrams of Rn and Cs. The metastable Rn state at -3.9 eV
is populated by resonant charge exchange with Cs. It is then excited by a laser followed by branched fluorescence decay. Also
shown is collisional ionization of the metastable state (compare
section 2.3.2b.)

$7p6p^5$ configuration, which are easily excited by a cw laser. In this way
spins, moments and charge radii of Rn isotopes could be measured for the
first time, ranging from ^{202}Rn to ^{222}Rn [33]. This example proves once more
how much charge exchange has enhanced the versatility of the method.

So far collinear laser spectroscopy has been applied to unstable isotopes of the following elements: Li, Rb, Sr, In, Sn, Xe, Cs, Ba, Sm, Dy,
Eu, Gd, Er, Ho, Yb, Hg, Tl, Pb, Rn, Fr, Ra. Experience tells us that it
may be applied to any element for which beams of good ion optical quality
and an intensity exceeding 10^4 to 10^7 particles/s are available. The lower
number applies to the simplest, the higher to very complex atomic spectra.

2.3.2 Variants of Collinear Laser Spectroscopy

Very far off stability, as well as for refractory elements, present
on-line mass-separators cannot provide the intensity required for the standard collinear technique. Most of the methodological development in recent
years has concentrated, therefore, on the increase of sensitivity of on-line
optical methods. Three out of the four variants of collinear laser spectroscopy which are discussed in the following aim at that end.

2.3.2.a Fluorescent Atom Coincidence Spectroscopy (FACS)

In the standard scheme the sensitivity is limited primarily by background from stray laser light, radioactivity, photomultiplier noise, etc.,
which add up to typically 10000/s. A group at the Daresbury on-line separator recently succeeded in suppressing this background by tagging true
fluorescence photons with the coincidence signal of the incoming beam particle which was detected downstream by a channeltron (see Fig. 13a). In
this way, they were able to measure the resonance transition of light Sr
isotopes at a count rate as less as 60 particles/s[34,35]. Fig. 13b shows

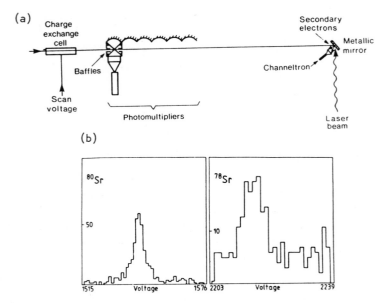

Fig. 13. a) Set-up for photon-atom coincidence detection in collinear laser spectroscopy[34].

b) Number of photon-atom coincidences from a fast beam of Sr atom as a function of the voltage scanning through the resonance[35].

the coincidence count rates for ^{80}Sr and ^{78}Sr as a function of the scanning voltage.

This elegant method requires a very clean beam, of course. Contamination by other isotopes or isobars will bury the signal in the background from accidental coincidences, especially since the coincidence window has to be as wide as ≈ 100 ns, because that is the time-of-flight of the particle through the observation region of the photomultiplier. In the Daresbury experiment this difficulty was overcome by carefully conditioning the surface ionization source. A possibility of getting clean beams will be offered by laser ion sources in the near future using the RIS technique. In a pilot experiment on stable Sr, a total ionization efficiency of 10% was recently reached[36]. Since these laser ion sources are pulsed, they will allow triple coincidence between the times of ion birth (t_0), of fluorescence detection (t_1), and of ion stopping (t_2). The foregoing discussion evidences the great impact with the development of new, sophisticated ion source separator combinations will have on the progress of on-line optical spectroscopy in the future.

2.3.2.b Detection by Collisional Ionization and Charge Exchange

As pointed out earlier, detection efficiency as well as background rejection may be improved by shifting the signal from photon to particle counting. In the scheme under discussion here, this is achieved by optical pumping into another electronic state of different binding energy. The OP effect is then sensed by the energy dependence of the collisional ionization or charge exchange cross-section. Such a scheme has been developed by Neugart et al.[37], and is explained in Fig. 12 for the case of Rn. The metastable $7s(3/2)_2$ state of Rn is populated by charge exchange with Cs. This state is then excited by a laser to a level of the 7p configuration, the decay of which branches into the tightly bound ground state by a cascade. Downstream of the OP region the beam passes another gas filled cell (containing preferably an electronegative gas), in which the remaining metastable fraction of the beam is ionized by collisions (see Fig. 14a). The reionized fraction is finally deflected onto a detector giving the signal. The pilot experiment was actually performed on the analogue case of a stable Kr beam, for which the reionization signal is shown in Fig. 14b in comparison with the standard fluorescence signal. The former, which has the character of a flop out signal, is a factor of 1000 stronger[37]. Quite recently the technique has been applied at ISOLDE to chains of Xe ($140 \leq A \leq 146$) and Rn ($223 \leq A \leq 226$)[38].

A similar idea was realized independently by Silverans et al. in a reversed sense[39]. In their scheme Sr^+ ions were pumped from the 5s ground state via excitation to 6p and subsequent branching into a metastable 4d state (compare the basic level diagram in Fig. 2). The pumping effect was monitored by the difference in neutralization probability of the two ionic states in a subsequent charge exchange cell. This technique was applied recently at the collinear set-up at ISOLDE and permitted measuring ^{100}Sr far out on the wing of the yield curve[40].

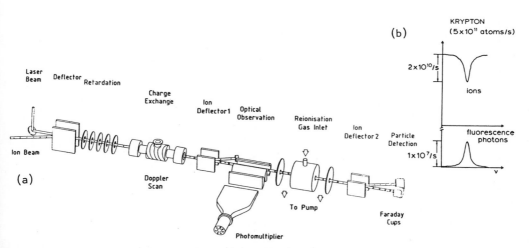

Fig. 14. a) Set-up for collisional ionization detection in collinear laser spectroscopy
b) Resonance signals from a beam of $5 \cdot 10^{11}$ krypton atoms/s; top: reionized fraction of beam; bottom: standard fluorescence signal[37].

2.3.2.c Collinear Laser RADOP

This experiment transfers the ideas of RADOP (see e.g. reference 2 and references therein) to collinear fast-beam laser spectroscopy (see Fig. 15a). A circularly polarized laser polarizes a fast beam of alkali atoms in flight by optical pumping. The beam is stopped in a suitable matrix exposed to an external field B. If relaxation times in this matrix are longer or at least comparable to the nuclear lifetime, the nuclear polarization P_I can be detected by the decay asymmetry measured by a pair of β-telescopes at 0° and 180° with respect to the field axis. A pilot experiment was conducted on some neutron-rich Rb isotopes at an on-line facility at the Mainz Triga reactor[41], followed by measurements on short-lived Li isotopes at ISOLDE. Fig. 15b shows asymmetry signals of ^{9}Li and ^{11}Li. The signal is small, of the order of 1% only, although a polarization P_I = 30% was attained in the beam, as could be verified by fluorescence signals from stable ^{7}Li in the same apparatus. The ^{11}Li current was 600 atoms/s. Spin and magnetic moment were determined[42].

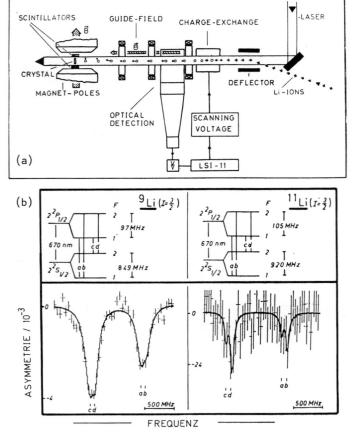

Fig. 15. a) Set-up for collinear laser RADOP experiment.
b) β-asymmetry signals from ^{9}Li and ^{11}Li when scanning the hfs of the D_1-resonance line (bottom); corresponding level schemes (top)[42].

2.3.2.d g_I-Measurement by Spin Rotation

This variant of collinear laser spectroscopy also makes use of spin polarization of a fast beam induced by optical pumping (see Fig. 16a). After being neutralized in a charge exchange cell the beam passes the pumping zone I, which is followed by a region of length L of a perpendicular magnetic field H_o in which the oriented spins precess by an angle $\theta = \gamma H_o L/v$, where v is the velocity of the beam. Note that H_o is strong enough to decouple the atoms by Zeeman splitting from the interaction with the laser during the precession phase. The precession angle can be measured downstream in the field free zone III by the modulation of the fluorescence intensity as function of H_o, because excitation by a polarized laser beam strongly depends on the angle of orientation of the ground state spin. If the atom in question has a diamagnetic ground state, as in the case of the alkaline earths, it is the nuclear spin which precesses and the measurement of θ determines directly g_I. The method described was developed by Vialle and collaborators[43]. The method has since been applied to Ra at the collinear set-up at ISOLDE. It yielded the first direct g_I-measurements for some isotopes of this element, namely [213]Ra and [225]Ra [44].

2.4 Summary of the Experimental Work

In 20 years of on-line optical spectroscopy, the methods have shifted radically from using traditional light sources to the use of tunable lasers. In fact, some of the new techniques, such as resonance ionization and collinear laser spectroscopy, are absolutely impossible without lasers.

At present, methodological work is progressing very efficiently towards higher sensitivity, pushing the exploration of nuclear structure by optical means further off stability and to refractory elements. In parallel, new schemes of ion souce and separation techniques are being installed which enrich the scope of separable elements and correspond better to the demands of on-line optical methods. It is foreseeable that laser spectroscopy of ions, stored in traps, will also soon have a chance in on-line

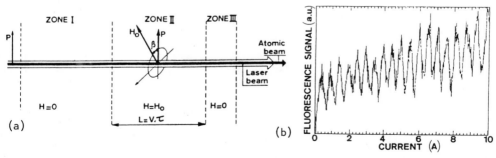

Fig. 16. a) Set-up for observation of spin precession in collinear laser spectroscopy[45]; I = pumping zone, II = precession zone, III = detection zone.

b) Fluorescence intensity of [213]Ra observed in zone III as a function of field H_o in zone II[44].

experiments. In view of the many vigorous activities, one may expect that our knowledge of spins, moments, and charge radii of nuclear ground states will be essentially completed in the next decade.

3. REMARKS ON THE INTERPRETATION OF RESULTS FROM ON-LINE OPTICAL SPECTROSCOPY

The stable nuclei form a rather narrow valley which covers only a small fraction of the total chart of bound nuclei. As such, it is not fully representative for all the phenomena of nuclear structure. The limitation to stable nuclei inhibits in particular a systematic exploration of nuclear structure as a function on N and Z, without which its complex phenomena cannot be understood. In the frame of this lecture we can touch upon a few selective questions only; for a more complete interpretation of results as well as of the experimental techniques themsemlves see, e.g., ref.[46].

3.1 Checking the Odd Group Coupling Model for Nuclear Moments

The kind of systematics obtained nowadays is demonstrated by Figs. 17 and 18, which show the series of magnetic dipole and electric quadrupole moments measured in a series of In-isotopes by collinear laser spectroscopy[30]. Selected are only those which have in addition to an odd proton number (Z = 49) also an odd neutro number. Most of them have rather long-lived isomeric states which could be measured in addition to the ground state by the present on-line technique. Since the moments of most of the relevant states in neighbouring isotopes and isotones (in Cd and Sn) are known as well, this total set of data offers an excellent chance for systematically checking the odd group coupling model. It predicts $\mu_{I,odd,odd}$ to be the vector sum of the moments of the odd proton group (p) in the adjacent odd-even isotope (of In) and of the odd neutron group (n) in the adjacent even-odd isotone (of Cd or Sn), provided the configurations are the same. The relation reads

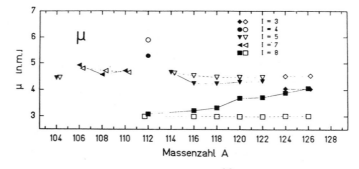

Fig. 17. μ_I-values of odd-odd In-isotopes[30]. Equal spins are connected by lines. Solid symbols represent experimental, open ones semi-empirical values.

$$\mu_{I,odd,odd} = \frac{I}{2}\ g_p + g_n + (g_p - g_n)\ \left|\ \frac{I_p(I_p+1) - I_n(I_n+1)}{I(I+1)}\ \right|\ . \tag{14}$$

These semi-empirical moments are plotted together with the experimental ones in Fig. 17[30]. For I < 8 the agreement varies from fair (I = 3) to very good (I = 5). For the discrepancy at I = 8 see below.

The odd group coupling model may be applied as well to spectroscopic quadrupole moments, although the quadrupole case has hardly been discussed in the literature, due to the lack of experimental examples. The odd group moments $Q_{s,p}$ and $Q_{s,n}$ are added with tensor coupling coefficients q_1, q_2, defined in ref. 47, to the resulting moment

$$Q_{s,odd,odd} = q_1\ Q_{s,p} + q_2\ Q_{s,n}. \tag{15}$$

The perfect agreement of these semi-empirical moments with the measured ones shown in Fig. 18 is a great surprise, indeed[30]. Only the I = 8 states form an exception again. (The moments of the latter have been taken from neighbouring Sn-isotones. The discrepancy for heavier mass numbers clearly indicates primarily that the structure of these states changes between indium and tin. Note that tin has a closed proton shell with Z = 50).

3.2 Decomposition of ms Charge Radii into Volume and Shape Effect

Since the field effect of the isotope shift measures the change of the mean-squared charge distribution, one expects it to be governed by nuclear gross properties rather than by fine details of the single-particle structure. According to any nuclear model with satured short range binding, the nuclear volume should increase proportional to the mass number and hence $\langle r^2 \rangle$ in proportion to $A^{2/3}$. From a brief look at the plots of isotope shifts below one learns, on the other hand, that $\delta \langle r^2 \rangle$ is not

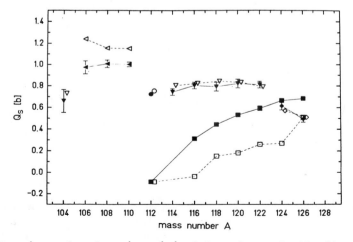

Fig. 18. Experimental and semi-empirical Q_s values of odd-odd In isotopes. Symbols are defined as in Fig. 17[30].

359

a smooth function of A. Obviously this cannot be explained by a liquid drop estimate of the change of nuclear volume alone, but one has to consider the influence of nuclear shape in addition; the latter has a strong and sometimes even abrupt A-dependence as a consequence of shell structure.

The assumption of a uniformly charge, sperical liquid drop with radius

$$R_o = r_o A^{1/3} \sim 1.2 \text{ fm } A^{1/3} \tag{16}$$

leads to the so-called standard or uniform shift

$$\delta\langle r^2\rangle^{AA'}_{unif} = 2/5 \ r_o^2 \ A^{-1/3} \ (A-A') \tag{17}$$

In the average, however, the isotope shift is about a factor of 2 smaller than predicted by eq. (17). This deficiency could be removed by the more realistic "droplet model" by Myers and Swiatcki[48,49] which also accounts very well for the nuclear binding energy. This success is mainly due to an additional degree of freedom of this model, namely that proton and neutron distributions are allowed to vary from each other. In particular, the model includes the build up of a "neutron skin" with increasing neutron excess.

Unless a nucleus has closed proton and closed neutron shells, its shape differs more or less from sphericity, which may be described by an angular dependence of the radius

$$R_d = R_o \ n \ (1 + \sum_i \beta_i \ Y_{io} \ (\theta)). \tag{18}$$

The deformation is described by the spherical harmonics Y_{io} with deformation parameters β_i; the normalization factor n preserves the nuclear volume. The lowest order contribution of β_i to $\langle r^2\rangle$ is given by

$$\langle r^2\rangle_d = \langle r^2\rangle_s \ \left[1 + \frac{5}{4\pi} \sum_i \langle\beta_i^2\rangle\right] \tag{19}$$

and hence for the isotope shift to first-order

$$\delta\langle r^2\rangle_d^{AA'} = \delta\langle r^2\rangle_s^{AA'} + \langle r^2\rangle_s \ \frac{5}{4\pi} \sum_i \delta\langle\beta_i^2\rangle^{AA'}, \tag{20}$$

where the index s refers to a spherical and d to a deformed nucleus. In eq. (20) the first term now corresponds to the change in nuclear charge volume (described by the spherical droplet model[49], for instance), and the second term represents the effect of a change of deformation on the isotope shift. Eq. (20) is called the two-parameter model, in particular, if one constrains the deformation to the leading quadrupole term β_2.

Fig. 19 shows the characteristic plot of $\delta\langle r^2\rangle$, as observed when a closed neutron shell (here at N = 82) is crossed. The plot covers the elements Xe^{50}, Cs^{51}, and Ba^{25}. Very characteristic is the kink at the magic number N = 82, which is one of the strongest and most persistent shell effects known in nuclear physics. It may be explained by comparison with a model which ignores shell effects such as the spherical droplet model, the slope of which is shown by the straight lines in Fig. 19 plotted as equi-deformation lines. These lines represent the pure volume effect at constant nuclear shape. One sees that the experimental line crosses the equi-deformation lines almost symmetrically on both sides of the magic number, so that the deformation increases steadily with distance from the magic number. Numerical calculations of $\delta \langle r^2 \rangle$ by advanced theoretical models including deformation are still unable to reproduce this effect quantitatively, as shown in the figure by the density-dependent, deformed Hartree-Fock calculation (DDHF)[52].

Fig. 20 shows the region around a closed proton shell (Z = 50) where the neutron covers a major part of the neutron shell $50 \leq N \leq 82$ and especially its center around N = 66. The isotope shifts fit strikingly well to parabolas with a slop at midshell corresponding to the droplet model. If one thus attributes the quadratic term to deformation, then the figures show that it maximizes at midshell and decreases towards both shell closures. For Cd, one derives from this data, for instance, a quadrupole deformation $\beta_2 = 0.29$ which is twice as large as expected from the quadrupole strength BE2 of nuclear gamma transitions. So far, only qualitative arguments for the discrepancy can be given.

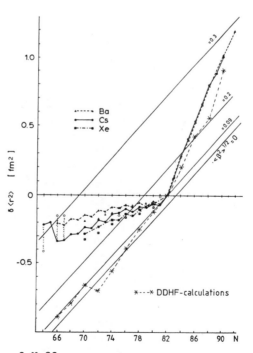

Fig. 19. Plots of $\delta \langle r^2 \rangle^{N,82}$ for Xe, Cs and Ba. (The open circles refer to isomers of Cs). Also shown are equi-deformation lines according to the droplet model and the result of a density-dependent Hartree-Fock calculation (stars)[53,52]

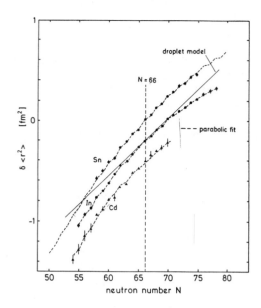

Fig. 20. Plots of $\delta\langle r^2\rangle$ for Cd^6, In^{30}, Sn^{54}. The Cd and In curves have been lowered at the reference point N = 66 by 0.4 fm^2 and 0.2 fm^2, respectively, for convenience. The broken lines are fitted parabolas with odd-even staggering. Also shown is the droplet line.

Whereas the isotope shift around the magic number Z = 50 was essentially quite regular, one observes extraordinary irregularities near the closed proton shell Z = 82 for very neutron deficient Au^{22} and especially for Hg isotopes (see reference 55 and references therein), (see Fig. 21). It occurs in the middle of the neutron shell $82 \leq N \leq 126$. In mercury it appears predominantly as an odd-even staggering, where the odd isotopes have a much larger $\langle r^2\rangle$ then their even neighbours, infering that the former ones are deformed, whereas the latter have essentially spherical shape. In ^{185}Hg one observes the phenomenon of a shape coexistence, since one finds a spherical isomer very close in energy to the deformed nuclear ground state. These observations have been the subject of many further experimental and theoretical investigations on nuclear spectra and structure of nuclei in this region. The huge odd-even staggering, e.g., is now understood (at least qualitatively) as due to the pairing effect which stabilizes the spherical structure of even isotopes and forms a critical balance with the deforming forces originating from the open neutron shell.

4. CONCLUDING REMARKS

What has been achieved by optical spectroscopy far off stability and what is still to be done? Charge radii, spins and moments have been measured along extended isotopic chains of about 20 elements covering about 400 nuclear ground states and isomers. Seen from a numerical standpoint of view we are not yet midway.

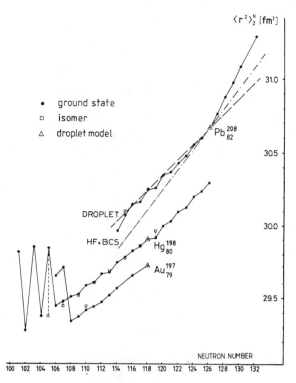

Fig. 21. Change of $\langle r^2 \rangle$ in the isotopic sequences of Au^{22}, Hg^{55}, Pb^{56}. The scales have been fixed at ^{208}Pb, ^{198}Hg and ^{197}Au by the droplet model (for simplicity). Isomers are indicated by open squares.

On-line laser spectroscopy, in combination with sophisticated isotope production and separation and signal detection techniques, will push the frontiers to higher sensitivity and enable measurements on refractory elements. The new techniques will enrich the scope of experiments at mass separators on line with proton and heavy ion accelerators. At the same time, they will have a large impact on the general problem of trace analysis in other fields of science and technology.

REFERENCES

1. H. L. Ravn, Phys. Rep., 54:201 (1979)
2. G. Huber, J. Bonn, H. J. Kluge, E. W. Otten, Z. Phys., A276:187 (1976)
3. G. Huber, C. Thibault, R. Klapisch, H. T. Duong, J. L. Vialle, J. Pinard, P. Juncar, P. Jaquinot, Phys. Rev. Lett., 34:1209 (1975)

4. T. Kühl, P. Dabkiewicz, C. Duke, H. Fischer, H. J. Kluge, H. Kremmling and E. W. Otten, Phys. Rev. Lett., 39:180 (1977)

5. P. Dabkiewicz, C. Duke, H. Fischer, T. Kühl, H. J. Kluge, H. Kremmling, E. W. Otten and H. Schüssler, J. Phys. Soc. Japan, 44, Suppl.:503 (1978)

6. F. Buchinger, P. Dabkiewicz, H. J. Kluge, A. C. Mueller and E. W. Otten, subm. to Nucl. Phys. A

7. P. Jaquinot and R. Klapisch, Rep. Progr . Phys., 42:773 (1979)

8. H. J. Kluge, in: "Progress in Atomic Spectroscopy", part B, W. Hanle and H. Kleinpoppen ed., Plenum Publishing Corporation, New York (1979) p. 727

9. E. W. Otten, Nucl. Phys., A354:471c (1981)

10. W. H. King, "Isotope Shifts in Atomic Spectra", Plenum Press, New York (1984)

11. H. Kopfermann, "Nuclear Moments", Academic Press, New York (1958)

12. I. Lindgren and A. Rose, Case Stud. in At. Phys., 4:93 (1974)

13. W. Neuhauser, M. Hohenstatt, P. E. Toschek, H. G. Dehmelt, Phys. Rev. Lett., 41:233 (1978)

14. K. Bekk, A. Andl, S. Göring, A. Hanser, G. Nowicki, H. Rebel; G. Schatz, Z. Phys., A291:219 (1979)

15. R. C. Thompson, M. Anselment, K. Bekk, S. Göring, A. Hanser, G. Meisel, H. Rebel, G. Schatz and B. A. Brown, J. Phys., G9:443 (1983)

16. V. S. Letokhov, "Non-linear Laser Chemistry", Springer Series Chem. Phys. 22, Springer Verlag, Heidelberg Berlin (1984)

17. G. S. Hurst, M. G. Payne, S. D. Kramer, J. P. Young, Rev. Mod. Phys., 51:767 (1979)

18. G. Bollen, A. Dohn, H. G. Kluge, U. Krönert and K. Wallmeroth, Inst. Phys. Conf. Ser. No. 84: Sect. 8:285 (1986)

19. J. D. Alkhazov, A. E. Barzakh, E. I. Berlovich, V. P. Denisov, A. J. Dernyatin, V. S. Ivanov, A. N. Zherikhin, O. N. Kompanets, V. S. Letokhov, V. I. Mishin and V. N. Fedoseyen, JETP Lett.,37:274 (1983)

20. V. N. Fedoyesev, V. S. Letokhov, V. I. Mishin; G. D. Alkhazov, A. E. Barzakh, V. P. Denisov, A. G. Dernyatin and V. S. Ivanov, Opt. Comm., 52:24 (1984)

21. V. S. Letokhov, Moscow, private communication

22. K. Wallmeroth, G. Bollen, M. J. G. Borge, J. Campos, A. Dohn, P. Egelhof, J. Grüner, H. J. Kluge, U. Krönert, F. Lindenlauf, R. B. Moore, A. Rodriguez, A. Venugopalan and J. Wood, Hyp. Int. in print

23. S. L. Kaufmann, Opt. Comm., 17:309 (1976)

24. W. H. King, G. A. Ruff, W. E. Lamb jr. and J. J. Spezewski, Phys. Rev. Lett., 36:1488 (1976)

25. A. C. Mueller, F. Buchinger, W. Klempt, E. W. Otten, R. Neugart, C. Ekström and J. Heinemeier, Nucl. Phys., A403: 234 (1983)

26. K. R. Anton, S. L. Kaufman, W. Klempt, G. Moruzzi, R. Neugart, E. W. Otten and B. Schinzler, Phys. Rev. Lett., 40:642 (1978)

27. F. Scheck, Phys. Rep., 44:187 (1978)

28. K. Wendt, doctoral thesis, Mainz (1985)

29. R. Neugart, in:"Lasers in Nucl. Physics", C. E. Bemis jr. and H. K. Carter ed., Harwood Academic Publishers 3, New York (1982), p.231

30. J. Eberz, U. Dinger, G. Huber; H. Lochmann, R. Menges, R. Neugart, R. Kirchner, O. Klepper, T. Kühl; D. Marx, G. Ulm and K. Wendt, Nucl. Phys., A464:9 (1987)

31. T. Meier, H. Hühnermann and H. Wagner, Opt. Comm., 20:397 (1977)

32. K. Dörschel, W. Hedderich, H. Hühnermann, E. W. Peau, H. Wagner, G. D. Alkhazov, E. Y. Berlovich, V. P. Denisov, V. N. Panteleev, A. G. Polyakov, Z. Phys., A317:233 (1984)

33. W. Borchers, R. Neugart, E. W. Otten, H. T. Duong, G. Ulm and K. Wendt, Hyp Int. in print and to be published

34. D. A. Eastham, P. M. Walker, J. R. H. Smith, J. A. R. Griffith, D. E. Evans, S. A. Wells, M. J. Fawcett and I. S. Grant, J. Phys., G12:205 (1986)

35. D. A. Eastham, P. M. Walker, J. R. H. Smith, D. D. Warner, J. A. R. Griffith, D. E. Evans, S. A. Wells, M. J. Fawcett and I. S. Grant, subm. to Phys. Rev. Lett.

36. S. V. Andreev, V. I. Mishin and V. S. Letokhov, Opt. Comm., 57:317 (1986)

37. R. Neugart, W. Klempt and K. Wendt, Nucl. Instr. Meth., B17:354 (1986)

38. R. Neugart, private communication and to be published

39. R. F. Silverans, G. Borghs, P. de Bishop and M. van Hove, Hyp. Int., 24:181 (1985)

40. R. F. Silverans, private communication and R. F. Silverans, P. Lievens, F. Buchinger, E. B. Ramsay, E. Arnold, W. Neu, G. Ulm and K. Wendt, to be published

41. J. Bonn, Hyp. Int., 22:57 (1985)

42. E. Arnold, thesis, Mainz (1986) and E. Arnold, J. Bonn, R. Gegenwart, W. Neu, R. Neugart, E. W. Otten, G. Ulm and K. Wendt, to be published

43. M. Carre, J. Lerme and J. L. Vialle, J. Phys., B19:2853 (1986)

44. E. Arnold, W. Borchers, M. Carre; H. Y. Duong, P. Juncar, J. Lerme, S. Liberman, W. Neu, R. Neugart, E. W. Otten, M. Pellarin, J. Pinard, G. Ulm, J. L. Vialle and K. Wendt, subm. to Phys. Rev. Lett.

45. N. Bendali, H. T. Duong, P. Juncar, S. Liberman, J. Pinard, J. M. Saint-Jalm, J. L. Vialle, S. Büttgenbach, C. Thibault, F. Touchard A. Pesnelle, A. C. Mueller, C. R. Acad. Sc., 229:1157 (1984)

46. E. W. Otten in:"Treatise on Heavy Ion Physics", Vol. 8, ed. A. Bromley, Plenum Press, New York (1987)

47. A. Bohr and B. R. Mottelson "Nuclear Structure", Vol. 1, Benjamin, New York (1969)

48. W. D. Myers and W. J. Swiatecki, Ann. Phys., N.Y., 55:395 (1969)

49. W. D. Myers and K. H. Schmidt, Nucl. Phys., A410:61 (1983)

50. W. Fischer, H. Hühnermann, G. Krömer and H.J. Schäfer, Z. Phys.,
51. 270:113 (1974)

51. A. Coc, C. Thibault, F. Touchard, H. T. Duong, P. Juncar, S. Liberman, J. Pinard, M. Carre, J. Lerme, J. L. Vialle S. Büttgenbach, A. C. Mueller and A. Pesnelle, subm. to Nucl. Phys. A

52. M. Epherre, G. Audi and X. Campi, Proc. 4th. Int. Conf. on Nuclei far from Stability, Helsingör 1981, CERN Rep. 81-09:62 (1981)

53. X. Campi, H. Flocard, A. Kerman and S. Kooni, Nucl. Phys., A251:193 (1975)

54. M. Anselment, W. Faubel, S. Göring, A. Hanser, G. Meise, H. Rebel and G. Schatz, Nucl. Phys., A451:471 (1986)

55. G. Ulm, S. K. Bhattacherjee, P. Dabkiewicz, G. Huber, H. J. Kluge, T. Kühl, H. Lochmann, E. W. Otten, K. Wendt, S. A. Ahmad, W. Klempt and R. Neugart, Z. Phys., A325: 247 (1986)

56. M. Anselment, W. Faubel, S. Göring, A. Hanser, G. Meisel, H. Rebel, and G. Schätz, Nucl. Phys., A451:471 (1986)

APPLICATION OF LASER COOLING TO THE ATOMIC FREQUENCY STANDARDS

F. Strumia

Dipartimento di Fisica - Università di Pisa
P.za Tarricelli, 2 - 56100 PISA
and INFN Sezione di Pisa - CISM Unità di Pisa

INTRODUCTION

The Atomic Frequency Standards (AFS) are the most precise devices ever built and have many important applications both in science and technology. In fact the development of more and more precise time/frequency standards has been stimulated and supported by the need of any civilization to improve the timekeeping and the navigation systems. The present sophisticated navigation systems are based on a set of spaceborne AFS. As a consequence several countries (USA, Canada, FRG , France, UK, Italy, URSS, Japan and China) support metrological national laboratories dedicated to the maintaining and development of time and frequency standards.

The measurement and intercomparison of frequencies is, at least in principle, a measurement of unlimited precision, the error being ± 1 count independently of the number of the counts. The final precision is therefore limited only by the stability of the master oscillator. The development of very fast diodes (MIM diodes) has permitted in the recent years the frequency synthesis up to the visible, and the direct frequency measurement of visible and near infrared actively stabilized lasers by heterodyne comparison with the microwave standards. Contrary to the wavelength measurements, the frequency measurements are not limited by diffraction and the BIPM decided at the 17th CGPM meeting on October 20th, 1983 to adopt a new definition of the meter which considers the velocity of light as a fixed number: "The meter is the length of the path travelled by light in vacuum during a time interval of 1/299792458 of a second". With this decision the two most important units, time and length, are referred to the ground state hyperfine transition F=4, $m_F=0 \leftrightarrow F=3$, $m_F=0$ of ^{133}Cs, whose central frequency is defined as

$$\nu_{Cs} = 9\ 192\ 631\ 770.000\ Hz \tag{1}$$

in absence of perturbations (13th General Conf. on Weights and Measures - 1967). Unfortunately the precision and resolution of the frequency measurements in the microwave region is not yet full extended to the visible region because of the phase noise introduced in the high order multiplication process and of the non adequate reproducibility of the reference oscillator in the visible and near infrared.

Today devices permit a frequency multiplication and phase locking of oscillators with the stability and resolution of the Cs primary standard only up to about 10^{13} Hz[1]. This confines the development of new kinds of AFS of improved performances to the MW and FIR (Far-infrared) spectral region. It is worth noting that the Cs AFS has been extensively studied during the last 30 years. It has reached a mature state, no significant improvement has been obtained in the past ten years and none is reasonably to be expected for the furure, at least for the classical design of Stern-Gerlach states selection and Ramsey interrogation scheme on a thermal beam.

The purpose of the present paper is to present the status and the perspective of AFS based on fine structure transitions observed in thermal beam of metastable Mg and Ca atoms, which are the bes candidates for AFS of performances better than that of Cs[2]. Recently the velocity reduction in an atomic beam has been demonstrated by resonant absorption of a counterpropagating laser beam. The impact of this new technique on the improvement of Mg, Ca, and Cs AFS will be discussed.

THE ATOMIC FREQUENCY STANDARDS

An atomic frequency standard consists of a high quality quartz oscillator phase locked to an atomic transition of particular insensitivity to external perturbations, with a very high $Q=\nu/\Delta\nu$ and a good signal to noise ratio (S/N) in the detection system as shown in Fig. 1. The above requirements greatly restrict the number of the atomic or molecular transitions suitable as a reference for an atomic frequency standard. The quality of a frequency standard is defined by its <u>Precision</u>, <u>Reproducibility, and Stability</u>. The Stability, which expresses the mean square deviation of the standard frequency over an observation time τ, is a function of τ^n, n depending on the prevalent kind of noise (for the shot or white noise n=-1/2, for flicker or 1/f noise n=0). The accuracy is the capability to agree with the master laboratory standard, and the reproducibility reflects the degree to which a set of standards will produce the same output frequency and the degree to which the ideal frequency of the reference atomic line is reproduced. With respect to reproducibility the AFS are distinguished between "primary" and "secondary" standards. The first are so called because the atomic reference sample is strongly protected against perturbations, in particular against collisions with other atoms and/or walls. The last requirement is fully satisfied by using atomic beams as in the case of the Cs AFS. On the contrary, other frequently used AFS, like the optically pumped Rb vapor or the Hydrogen maser, are not primary standards even if the latter has a better short term stability with respect to Cs. Secondary AFS can be preferred in the applications because of their lower cost, dimensions, weight and power consumption, but they must be calibrated more or less frequently against the Cs primary AFS. In conclusion an AFS referred to a transition observed in an atomic beam is of particular interest for precision and re-

producibility because the reference system is immune from environment perturbations to the maximum degree. In this case there are however limitations
in the sample size and in the linewidth of the reference transition since
the interrogation time depends on the beam velocity and the interrogation
length: $\Delta t = L/v$

The stability, when measured following a particular procedure, is
known as the Allan variance $\sigma(\tau)$ and is given, in the case of shot noise,
by

$$\sigma(\tau) = \frac{K}{Q \ S/N \ \sqrt{\tau}} \tag{2}$$

where K is a constant of the order of unity and depends on the actual physical structure of the standard. As shown in Fig. 1, the beam preparation
and interrogation scheme is responsible for the reference line Q, the
system for detecting the transition is responsible for the largest part of
the noise, while the oscillator correction time constant is adjusted by the
phase lock electronics. The larger Q and/or S/N, the shorter is the time
necessary for reaching a given stability. It is then convenient to use a
high frequency transition as reference, since Δv (and $Q = v/\Delta v$) is limited
by the interrogation time τ_i. Eq. 2 shows that stability can be improved
also by using longer and longer integration times as a consequence of the
white noise averaging. However, for very long time ($\tau \cong 10^4 \sim 10^5$ s for AFS)
the 1/f noise becomes prevalent and the Allan variance becomes independent
of τ, as shown in Fig. 2. For even longer integration times ($\tau > 10^6$ s)
other kinds of noise (as the random walk noise) will start to degrade the
stability. As an experimental result the time dependence of the Allan
variance can be written as

$$\sigma(\tau) \propto (b^2 \tau^{-1} + c^2 + d^2 \tau)^{1/2} \tag{3}$$

where the relative values of the b, c, and d constants depend on the specific AFS, but the general trends are that shown in Fig. 2.

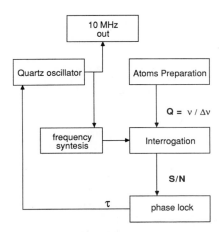

Fig. 1. Block scheme of an atomic frequency standard.

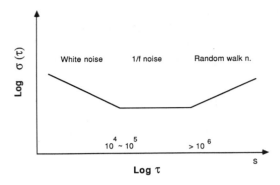

Fig. 2. Stability dependence (Allan variance) on the interrogation time
of an atomic frequency standard.

The Cs Atomic Frequency Standard

The most convenient design of the Cs AFS is well established, following about 40 years of development[3,4,5,6]. The atomic beam is obtained
by thermal effusion from a thermostatized oven and, after collimation, the
atoms in the F=4, m_F=0 hyperfine level are isolated by using a Stern-Gerlach
magnet. The B field must be strong enough to break the coupling between the
nuclear and electronic dipole moments: a few kGauss. The hyperfine transition is induced by using the Ramsey technique of two separate oscillating
fields[6] so that an interaction length up to a few meters can be used. The
atoms transferred into the F=3, m_F=0 hyperfine level are then selected by
means of a second Stern-Gerlach magnet and detected by ionization on a hot
wire with about 100% efficiency. It is also a common practice to select
atoms with a velocity lower than the average of the thermal distribution
by a proper positioning of the Stern-Gerlach magnets and the collimating
slits. Since, moreover, Cs has a relatively high vapor pressure (Cs melts
at 306° K), a line Q of the order of $\cong 10^7 \cong 10^8$ can be obtained. Among the
disadvantages of Cs there is the large number of hyperfine sublevels, 16,
which reduces the number of useful atoms and imposes a tight beam collimation with a further significant flux reduction. The performances of the
Cs AFS have been evaluated[4,5,20] and are summarized in Table I.

The Submillimeter AFS: Magnesium and Calcium

As it was pointed out, the reference line Q in an AFS based on an
atomic beam apparatus is given by

$$Q = \nu/\Delta\nu = k \, \nu \, L/v \tag{4}$$

where the constant k is of the order of the unity and depends on the actual
intensity distribution of the microwave field along the beam. In order to
increase the Q value several approaches are possible. The interaction length
L can be considered, which ranges from a few cm in commercial standards to
more than 3 m in the NBS6. A larger L value is not convenient for several
practical reasons. For example, it will cause a reduction of the beam intensity and hence of the S/N ratio (Eq. 2), and the shielding of the interrogation region from the external magnetic field and its fluctuation

370

Table I. Characteristics of cesium atomic beam frequency standards
(from published data)

	Commercial AFS	NBS-6	NRC-V
Short term stability	1.5×10^{-11}	5×10^{-13}	3×10^{-12}
Long term stability	3×10^{-13} (6 months) 3×10^{-12} (5 years)	1×10^{-14} (1 week)	1.5×10^{-14}
Accuracy	7×10^{-12}	8×10^{-14}	1.5×10^{-13}
Volume	$.03 \ m^3$	$5 \ m^3$	
Interaction length	7.5 cm	375 cm	213 cm
Microwave Q	1×10^7	2.8×10^8	1.5×10^8
Mean Atomic Velocity	100 m/s	190 m/s	260 m/s

will become difficult. In fact, being the reference line of the kind $m=0 \leftrightarrow m=0$, there is a residual quadratic Zeeman effect (higher order terms are negligible)

$$\nu = \nu_0 + \alpha \ B^2 \tag{5}$$

where $\alpha = 427$ Hz/Gauss^2 in the case of Cs : a number not sufficiently small for metrological purposes. Since L, in conclusion, is limited to the order of 1m, the Q can be substantially increased only by increasing the frequency of the reference transition and/or reducing the atomic velocity in the beam.

Let us first consider the velocity reduction: in an effusive thermal beam the velocity distribution is expected to be given by

$$f(x) = 2 \ x^3 \ exp(-x^2) \tag{6}$$

where $x = v/\beta$ and

$$\beta = 129 \sqrt{\frac{T}{M}} \ m/s$$

where T is the oven temperature in K, and M the atomic weight. The most probable velocity $(v^M=(3/2)^{1/2}\beta = 158(T/M)^{1/2})$ is of the order of a few hundredths of m/s for any kind of atoms. In the real beams the very slow and fast atoms are kicked off the beam by collisions, otherwise Eq. 6 is a very good description of the velocity distribution of a thermal beam. It is possible to select the low velocity part of the beam by a proper use of the Stern-Gerlach magnets or by using rotating wheels. However the flux decrease is dramatic as a consequence of the cubic term in Eq. 6.

In conclusion also the beam velocity v is to be considered a fixed parameter in absence of laser cooling. The velocity manipulation by laser light will be considered in the next section. Here we will discuss the possible use of atomic transitions at a frequency substantially higher than that of Cs.

The problem of the maximum multiplication of the quartz oscillator frequency without an intolerable spectral degradation has been carefully investigated both theoretically[7] and experimentally[1,8-10]. The present situation is that a frequency synthesis suitable for AFS can be realized up to about 1 THz by phase locking a Carcinotron and up to about 6 THz if the phase locking of optically pumped Far-infrared (FIR) lasers is also considered. In the latter case only a comb of nearly fixed frequencies is presently available. No substantial improvements can be foreseen for the next few years.

In looking for reference transitions at frequencies significantly higher than that of Cs, it is worth noting that the hyperfine structure splittings are restricted to the microwave region due to the magnitude of the nuclear magnetic moment. Transitions suitable for AFS in the submillimeter region can be found by considering allowed magnetic dipole transitions between fine structure levels in the fundamental or metastable states. In this case the number of the magnetic sublevels is reduced if nuclei with I=0 are used. The absence of hyperfine structure results also in the important advantage that the α constant of the residual quadratic dependence from the magnetic field is reduced by about four orders of magnitude (α is inversely proportional to the square of coupling constant).

Unfortunately, the Stern-Gerlach magnetic deflection is no more useful for the selection of the atoms in a $m_J=0$ sublevel for energy splittings in the submillimeter region. In fact the deflection angle θ is given by

$$\theta = \frac{L}{m\,v^2}\,\mu_{eff}\,\frac{\partial B}{\partial x} \tag{8}$$

and it is necessary for the magnetic field B to be large enough to break the multiplet structure in order to have $\mu_{eff} \neq 0$ for the $m_J = 0$ levels.

In the case of hyperfine structure

$$B > \mu_N/\mu_0 \cdot B(0) \cong 0.1 \text{ T} \tag{9}$$

where μ_N is the nuclear magnetic dipole moment, μ_0 the Bohr magneton, and B(0) the magnetic field generated in the atom by the electrons motion. The field and gradient required by eqs. 8 and 9 can be produced and shielded from the interrogation region. On the contrary the field necessary in the case of a fine structure splitting in the submillimeter region must be of the order of

$$B \geq B(0) > 10^2 \text{ T} \tag{10}$$

and can not be realized with the present technology.

Two alternative ways can be considered in order to obtain a significant population difference between the $m_J = 0$ levels:

(i) optical pumping, possibly by using resonant laser radiation;
(ii) the use of metastable levels with a different lifetime in order
      to realize a temporal state selection.

Furthermore a sensitive method must be available for the detection of the
reference transition with a good S/N ratio. This constrains were considered
in ref. 2, and it was demonstrated that only two elements , Mg and Ca, have
the required spectroscopic properties to be considered for a possible AFS.
In fact the metastable triplet states $^3P_0$, $^3P_1$, and $^3P_2$ and the associated
magnetic dipole transitions at 0.6 and 1.2 THz for Mg and 1.5 and 3.2 for
Ca respectively have a set of properties that are of particular interest.
A scheme of the lowest energy level is given in Fig. 3 and 4 for Mg and
Ca respectively. The radiative decay of the ^3P levels towards the ground
$^1S_0$ state is strongly forbidden with the exception of the $^3P_1$ level which
has a lifetime of about 10^{-3} s as a consequence of a small mixing with the
singlet $^1P_1$ state of the same configuration. The best available numerical
data concerning this levels are given in Table II.

It is then possible to obtain a time of flight states selection and
the SMM transition $^3P_0 - ^3P_1$ can be monitored by the fluorescence in the
visible from $^3P_1 - ^1S_0$ transition. A scheme of such a standard is shown in
Fig. 5. A beam of metastable atoms can be easily obtained with high efficiency (30≅50%) by electron bombardment in a magnetically confined plasma[14].
The ^3P sublevels are then equally populated, but the atoms in the $^3P_1$ state

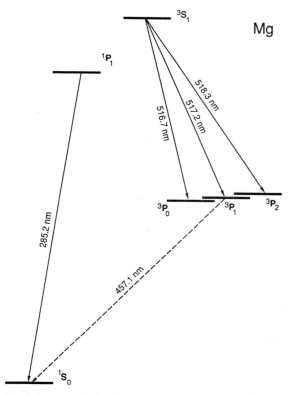

Fig. 3. Scheme of the lowest energy Magnesium levels.

	Mg	Ca
ν_{01} (MHz)	601 277.157 860(20)[a]	1 563 630.(150)[b]
ν_{12} (MHz)	1 220 575.1(33)[c]	3 174 230.(100)[b]
g_1(exp)	1.501111(16)[c]	1.5010834(21)[d]
g_2(exp)	1.501102(16)[c]	1.5011313(80)[d]
g_1(th)	1.501158	1.50108
g_2(th)	1.501160	1.501160
α_{01} (Hz/gauss^2)	3.8093[d]	1.4654.[d]
$\tau(^3P_1)$ (ms)	2.4(2)[e]	0.55(4)[2]
		0.50(4)[f]
$\tau(^1D_2)$ (ms)		2.3(5)[f]
$g(^1D_2)$		0.999 95(6)[g]
A_{21} (s^{-1})[h]	0.910x10^{-6}	16.0x10^{-6}
A_{10} (s^{-1})[h]	0.145x10^{-6}	2.57x10^{-6}

(a) ref. 18; (b) ref. 41; (c) ref. 42; (d) ref. 21; (e) ref. 11; (f) ref. 39, (g) ref. 43; (h) ref. 44.

start to decay to the ground state by emitting blue light at 475 nm (Mg), a radiation that can be detected with photomultipliers of quantum efficiency larger than 20%. The $^3P_1$ state population is reduced to 1/e at a distance $L \cong 2m$ from the excitation point[11]. The $^3P_0-^3P$ transition can be induced by using a folded open mirrors cavity as shown in Fig. 5. In this way two separate interaction regions of about 1.5 cm and separated by a distance L' are obtained with two waists in correspondence of the interaction zones. In order to observe an optical Ramsey lineshape it is necessary for the atoms to cross zones of equal phase in both interaction regions. As a consequence of the beam cross section and the unavoidable beam divergence the above requirement is not satisfied and the interference signal is averaged out. A possible solution is to use three equally spaced oscillating fields, but a more practical solution in the submillimeter region consists in the use of a space selector made of a grid of $\lambda/2$ period and mounted on a translator for a precision alignment with respect to the standing wave antinodes as shown in Fig. 5[13]. The $^3P_0-^3P_1$ transition is observed as an increase of the fluorescence light emitted by the beam atoms. In the case of Mg this background fluorescence signal can be larger than the signal and contribute significantly to the reduction of the S/N ratio.

In 1983 the first submillimeter source of coherent radiation with a

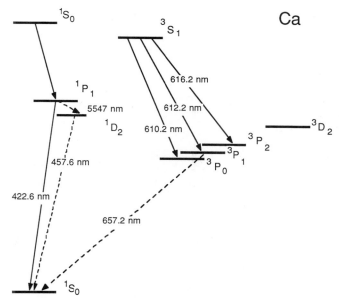

Fig. 4. Scheme of the lowest energy Calcium levels.

sufficient power and tunable around 601 GHz became available. It consists
of a carcinotron phase locked to a high stability and spectral purity quartz
oscillator[1]. The first observation of the $^3P_0-^3P_1$ Mg line was soon reported[14]
by using a single oscillating region and the first accurate frequency measurements obtained both for the most abundant isotope ^{24}Mg(78.6%, I=0) and the
less common ^{26}Mg(11.3%, I=0), and ^{25}Mg(10.1%, I=5/2) isotopes[15,16]. The
observed linewidth of 77 kHz was in accordance with the interrogation length
of only 0.5 cm and the experimental result that the atoms with lower velocity
are preferred for both excitation and interrogation processes.

In spite of this large linewidth, a first operation as an AFS was realized in 1984[17] by activating the phase loop control of the quartz oscillator,
and a stability $\sigma \cong 5 \times 10^{-10} \tau^{-1/2}$ was obtained in agreement with the eq.(2)
for the observed S/N ratio of $\cong 240$ for $\tau = 1s$ and a form factor $K \cong 1$.

The low value of the S/N ratio (at least for an AFS) is a consequence
of the limited power available at 601 GHz and of the large background signal.
In fact a S/N ratio up to $10^4 \cong 10^5$ is expected in more favorable experimental
conditions. Let L_b be the distance between the position of the metastable
states excitation and that of the detection and L_s the distance between
interrogation and detection. The background signal is then given by

$$B = N \ P \ \Omega \ Q \ a \ \Delta L \ exp(-aL_b) \qquad (11)$$

where N is the flux of atoms in the $^3P_1$, Ω the light collection solid angle,
Q the photomultiplier quantum efficiency, ΔL the observation length, $a=1/v\tau$
the average decay lenght, and the coefficient P<1 takes into account a
possible depopulation by optical pumping (in the present case P=1). The
signal is given by

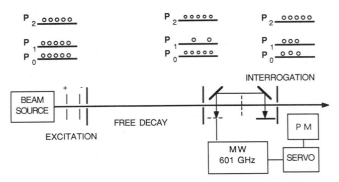

Fig. 5. Scheme of a submillimeter atomic frequency standard based on a Mg atomic beam.

$$S = N/3 \; P'\Omega QT_r \; a \; \Delta L \; \exp(-aL_s) \tag{12}$$

where an equal excitation probability for the metastable sublevels has been assumed, $P'>1$ takes in account a possible optical pumping, and T_r is the fraction of atoms transferred from the $^3P_0$ to the $^3P_1$ level into the interrogation region ($T_r \cong 1.5 \times 10^{-3} \times I(mW/cm^2)$ for an interrogation length of 0.5 cm). Since $N = (S + B)^{1/2}$ when the white noise is dominant, we obtain

$$S/B = Tr/3 \; P'/P \; \exp(a(L_b - L_s)) \tag{13}$$

and

$$\frac{S}{N} = \frac{S}{\sqrt{S+B}} = \frac{\sqrt{N \Omega QP' \, a \, \Delta L}}{3} \; \frac{T_r \exp(-aL_s)}{\{P \exp(-aL_b) + P'T_r/3 \exp(-aL_s)\}^{1/2}} \tag{14}$$

The value of $T_r \cong 1.3 \times 10^{-2}$ is obtained from eq. (13) and the experimentally observed value $S/B = 0.81 \times 10^{-2}$, a result in agreement with the available power. The expected value $S/N \cong 300$ can then be obtained from eq. (14); again in good agreement with the experimentally observed value of 240. From the above discussion it follows that a large improvement is still possible. As an example a $S/N \cong 3 \times 10^4$ can be predicted if $P=1/5$ and $T_r = 1/2$, with a corresponding stability increase. The optical pumping effect of the order of 1/5 is estimated from the power output of a Mg spectral lamp and the efficiency of birefringent filters especially designed for the separation of the $^3S_1 \leftrightarrow {}^3P_1$ line from the $^3S_1 \leftrightarrow {}^3P_0$ and $^3S_1 \leftrightarrow {}^3P_2$ lines (Fig. 3). A larger value of T_r can be obtained by using a resonant cavity of large r finess and/or by using a more powerful carcinotron (a power increase of about 5 is already possible according to the manufacturer). In conclusion it is reasonable to forecast a S/N ratio improvement of one or two orders of magnitude.

As for the line Q, a factor larger than 10^3 was to be gained, since the natural linewidth of the $^3P_1$ level is 65(3) Hz. A first significant improvement was obtained in 1986 with the successful realization of the optical Ramsey interrogation scheme[18]. A linewidth of 1.2 kHz ($Q \cong 5 \times 10^8$) and a S/N=250 was demonstrated with a distance of 30 cm between the two oscil-

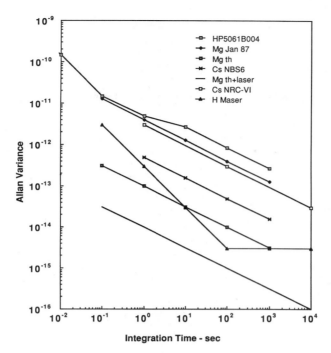

Fig. 6. Comparison of the frequency stability of high quality AFS. "Mg Jan'87" is the present stability of the Magnesium AFS; "Mg th" is expected stability without laser cooling; and "Mg th+laser" is expected stability with laser cooling.

lating fields. The stability was measured by comparison with a commercial high performance Cs AFS and was estimated to be $\sigma(\tau) = 8 \times 10^{-12} \times \tau^{-1/2}$, again in agreement with eq. (2) and a form factor $K \cong 1$. More recently[19] a S/N=500 has been obtained increasing the stability to

$$\sigma (\tau) = 4 \times 10^{-12} \times \tau^{-1/2} \tag{15}$$

This short term stability is better than that of the present commercial Cs AFS and inferior only to that of the large laboratory Cs AFS of the National Bureau of Standards - USA and National Research Council - Canada (Fig. 6)[5,6,20]. On the contrary, the Q is already larger than that of all the present Cs AFS, and can be easily further increased.

The future developments of a Mg standard based on a thermal beam can be forecast as:

(i) a S/N increase of about a factor 5 by improving the depopulation of the $^3P_1$ level with optical pumping as shown in the scheme of Fig. 7, and by using a polarizer in front of the photomultiplier in order to eliminate the fluorescence light from the $\Delta M = \pm 1$ transitions;

(ii) a further S/N increase of about $\cong 10$ or 20 by increasing the intensity of the interrogating field;

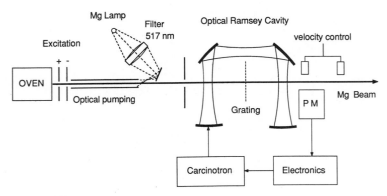

Fig. 7. Scheme of the experimental apparatus of an AFS based on a Mg thermal beam. The optical pumping out of the $^3P_1$ level is an option to increase the S/N ratio.

(iii) a Q increase to $\cong 10^9$ or 1.5×10^9 by using a longer interaction length.

In conclusion a Mg AFS realized by following the scheme of Fig. 7 will reach a short term stability of

$$\sigma(\tau) \leq 5 \times 10^{-14} \times \tau^{-1/2} \qquad (16)$$

and a precision and reproducibility of the order of $10^{-12} \cong 10^{-13}$. The most significant limitations will be a consequence of the second order Doppler effect which causes a red shift

$$\frac{d\nu}{\nu} = -\frac{2v\,dv}{c^2} - 1.5 \times 10^{-14}\,dv \ (m/s) \qquad (17)$$

and of the phase shift in the Optical Ramsey interrogation as a consequence of imperfections and thermal fluctuations.

There is also a convenience to use the $^3P_2 - ^3P_1$ transition as a reference if enough power will become available in the future at 1220 GHz. In such a case the line Q should be doubled in the same experimental apparatus as a consequence of the frequency increase. Also the S/N ratio should be improved because the Landè factors g_2 in the level $^3P_2$ and g_1 in the $^3P_1$ level respectively are nearly equal and the transitions J=2, M=-1—J=1, M=-1 and 2, +1—1, +1 are degenerate with 2,0—1,0 transition within the natural linewidth for a magnetic field intensity of about 1 Gauss or less. The signal S will then be larger than in the case of the 0,0—1,0 transition since three time more atoms will be involved and the transition probability is 8 times larger (Table II). The Zeeman effect in the above transitions can be calculated up to the second order (the third order terms are identically zero) to be

$$\nu(0,0\leftrightarrow1,0) = \nu_1 + (\frac{4}{3}\varepsilon_1 - \frac{1}{3}\varepsilon_2 + \frac{2}{15}w)\,B^2 \qquad (18)$$

$$\nu(2,0\leftrightarrow1,0) = \nu_2 + (-\frac{2}{3}\varepsilon_1 + \frac{2}{3}\varepsilon_2 - \frac{1}{5}w)\ B^2 \qquad (19)$$

$$\nu(2,-1\leftrightarrow1,-1) = \nu_2 - (g_2 - g_1)\mu_0\ B + \frac{1}{2}\varepsilon_2\ B^2 \qquad (20)$$

$$\nu(2,\ 1\leftrightarrow1,\ 1) = \nu_2 + (g_2 - g_1)\mu_0\ B + \frac{1}{2}\varepsilon_2\ B^2 \qquad (21)$$

where ε_1 and ε_2 are the second order expansion coefficients and w is the diamagnetic quadratic effect[21]. The following numerical values can be calculated for the Mg atom

$$(g_2 - g_1)\mu_0 \cong 10(8)\ \ Hz/Gauss \qquad (22)$$

$$\varepsilon_1 = \frac{\mu_0^2}{E_1 - E_0} = 3.2579215\ \ (Hz/Gauss) \qquad (23)$$

$$\varepsilon_2 = \frac{\mu_0^2}{E_2 - E_1} = 1.6049105\ \ (Hz/Gauss) \qquad (24)$$

$$w = \frac{e^2 <r^2>}{8m} \cong 0.003\ (2)\ \ (Hz/Gauss) \qquad (25)$$

and the $\Delta M=0$ transitions have the frequencies

$$\nu(0,0\rightarrow1,0) = \nu_1 + (3.808925 + 0.0004)B^2 \qquad (Hz/Gauss) \qquad (26)$$

$$\nu(2,0\rightarrow1,0) = \nu_2 + (-1.102007 - 0.0006)B^2 \qquad (Hz/Gauss) \qquad (27)$$

$$\nu(2,1\rightarrow1,1) = \nu_2 + 10\ B + 0.802455\ B^2 \qquad (Hz/Gauss) \qquad (28)$$

$$\nu(2,1\rightarrow1,1) = \nu_2 - 10\ B + 0.802455\ B^2 \qquad (Hz/Gauss) \qquad (29)$$

For a magnetic field of small intensity (≤1 gauss) the three $\Delta M=0$ Zeeman components of the $^3P_2 - ^3P_1$ line are well within the natural linewidth (65(3) Hz) and can be considered as a single line of frequency

$$\nu(2\rightarrow1) = \nu_2 + (0.04067 - 0.00080)\ B^2 = \nu_2\ (1 + 3.37 \times 10^{-14})\ B^2 \quad (30)$$

As a consequence the $^3P_2 - ^3P_1$ line not only will have a larger Q and intensity than the $^3P_0 - ^3P_1$, but also a much smaller dependence on the residual magnetic field. This would promise an AFS of better stability, precision, and reproducibility with respect to one referred to the $^3P_0 - ^3P_1$ line and stimulate the realization of a suitable source at 1220 GHz.

Let us now mention the advantages of an AFS based on the Ca lines. In this case $^3P_1$ state lifetime is shorter than that of Mg and the maximum theoretical Q is about two times smaller, however the shorter decay length would give a better S/N ratio in a more compact device in the case of thermal beams.

SMALL EFFECTS RESPONSIBLE FOR PRECISION AND REPRODUCIBILITY

The small perturbations that are esponsible for the maximum attainable
precision and reproducibility in the passive primary AFS will be briefly
discussed in this section. The numerical values summarized in Table III for
the best available Cs beam AFS represent the results of more than three
decads of continuous experimental and theoretical efforts. Several other
effects have been also considered, like, for example, the Bloch–Siegert
shift. However their contributions resulted much less important than that
reported in Table III and do not affect the present performances of the
AFS. The relevance of many of the effects considered in the table is a
function of the atoms velocity, mainly as a consequence of the Q dependence
upon the interrogation time, and is shown in the second column.

A first set of perturbations of Table III is expected to be much smaller
in Mg and Ca than in Cs as a consequence of a more favorable levels structure and the much higher transition frequency. In fact the shift caused by
the residual B field and its inhomogeneities and fluctuations is reduced by
about four orders of magnitude, as discussed in the previous section, and
can be disregarded.

Fluctuations in amplitude and direction of the magnetic field in the
interrogation region may also induce Majorana transitions from m=0 to the
m= 1 levels, and these transitions may pull the reference line $(0,0) \rightarrow (1,0)$
if they are asymmetric. This effect is smaller in Mg because the Zeeman
splitting exists only in the $^3P_1$ state and is about five time larger reducing in correspondence the Majorana transition probability in presence of
the same magnetic field inhomogeneity.

The tail pulling is a similar effect, which will follow from unbalanced
$(0,0) \rightarrow (1,\pm1)$ transitions induced at the frequency of the $(0,0) \rightarrow (1,0)$ line.
This effect is smaller than in Cs not only because of a larger Zeeman effect
but also because in Mg there is only one $\Delta m=0$ transition and the other
Zeeman transitions are $\Delta m=\pm1$ and will require an oscillating field of orthogonal polarization.

Majorana and tail pulling effects will also be reduced by the increased
value of the line Q, and, in conclusion, are expected to be negligible in
the Mg AFS.

The black body shift follows from the perturbation of the levels energy
by the black body spectrum in the interrogation region. It can be calculated
as a function of the temperature[22] and the reference frequency corrected
without problems.

The cavity frequency pulling is a shift caused by the mistuning $\Delta v_c = v_c - v_0$ of the resonant cavity frequency and is given by

$$\frac{\Delta v}{v_0} = \left(\frac{Q_c}{Q}\right)^2 \frac{\Delta v_c}{v_0} \tag{31}$$

and is reduced by increasing the Q of the reference line and the frequency

Table III. Accuracy and precision limitations of the NBS-6 (a)
and NRC-CsVI (b) Cs atomic beam frequency standards
(the data are in units of 10^{-14}). The dependence of
the effects on the atoms velocity is also shown.

Effect	Velocity dependence	Value NBS	Uncertainty NBS	Uncertainty NRC
1 B-field		5335	3	5
2 B-field inhomogeneity			0.2	2
3 Majorana effect	v^2		0.3	
4 tail pulling	v^4		2	
5 blackbody shift		1.7	0.0	
6 cavity pulling	v^2		0.1	
7 second order Doppler	v^2	26	1.0	2
8 cavity phase shift	v	36	8.0	3
9 uncertainty in phase due to n(v)	v		1.0	
10 RF spectrum	v^2		1.0	2
11 amplifier offset	v		1.0	
12 second harmonic distortion	v		2.0	

(a) ref. 4; (b) ref. 20.

of the reference transition. Reasonable numbers for a Mg AFS are $Q_c \cong 10^9$ and $\Delta\nu_c \cong 10^5$, and the relative frequency pulling will be in the $10^{-15} \cong 10^{-16}$ range.

The second order Doppler effect (eq.17) is more important in a Mg or Ca AFS than in the Cs AFS due to the larger velocity in the atomic beam. The only way to reduce this effect is to decrease the velocity and/or to increase its control.

The cavity phase shift is an effect that appears if a phase difference exists between the interrogatin fields: the Ramsey resonance curve will be asymmetrically distorted with a resulting frequency error. In the case of the conventional Ramsey interrogation with two separate fields the relative frequency shift results

$$\frac{\Delta\nu}{\nu} = \frac{\Delta\Phi}{\pi Q} \qquad (32)$$

where the $\Delta\Phi$ is the phase difference. This effect has not yet been satisfactorily studied in the case of open mirrors cavities. It will depend on optics quality, cavity symmetry, coupling and intracavity losses, mechanical and thermal stability etc. It is expected to be an important source of precision limitation also for the Mg AFS, even if the cavity phase shift effect is reduced by increasing the Q,.

The RF spectrum frequency shift follows from the presence of asymmetrical sidebands with respect to the carrier frequency. The shift from a single sideband is given by[6]

$$\frac{\Delta \nu}{\nu} = \frac{A_s}{A_0} = \frac{\delta \nu^2}{\nu_0 (\nu - \nu_s)} \tag{33}$$

where $\delta \nu$ is the transition linewidth, then proportional to v, ν_s the sideband frequency, and A_s its intensity. Also this effect is expected to be smaller in the Mg AFS. The last two effects of Table III depend primarily on the quality of the electronic apparatus and in principle are expected to be less important by increasing the Q.

In conclusion the accuracy and reproducibility of a submillimeter AFS is expected to be of the order of one part in 10^{-13}. The most important sources of error being the second order Doppler effect and, probably, the cavity phase shift. It is however worth noting that many effects that plague the primary Cs AFS are here much less important.

The benefit of reducing the beam velocity below the limit of the thermal beams is fully evident in table III and worth to do in the case of Ms and Ca. The accuracy and reproducibility will probably be increased to less than one part in 10^{-15}. In the case of Cs, on the contrary, the larger residual Zeeman effect is responsible of velocity independent frequency fluctuations, that will reduce the benefits of a deceleration.

LASER DECELERATION AND COLLIMATION

Laser deceleration of atomic beams

As discussed in the previous sections linewidth of 1.2 kHz was obtained in the case of the Mg thermal beam and an interrogation length of 30 cm. This length must be increased up to the impractical value of 6m in order to reach the minimum linewidth (65 Hz) allowed by the natural lifetime of the $^3P_1$ level. The same result could also be obtained with L=30 cm by reducing about twenty times the mean atomic velocity. This solution is not possible by using velocity selectors because of the enormous reduction of the beam intensity[2]. On the contrary, the laser deceleration can be a convenient solution as discussed below. The manipulation of the velocity of the beam atoms by absorption of resonant radiation was successfully demonstrated in 1933[23] as a deflection of Na atoms irradiated with the D lines photons from a spectral lamp. In fact, each time a photon is absorbed, its momentum is transferred to the atom, whose velctiy in the direction opposite to that of the impinging photon is reduced by the amount

382

$$\Delta v = h/\lambda M \tag{34}$$

where M is the atom mass. For the Na D lines it results $\Delta v = 2.3$ cm/s, a very small quantity compared to the average velocity of a thermal beam ($v \cong 7 \times 10^4$ cm/s). As a consequence spectral lamps with their maximum excitation rate of about 10^3 photons/s are not sufficient for obtain a significant deceleration of atomic beams. Only a laser source can have enough power density for practical applications[24].

Several techniques for realizing the laser deceleration (or laser cooling) have been proposed[25], but only the methods based on the resonant absorption have been successful up to now and will be discussed below.

In a two levels atom the maximum spontaneous emission rate R_M is reached when the transition is strongly saturated and, as a consequence, the levels equally populated

$$R_M = 1/2\tau \tag{35}$$

where τ is the spontaneous lifetime of the upper level. The spontaneously reemitted radiation is isotropic, its momentum is averaged out, and the atoms experience the acceleration

$$a_M = \frac{\Delta v}{2\tau} = \frac{h}{2 M \lambda \tau} \tag{36}$$

If the laser power is constant, an atom with an initial velocity v_0 is stopped after a minimum distance

$$L_m = \frac{v_0^2 \tau}{\Delta v} = \frac{v_0^2 M \lambda \tau}{h} \tag{37}$$

It follows that light weight atoms with a short lifetime and a short wavelength resonant transition are most easily decelerated. In case of a finite saturation, the spontaneous emission rate is given by

$$R(S) = \frac{1}{\tau} \frac{S}{1 + 2S} \tag{38}$$

where S is the saturation parameter. The previous equations become

$$a(S) = \frac{\Delta v}{\tau} \frac{S}{1 + 2S} \tag{39}$$

and

$$L(S) = L_m \left(1 + \frac{1}{2S}\right) \tag{40}$$

Also for a relatively small saturation the stopping length is of the same order of L_m, and a reasonably weak laser beam may be still useful. Meantime the momentum averaging of the spontaneously reemitted photons is not exactly zero. In fact, if N photons are scattered during the deceleration process, the transverse velocity of the atoms will execute a random walk of N steps, each of equal size Δv. Thus the final spread of the transverse velocity is approximately

$$\delta v \cong \Delta v \, N^{1/2} \qquad (41)$$

The actual excitation rate E and the saturation parameter are given by the equations

$$E = B_{12} \frac{I(\nu)}{c} = \frac{\lambda^2}{2\pi} \frac{g_2}{g_1} \frac{I(\nu)}{h\nu} \frac{\gamma^2}{(\nu - \nu_0)^2 + \gamma^2} \qquad (42)$$

$$S = \frac{B_{21}}{c \, A_{21}} \frac{I(\nu)}{} \left(1 + \frac{g_2}{g_1} \right) = \tau E \left(1 + \frac{g_2}{g_1} \right) \qquad (43)$$

where γ is the natural HWHM of the transition frequency, g_1 and g_2 are the degeneracy of the lower and upper level respectively, and $I(\nu)$ the intensity of the laser beam. In the case of Mg and Ca ${}^1S_0 - {}^1P_1$ transition it is convenient to have the quantization axis along the beam, since only the m=0 → m=+1 (or -1) transition is excited when the laser beam is circularly polarized and then is $g_1/g_2 = 1$.

Unfortunately, if the laser beam has a fixed frequency, only a small fraction of the atoms in the thermal velocity distribution can absorb the laser radiation. It consists of those atoms that have a velocity component along the beam that compensates the laser offset by means of the Doppler effect. As soon the deceleration process starts, the velocity is reduced and the atoms are Doppler shifted out of resonance. No more absorptions are possible, and the velocity distribution is only modified as shown in Fig. 8. Three methods have been proposed and experimentally demonstrated for avoiding the limitation on cooling imposed by the Doppler effect. The first was realized at Institute of Spectroscopy - Moscow[26]. It consists in increasing the laser power in order to have a very large saturation to increase the homogeneous linewidth ($\gamma = \gamma_0 \sqrt{(1+S)}$), and, in addition, to focusing the laser beam in such a way that the intensity increases as the atoms propagate toward it, giving increased saturation broadening as the atomic velocity is shifted farther from resonance. By this method a continous beam of cooled atoms can be obtained. Unfortunately it would require a presently unavailable laser power in the case of Mg and Ca.

The second method was also proposed by V.S. Letokhov et al[27] and consists in sweeping the laser frequency to keep it resonant with the Doppler shifted atoms during the deceleration. As a consequence all the atoms with a velocity slower than the initially resonant velocity v_i, are swept into a narrow velocity group around the final velociti v_f. Here $v_i - v_f = \lambda \Delta \nu$, where $\Delta \nu$ is the frequency tuning range of the laser. The method was fully demonstrated at JILA in the case of Na[28], and Cs[29]. Unfortunately, it will produce only pulsed cooled beam and, in addition, it will not realize a spatial compression if the velocities in the sense that atoms of a partic-

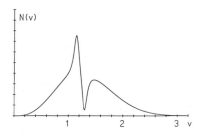

Fig. 8. Atomic velocity distribution in a thermal beam after deceleration with a laser beam of fixed frequency.

ular velocity appear at a particular time but spread out in space. Moreover commercially available Dye lasers do not scan rapidly enough, nor are suitable frequency modulator available. Semiconductor diode lasers are easier to operate, but their emission frequencies restrict the application to Cs only.

The third successful method changes the frequency of the atoms rather than that of the laser. This can be obtained with the Zeeman frequency shift produced by a magnetic field of appropriate intensity along the beam. With reference to Fig. 9, an atom with a velocity v has its peak absorption at the frequency

$$\nu_a = \nu_L \left(1 + \frac{v}{c}\right) = (\nu_0 - \Delta\nu)\left(1 + \frac{v}{c}\right) \cong \nu_0\left(1 + \frac{v}{c}\right) - \Delta\nu \tag{44}$$

where ν_0 is the peak absorption frequency in absence of Doppler effect, and $\Delta\nu$ is the laser detuning with respect to ν_0. In the case of a ${}^1P_1$ state is

$$\nu_a = \nu_0 \pm \mu B \tag{45}$$

with μ = 1.399 MHz/gauss. The resonant condition along the beam is

$$\mu B(z) = \nu_0 \, v/c - \Delta\nu \tag{46}$$

with a final velocity v_f, when B=0,

$$v_f = \lambda \Delta\nu \tag{47}$$

For a linear Zeeman effect, as in Mg and Ca, and a constant deceleration a is

$$v(z) = (v_0^2 - 2 a z)^{1/2} \tag{48}$$

and

$$B(z) = \frac{v_0}{\mu\lambda} \sqrt{1 - \frac{2 a z}{v_0^2}} = B_0 \sqrt{1 - \frac{z}{L(S)}} \quad . \tag{49}$$

In fact the above equation is just giving the minimum value of the field.

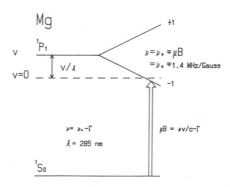

Fig. 9. Scheme of the Zeeman tuning of the laser cooling by means of an axial magnetic field.

Other field profiles are allowed but, since the acceleration has a finite value, there is also an upper limit on the field gradient, that is given by the derivative of eq. 48

$$\frac{dv}{dB} \cdot \frac{dB}{dz} \leq \frac{a}{v} \ \rightleftarrows \ \frac{dB}{dz} \leq \frac{a}{\mu \ \lambda \ v} \tag{50}$$

This restriction is not particularly stringent in the case of Mg and Ca. The method of Zeeman tuning will produce a continuous beam of cooled atoms by using a laser beam at a fixed frequency. It was experimentally demonstrated for the first time on a Na beam in the NBS laboratories[30], and its application to the case of Mg and Ca AFS is even more convenient as a consequence of the small saturation required ($S \cong 1$).

The numerical values of parameters involved in the laser cooling of Mg and Ca are given in Table IV, together with those of Na and Cs for comparison. As it can be seen less photons and a shorter distance are required in the case of Mg and Ca. Moreover, in spite of a laser beam in the ultraviolet and violet respectively, their atomic structure is much more convenient for laser cooling and AFS than that of Na and Cs. In fact:

i) – Na and Cs have a $^2S_{1/2}$ ground state which is split in two hfs sublevels. As a consequence a single frequency laser beam, coincident with a given hfs component, will cause a large optical pumping effect, the atoms will quickly accumulate in the other hfs sublevel and it will be impossible for them to scatter a sufficient number of photons to be significantly decelerated. To avoid the optical pumping effect it is necessary to use a two frequencies laser beam or two separate laser beams that are frequency tuned to excite the atoms from both the hfs level. It is also necessary to use a circularly polarized laser light to avoid Zeeman optical pumping and, as a consequence, the atoms are eventually cooled in the $m_F = F$ sublevel, a result not convenient in the case of AFS. On the contrary in Mg and Ca the ground state has J=0 and I=0: no optical pumping is possible, a single frequency laser beam is sufficient, and

Table IV. Numerical data on laser cooling (N is the number of photons
needed for cooling starting from the velocity v_0, and I_s is
the saturation intensity of the resonance position).

	^{23}Na	^{133}Cs	^{24}Mg	^{40}Ca
λ(nm)	589.0	852.1	285.2	422.6
τ(ns)	16	32	2.02	4.57
γ HWHM (MHz)	5	2.5	39	17.5
Δv (cm/s)	2.29	0.35	5.84	2.36
v_0 (m/s)	800	300	1000	800
N ($\times 10^3$)	35	85.7	17.1	33.9
L_m (cm)	45	82.3	3.46	12.4
v_t (m/s)	4.3	1.02	7.6	4.34
v_m (m/s)	0.41	0.12	1.13	0.41
I_s (mw/cm^2)	6.4	1.05	444	60

the circular polarization is used only to double the excitation
rate in presence of a magnetic field.

ii) - The Na and Cs ground state is strongly paramagnetic and the
cooled atoms are easily deflected by residual magnetic field
at the end of the coil. Mg and Ca have a completely diamagnetic
ground state.

iii) - in the case of Na and Cs γ is rather small (5 MHz and 2.5 MHz
respectively). The laser frequency stability and the magnetic
field gradients must be carefully controlled. This conditions
are an order of magnitude less stringent for Mg and Ca.

iiii) - In the case of Na and Cs the cooling laser beam interacts with
the ground state which is perturbed (light shift) by an amount
that will depend on the actual experimental apparatus. The
same level is also used for the reference transition of the
AFS and the fluctuations in frequency and intensity of the
laser beam will cause severe limitations in the performances
of the AFS. On the contrary, in the case of a Mg or Ca AFS the
reference transition is between the metastable triplet states
that are very far from resonance with the frequency of the
laser radiation.

In conclusion laser cooling is a much more convenient and simpler
method for improving Mg and Ca AFS than the Cs one. Of course a CW, frequency stabilized laser beam of sufficient power must be available at
285.2 nm and at 422.6 nm respectively for Mg and Ca. In the second case
a ring Dye laser using Stilben 3 pumped by UV ion lasers can give a laser
beam of about 100 mW in power and actively stabilized both in frequency

($\cong$1 MHz) and in power. In the Mg case a CW laser beam can be obtained only by frequency doubling the radiation at 570.4 nm obtained in a ring Dye laser by using Rhodamine 6G dye. Intracavity doubling is particularly efficient and is commercially available (Coherent 699-21-7500). We have obtained a laser beam of about 4$\cong$5 mW, single mode TEM$_{00}$ and with a r.m.s. linewidth of $\cong$2 MHz. With such a beam it will be possible to decelerate a Mg beam about 1.2 mm in diameter when S=1 in the near future.

Laser Collimation of atomic beams

As it was previously discussed the laser deceleration will increase the transverse velocity v_t, that cause a beam intensity reduction in the subsequent collimation needed in the interrogation region. In fact, even if the atoms are interrogated with a stationary electromagnetic field orthogonal to the mean beam direction, the transverse velocity component will increase the linewidth via the first order Doppler effect. In order to obtain the minimum linewidth expected from the interrogation time, the transverse velocity v_t must be smaller than

$$v_t \leq c/Q. \tag{51}$$

For Mg the maximum Q is $\cong 10^{10}$ and $v_t \leq 3$ cm/s. The collimation imposed by the optical Ramsey interrogation scheme is automatically sufficient to avoid a first order Doppler bradening but the intensity reduction will be large if the incoming beam is scarcely collimated.

Fortunately the laser radiation can also be used for the collimation of an atomic beam as it was proposed[31] and demonstrated recently[32]. If the atoms are irradiated orthogonally to the beam, and from opposite directions, with monochromatic radiation of frequency ν, they are subjected to a lateral acceleration, that, in the limit S<1, is[33]

$$a_t = \frac{h}{2 M \lambda \tau} \frac{S (\Lambda^- + \Lambda^+)}{1 + S (\Lambda^- + \Lambda^+)} \tag{52}$$

and

$$\Lambda^{+;-} = \frac{\gamma^2}{(\nu - \nu_0 \pm v_t/\lambda)^2 + \gamma^2} \tag{53}$$

where ν is the laser frequency. The acceleration is towards the beam axis if $\nu < \nu_0$ as shown in fig. 10. The maximum collimation is obtained when $\nu - \nu_0 = -\gamma$, is limited by the natural linewidth and is given by[31,34]

$$v_t^m = \sqrt{\frac{2 h \gamma}{M}} \tag{54}$$

The minimum transverse velocity will be 1.14 m/s and 0.41 m/s respectively for Mg and Ca.

388

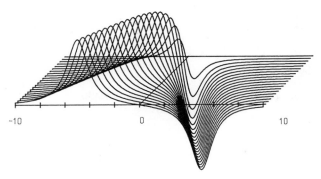

Fig. 10. Collimating acceleration induced by two counterpropagating laser
beams orthogonal to the atomic beam as a function of the transverse velocity and for several laser detuning from the line center.
The acceleration is towards the beam axis when the detuning is
negative. The detunings are in units of γ, from -5γ to 0 in steps
of .25 γ. The saturation is assumed to be S=0.1. The horizontal
axis reports the transverse velocity in units of γ ($1\gamma \cong 6$ m/s).

Excitation of the metastable levels

At the end of the deceleration process, the atoms are left in the
ground state and must be excited into the metastable ^3P states. The excitation can be obtained by inelastic collisions with electrons accelerated
along the beam axis. In the case of a thermal beam the momentum exchange
with the colliding electrons has a negligible effect on the velocity
distribution. On the contrary the excitation probability is proportional
to the interaction time and, therefore is larger for the atoms with smaller
velocity

$$p = j_e \sigma L'/v \tag{55}$$

where j_e is the current density, σ the excitation cross section, and L' the
interaction length between atoms and electrons. This effect was observed
as a decrease in the average velocity of the metastable atoms when j_e is
decreased[11]. A further decrease of the average velocity was observed in the
metastable atoms that make the $^3P_0 - ^3P_1$ transition since also the transition rate is a velocity dependent process[17,35]. In the case of a cooled
beam the velocity spread is reduced and the previous effect is negligible.
On the contrary the momentum exchange with the colliding electrons is
important in this case, and the changes in the velocity distribution must
be considered in detail.

The velocity v_B of the center of mass of the atom-electron system is

$$v_B \cong v^1 \pm \frac{m}{M} \sqrt{\frac{2 \, e \, V}{m}} \tag{56}$$

where M and m are the masses of the atom and electron respectively (m/M =
2.26×10^{-5} in the case of Mg), v^1 is the atom velocity in the laboratory
frame, V is the electron acceleration voltage, and the $\pm$ refers to copro-

pagating and counterpropagating electrons respectively. After an inelastic collision the energy eV_e (V_e = 2.72 V and V_e = 1.89 V respectively for Mg and Ca) is transformed in internal energy and the moduls of the atom velocity in the center of mass is

$$v_c \cong \frac{m}{M} \sqrt{\frac{2e}{m}} (V - V_e) \cong 13.37 \sqrt{V - V_e} \quad m/s \quad (Mg) \tag{57}$$

and

$$v_c^1 = v_B + v_c \tag{58}$$

where the numerical value refers to the case of Mg (for Ca is 8.0 m/s) and the velocities combine as shown in Fig. 11. Along the beam axis we have

$$v_c^1 = v_B \pm v_c = v^1 + \frac{\sqrt{2\ m\ e}}{M} (\pm \sqrt{V} \pm \sqrt{V - V_e}). \tag{59}$$

When $V = V_e$ the collision is completely inelastic, the atom velocity is in the beam direction, and is increased of decreased of 22.0 m/s and 11.0 m/s respectively for Mg and Ca. This is obviously the most convenient situation, but, unfortunately, a spread in the electron velocities is unavoidable, and, according to Fig. 11, there is a deflection out from the beam axis. The maximum deflection angle θ^M is

$$\theta^M = \sin^{-1} \left(\frac{v}{v_B}\right) = \sin^{-1} \left[\frac{\sqrt{2me\ (V - V_e)}}{M v^1 \pm \sqrt{2meV}} \right] \tag{60}$$

where the ± refers to the copropagating and to the counterpropagating case respectively.

The actual velocity spread will also depend on the differential cross section $\sigma(\theta)$ which is expected to be large for small angles and much smaller for $\theta \cong 180°$. The only available data[36] were obtained with a minimum energy of 10 eV. It is however possible to conclude that the velocity along the beam will be

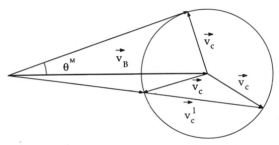

Fig. 11. Scheme of the velocity modification after an inelastic electron collision.

$$v_c^1 = v_B - v_c = v^1 + \frac{\sqrt{2\,m\,e}}{M}\,(+\sqrt{V} - \sqrt{V - V_e}) \qquad (61)$$

for a copropagating beam, and

$$v_c^1 = v_B + v_c = v^1 + \frac{\sqrt{2\,m\,e}}{M}\,(-\sqrt{V} + \sqrt{V - V_e}) \qquad (62)$$

for a counterpropagating beam so that the final velocity will be about the same in both cases. The dependence of θ^M from V and for practical values of v^1 is shown in Fig. 12 in the case of Mg. The final velocity along the beam axis is larger in the case of a copropagating electrons beam and this has been taken into account in the figure. Nevertheless, by comparison of near equivalent final velocity curves we can see that the lateral velocity spread is smaller in the case of a copropagating electrons beam, that is to be preferred in the experiments.

In view of this results it follows that a laser collimation after the deceleration is apparently of no particular convenience because the velocity spread caused by the excitation is in any case larger. Laser collimation before the deceleration shall be useful for counterbalancing the beam intensity losses caused by the geometrical collimation in the interrogation region.

Before concluding this section, it is worth considering also the purely elastic collisions, which do not contribute to the excitation but are responsible for a large velocity spread. The cross section for elastic collisions is expected to be about 10~15 times larger than the excitation cross section[36]. As a consequence the electron current density must be limited to about 100 mA/cm^2 to avoid that a large fraction of the metastable atoms would scattered out of the beam. This loss can be avoided by performing a laser collimation also in the excitation region.

Laser deceleration of Calcium atoms

The Ca atom has the resonance line $^1S_0 - {}^1P_1$ at a favorable wavelength, $\lambda = 422.6$ nm, where the stilben dye laser can provide more than 100 mW of output power. The Ca level scheme is given in Fig. 4. Unfortunately the lowest $^1D_2$ level is less energetic than the $^1P_1$ level to which it is connected by an allowed electric dipole transition in the middle infrared, $\lambda = 5547$ nm. As a consequence a population transfer from the $^1S_0$ level to the $^1D_2$ level is possible during the deceleration process (optical pumping). From the $^1D_2$ level the atoms can return to the ground state only via an electric quadrupole transition of very long lifetime and, as a consequence, are lost for the deceleration process. The amount of this unwanted effect will depend on the branching ratio R between the $^1P_1 - {}^1S_0$ and the $^1P_1 - {}^1D_2$ spontaneous transition probabilities. Since about 10^4 spontaneous emission are necessary for the deceleration, the optical pumping effect must be carefully considered.

A first simple solution for realizing a successful laser cooling can

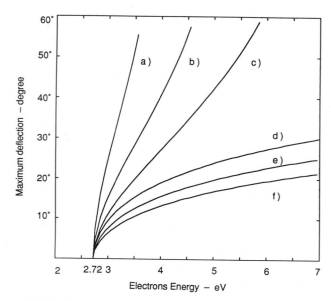

Fig. 12. Maximum deflection angle after inelastic collisions with counterpropagating (curves a, b, and c) and copropagating (curves d, e,
and f) electrons as a function of the electrons energy. The initial
atomic velocity in the laboratory frame is assumed to be 40, 50,
and 60 m/s for the curves a, b, and c respectively, and 20, 30,
and 40 m/s for the curves d, e, and f respectively.

be the use of a second laser beam for pumping back the atoms to the $^1S_0$ or
$^1P_1$ levels. A dye laser tuned at 671.1 nm can excite out from the $^1D_2$ level
toward upper levels and than to the ground state. An even more convenient
and practical solution is to use a laser tuned directly on the $^1D_2 - {}^1P_1$
line at 5547 nm. The feasibility of such a laser, also with operation in
CW regime, has been demonstrated with a Ca discharge in a special gas
mixture[37].

It is however important to determine the best value of R and to calculate its influence on the cooling process before increasing the complexity
of the experimental apparatus. A rough calculation of the $|<^1P_1|r|^1D_2>|$
integral would give $R \cong 10^4$, a value to small to avoid the use of a second
laser beam. More accurate calculations have shown that the radial integral
value is smaller than expected and the branching ratio R can be larger by
an order of magnitude. By considering all the available data[11,38,39], R can
be estimated to be in the range $0.9 \times 10^5 < R < 5 \times 10^5$ with a most probable
value of

$$R = 1.5 \times 10^5$$

This value is also supported by the recent experimental measurement
of the corresponding transition in the Sr atom[40]. This large R value, which
is much more favorable for laser cooling, corresponds to a transition
probability of the same order of the spin forbidden ($\Delta S \neq 0$) intercombina-
tion lines, that must also be considered. The most probable values of

transition rates of this lines are shown in Fig. 13. The $^1P_1$ state has two more weak decay modes toward the $^3D_1$ and $^3D_2$ states, while the most probable decay mode of the $^1D_2$ state is toward the metastable $^3P_1$ and $^3P_2$ states. Only 10% of the atoms in the $^1D_2$ state return directly to the ground state emitting photons at 457 nm of the electric quadrupole transition; 69% decay into the $^3P_1$ level and then into the ground state emitting the red inter-combinating line at 657 nm; finally 21% decay into the very long lifetime $^3P_2$ level. This theoretical prediction is in agreement with the experimental observation that the blue fluorescence at 457 nm is much weaker than the red fluorescence at 657 nm in a beam of metastable Ca in spite of an experimental lifetime of $2.3(5) \times 10^{-3}$ s[40].

In conclusion only about 10^{-5} of the atoms excited into the $^1P_1$ state do not return directly to the ground state. About 19% of this small fraction is eventually trapped in the $^3P_2$ and $^3P_0$ metastable states, the remaining 81% returns to the ground state in about 3 ms, a time longer than the deceleration time.

From the above data it is possible to calculate the fraction of Ca that can be completely decelerated by using only a resonant laser beam at 422 nm. The probability for an atom to be excited n times is

$$P_s = \left[\frac{R}{R+1}\right]^n \cong \exp\left[-n/R\right] \tag{63}$$

and the probability of not decay in the ground state at the n+1 excitation is

$$P_D = \frac{1}{R+1}\left[\frac{R}{R+1}\right]^n \cong \frac{1}{R}\exp\left[-n/R\right] \tag{64}$$

If $R \to \infty$, the fraction of atom decelerated from the initial velocity x_0 to the final velocity x_f is

$$N(x_0, x_f) = \int_{x_f}^{x_0} N(x)\, dx \tag{65}$$

where $x = v/\beta$. Otherwise

$$N(x_0, x_f, R) = \int_{x_f}^{x_0} N(x)\, \exp\left[-K(x - x_f)\right] dx \tag{66}$$

where $K = \beta/(R\Delta v)$. The integrals are shown in Fig. 14 as a function of x and for $R = \infty$, 5×10^5, 1.5×10^5, 0.9×10^5 respectively, and T=960 K° (β =632 m/s).

Also the velocity distribution of the atom lost during the deceleration can be calculated: the probability that an atom with initial velocity v' will be lost at a velocity v is

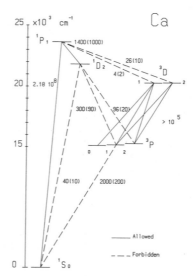

Fig. 13. Levels scheme and probability, in units of s^{-1}, of the allowed
and forbidden radiative transitions from the $^1P_1$ and $^1D_2$ levels
of the calcium.

$$P\ (D,\ v',\ v)\ =\ \frac{1}{R+1}\ \exp\left[-\ \frac{v'\ -\ v}{R\ \Delta v}\right] \tag{67}$$

and the velocity distribution for an initial velocity x_0 is

$$N^D\ (R,\ x_0,\ x)\ =\ K\ \int_x^{x_0}\ N\ (x')\ \exp\left[-\ K\ (x'\ -\ x)\right]\ dx' \tag{68}$$

and is shown in Fig. 14 for the above R values.

In conclusion a sufficient number of Ca atoms could be decelerated by using only one laser beam and without increasing the complexity of the experimental apparatus.

Benefits from laser deceleration and collimation

The laser deceleration will be very important for reaching the maximum available Q and for drastically reducing the small perturbations that are limiting the accuracy and reproducibility of the AFS as discussed in the previous sections. However other practical benefits are obtained from the laser cooling:

i) the decay length from the $^3P_1$ state is reduced proportionally to the velocity. As a consequence the background radiation in the detection region will be largely reduced, while the signal radiation will be almost completely emitted in front of the photomultiplier increasing the collection efficiency towards the unity. The combination of these effects will significantly increase the S/N ratio of a laser cooled AFS.

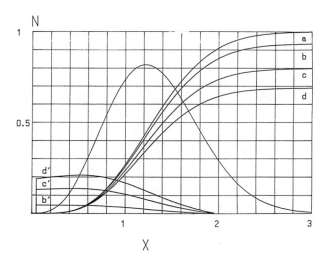

Fig. 14. Plots of the function $N(x)$ and of its integral (curve a) as a function of x. The function $N(x_0, x_f, R)$ is shown for $x_f = 0.05$ and $R = 5 \times 10^5$ (curve b), 1.5×10^5 (curve c), and 0.9×10^5 (curve d) respectively. The function $N^D (R, x_0, x)$ is also shown for $x_0 = 2$, $x_f = 0.005$, and the same values of R (curves b', c', and d').

ii) The submillimeter power necessary to interrogate the reference transition is reduced proportionally to v^2. As a consequence the presently available power at 601 GHz will become sufficient to reach the optimum excitation rate. Moreover the probability to have, in the near future, coherent sources for exciting also the other fine structure transitions in Mg and Ca is also increased.

These additional benefits can compensate the decrease in the atomic beam intensity as a consequence of the laser deceleration and of the metastable levels excitation process. A S/N ratio of $10^4 \cong 10^5$ can be forecast for a decelerated beam and the present experimental apparatus.

It is evident that with the temporal states selection method a popula-tion inversion is obtained between the $^3P_2$ and $^3P_1$ levels. Unfortunately the small signal gain of the $^3P_2 - ^3P_1$ transition is too small in the case of a thermal beam and maser action cannot be obtained. However the gain is proportional to $1/v^2$ until the interrogation line width is equal to the natural line width of the $^3P_1$ level. In the case of Ca the presently avail-able laser power can be of the order of a several hundred mW and the flux of the decelerated Ca atoms as large as $10^{12} \cong 10^{13}$ atoms/s. If the beam is also laser collimated, the population inversion between the $^3P_2$ and $^3P_1$ levels will be possibly large enough to reach the threshold condition for maser oscillation at 3.17 THz. Such a maser is expected to have an excellent stability with an output power of a few nW. In any case a very efficient laser deceleration and collimation will be necessary.

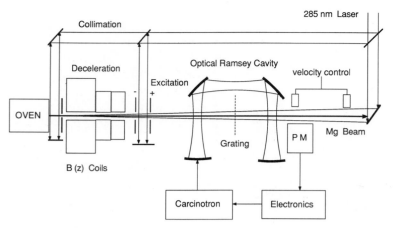

Fig. 15. Scheme of the experimental apparatus of an AFS based on a laser
cooled Mg beam. The velocity monitor signal can be used for the
long term stabilization of the laser frequency.

DESIGN OF A COOLED AFS AND EXPECTED PERFORMANCES

The scheme of an AFS using a beam of decelerated Mg atoms is shown in
Fig. 15. The previous analysis of the metastable states excitation by inelastic collision with low energy electrons suggest that the most convenient
final velocity should be in the 20≅40 m/s range, about 20 times less than
the average velocity of the thermal beam. It is worth noting that the natural
linewidth of the 285 nm Mg line is equivalent to a velocity of 12 m/s and,
as a consequence, a Mg beam cannot be decelerated to a velocity smaller
than about 15 m/s without the losses of the atoms that are stopped and sent
back toward the oven. The laser collimation, which require only a small
fraction of the available laser power, can be used in several places along
the beam in order to increase its intensity. A first collimation could be
performed at the oven exit before the deceleration region: the beam intensity
could then be increased up to about an order of magnitude before the Mg–Mg
elastic collisions become important. Second, the cooling laser beam could
have a convergence toward the oven to maintain the atomic beam divergence
within the limits of eq.(54). Third, the laser collimation will be certainly
important in the metastable staes excitation region, since it can compensate
the beam divergence increase due to the atom–electron elastic collisions.
The electrons current density could then be increased to obtain a metastable
excitation fraction of about 10≅30% as in the case of the thermal beams. The
presence of the decelerating laser beam in this region will also compensate
the longitudinal velocity variation from the elastic collisions. It is worth
noting that the same laser frequency offset can be used for both deceleration
and collimation.

A magnet for the Zeeman deceleration of both Mg and Ca has been realized
by assembling three coils encapsulated in a soft iron sheet. A soft iron
magnetic pole has been also placed at the coil entrance in order to increase
the magnetic field intensity and asymmetry. The measured field intensity

396

along the beam axis is shown in Fig. 16. The deceleration length will be 19 cm, in the Mg case, with a saturation parameter S=0.35 and an initial field strength B=0.376 T, corresponding to an initial velocity of 1490 m/s (x≅2). The presently available laser power of 5 mW at 285 nm will allow the deceleration of an atomic beam of 0.18 cm in diameter.

With a final velocity of about 30 m/s, an optical Ramsey interrogation region of 20≅30 cm in legnth is sufficient for reaching the maximum line Q allowed by the natural linewith of the $^3P_1$ level. A detection region of about 10 cm in legnth will also be sufficient for a photon collection efficiency ≥50%.

Moreover the nondestructive detection method of the reference transition by using the fluorescence photons emitted from the $^3P_1$ level has two important advantages compared to the Cs AFS. First the beam velocity can be continuously measured without interfering with the standard operation by using two additional detectors, which look to the beam fluorescence at two short regions l' separated by a distance L'. The velocity is then given by

$$v = \frac{L'}{\tau \ln (I_1/I_2)} \tag{69}$$

where I_1 and I_2 are the signals from the first and the second detector respectively. This signal can be used in a feedback to control the laser frequency in order to maintain the beam velocity constant. The frequency

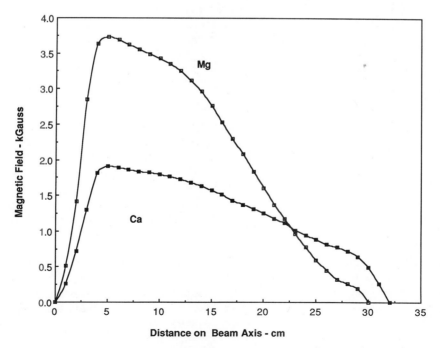

Fig. 16. Magnetic field intensity along the beam axis obtained with a three coils system.

fluctuations of the reference transition induced by the second order Doppler effect will then greatly reduced: as an example the frequency error will be only one part in 10^{-16} in the case of v=30 m/s stabilized within 1%.

The second advantage consists in the possibility to use the decelerated beam a second time by repeating the collimation excitation process after the detection region. A second AFS could then be realized on the same experimental apparatus. It will be partially correlated to the first only for the fluctuations in the beam velocity. Also the signal will be of course smaller by about one order of magnitude. However such a standard will be certainly useful for the measurement of the cooled AFS stability in consideration of the cost and complexity of the laser system.

In conclusion the total length of a laser cooled AFS will be about 1m. With a S/N ratio in the range 10^4-10^5, the expected stability is $\sigma(\tau=$ 1 s)$\cong 10^{-14}\cong 10^{-15}$ and will be increased by a factor three when it will become possible to use in the future the $^3P_2$ - $^3P_1$ transition at 1220 GHz as a reference. A comparison between the present and future stability of a Mg AFS and that of the Cs and H AFS is shown in Fig. 6. The Hewlett-Packard Cs atomic beam is also shown for comparison because it is the best commercial AFS presently available.

The expected precision and reproducibility can be inferred from the discussion of table III, where the dependence of the effects upon the beam velocity is given. In many cases the dependence follows from the decrease of the effect with the increase of the reference line Q. The contribution of the effects from 1 to 7 will become irrelevant, that from 10 to 12 depend primarily from the quality of the electronic components and will in any case be reduced in the future. The cavity phase shift will be probably the largest source of errors also for the Mg AFS. In any case it is reasonable to forecast an improvement of at least an order of magnitude of the stability and reproducibility with respect to the Cs AFS.

CONCLUSION

From the above discussions it is evident that the submillimeter AFS based on an atomic beam of Mg or Ca will be in the near future superior to the Cs AFS both in stability and in precision. The atomic structure of Mg and Ca is particularly favorable for the laser cooling, which is essential to reach the best expected performances. However two major technical problems are presently challenging the full development of the laser cooled AFS:

    i) the reliability and the long term performance of the dye laser. The present apparatus are also excessive in volume, power consumption, and cost. Improvements of the pump ion lasers and of the dye laser frequency doubling are desired and expected in the future.

   ii) The development of radiation sources above the THz with the required frequency, power, stability, and spectral purity.

The laser cooling can also be used on the Cs beam AFS. However in this case the expected advantages are less relevant than for Mg and Ca. In fact the frequency of the reference line is lower and the transition linewidth must be reduced below 1 Hz in order to reach a Q 10^{10}. The interrogation time will then be so long that the scattering with the background gas could become a serious limitation. It is worth noting that the Cs atoms have a very large spin-exchange cross section (2.20×10^{-14} cm^2)[45] and saturated vapor density (3×10^{10} atoms/cc at 21 °C).

Another kind of AFS that will offer a precision and reproducibility better than that of Cs, are the standards based on microwave transitions observed in stored ions. In this case the interrogation time can be very long; unfortunately the sample of the reference atoms must be very small and the S/N ratio poor. The stability will then be unsatisfactory for an AFS, and devices based on the ion storage will be probably useful only for the long term calibration of more stable AFS of different kind.

As a final remark it is worth noting that the present time scale is based only on the Cs clocks, and the long term precision is evaluated only by comparisons between Cs clocks. The development of clocks of different kind will give the opportunity to check if the present time scale is affected by some unexpected systematic drift.

REFERENCES

1. E. Bava, A. Godone, G. D. Rovera, Infrared Phys., 23, 157:160, (1983)
2. F. Strumia, Metrologia, 8, 85:90 (1972)
3. C. Audoin, The Cs Beam frequency standard etc., in: "Metrology and Fundamental Constants", A. Ferro Milone, P. Giacomo, and S. Leschiutta eds., North-Holland 1980, 169:259
4. L. Lewis, Prog. Quant. Electr., 8, 153 (1984)
5. S. R. Stein, Prog. Quant. Electr., 8, 129 (1984)
6. N. F. Ramsey, "Molecular Beams", Oxford University Press, (1956) and (1985)
7. F. L. Walls, A.De Marchi, IEEE Trans. Instrum. Meas., IM-24, 210:217 (1975)
8. A. Godone, A. De Marchi, E. Bava, "Proc. 33rd Ann. Symp. Freq. Control" Atlantic City, N.J., 498:503 (1979)
9. S. R. Stein, A. S. Risley, H. Van de Stadt, F. Strumia, Appl. Opt., 16, 1893:1896 (1977)
10. A. De Marchi, A. Godone, E. Bava, IEEE Trans. Instrum. Meas., IM-30, 132:138 (1981)
11. G. Giusfredi, P. Minguzzi, F. Strumia, M. Tonelli, Z. Physik, A274, 279:287 (1975)
12. A. De Marchi, E. Bava, A. Godone, G. Giusfredi, IEEE Trans. Instrum. Meas., IM-32, 191:197 (1983)
13. G. Kramer, J. Opt. Soc. Am., 68, 1634 (1978)
14. A. Godone, A. De Marchi, G. D. Rovera, E. Bava, G. Giusfredi, Phys. Rev., 28, 2562:2564 (1983)
15. E. Bava, A. Godone, A. De Marchi, G. D. Rovera, G. Giusfredi, Opt. Commun., 47, 193:195 (1983)

16. A. Godone, E. Bava, G. Giusfredi, Z. Physik, A318, 131:134 (1984)
17. A. Godone, E. Bava, A. De Marchi, G. D. Rovera, G. Giusfredi, IEEE Trans. Instrum. Meas., IM-34, 129:132 (1985)
18. A. Godone, E. Bava, G. Giusfredi, C. Novero, Yu-zhu Wang, Opt. Commun., 59, 263:265 (1983)
    E. Bava, A. Godone, G. Giusfredi, C. Novero, "Metrologia", in press
    E. Bava, A. Godone, G. Rietto, Appl. Phys., B41, 187:196 (1986)
19. E. Bava, A. Godone, Private Communication
20. A. G. Mungall, H. Daams, J. S. Boulanger, IEEE Trans. Instrum. Meas., IM-29, 291:297 (1980), and Metrologia, 17, 123:145 (1981)
    A.G. Mungall, C. C. Constain, IEEE Trans. Instrum. Meas., IM-32, 224:227 (1983)
21. N. Beverini, F. Strumia, Quaderni della Scuola Normale Superiore, Volume in honour of Adriano Gozzini, 361:173, Pisa 1987
22. E. Bava, A. De Marchi, A. Godone, Lett. Nuovo Cimento, 38, 107:110 (1983)
23. R. Frisch, Z. Physik, 86, 42 (1983)
24. T. W. Hänsch, A. W. Schawlow, Opt. Commun., 13, 68 (1975)
25. W. D. Philips, J. V. Prodan, H. J. Metcalf, J. Opt. Soc. Am., B2, 1751:1767 (1985)
    W. D. Philips, Ann. Phys. Fr., 10, 717:732 (1985)
26. V. I. Balykin, V. S. Letokhov, A. I. Sidorov, Opt. Commun., 49, 248 (1984) and Zh. Eksp. Teor. Fiz., 86, 2019 (1984)
27. V. S. Letokhov, V. G. Minogin, B. D. Pavlik, Opt. Commun., 19, 72 (1976)
28. W. Ertmer, R. Blatt, J. Hall, M. Zhu, Phys. Rev. Lett., 54, 996 (1985)
29. R. N. Watts, C. E. Wieman, Opt. Lett., 11, 291 (1986)
30. W. D. Philips, J. V. Prodan, H. J. Metcalf, A. Migdall, I. So, J. Dalibard, Phys. Rev. Lett., 54, 992 (1985)
31. V. I. Balykin, V. S. Letokhov, V. G. Minogin, T. V. Zueva, Appl. Phys., B35, 149:153 (1986)
32. V. I. Balykin, V. S. Letokhov, A. I. Sidorov, Pis'ma Zh. Eksp. Teor. Fiz., 40, 251 (1984)
    V.I. Balykin, V.S.Letokhov, V. G. Minogin, Yu . V. Rozhdestvensky, A. I. Sidorov, J. Opt. Soc. Am., B2, 1776 (1985)
33. V. G. Minogin, Opt. Lett., 10, 179 (1985)
    V. S. Letokhov, V. G. Minogin, Phys. Rep. 73, 1 (1981)
34. V. S. Letokhov, Comm. At. Mol. Phys., 19, 119 (1987)
35. E. Bava, A. Godone, G. Giusfredi, C. Novero, IEEE J. Quantum Electr., QE23, 455:457 (1987)
36. W. Williams, S. Trajmar, J. Phys. B, 11, 2021:2029 (1978)
37. V. M. Klimkin, S. S. Monastyrev, V. E. Prokopev, JETP Lett., 20,110, (1974)
    V. M. Klimkin, P. D. Kolbycheva, Sov. J. Quant. Electr., 7, 1037 (1978)
38. P. Hafner, W. H. E. Schwarz, J. Phys. B, 11, 2975 (1978)
    R. N. Diffenderfer, P. J. Dagdigian, D. R. Yarkony, J. Phys. B, 14, 21 (1981)
    C. W. Bauschlicher, S. R. Langhoff, H. Partridge, J. Phys. B, 18, 1523 (1985)
    K. Fukuda, K. Ueda, J. Phys. Chem., 86, 676 (1982)
    S. K. Peck, Optics Commun., 54, 12 (1985)
39. L. Pasternak, D. M. Silver, D. R. Yarkony, P. J. Dagdigian, J. Phys.

B, 13, 2231, 1523 (1980)

40. R. L. Hunter, W. A. Walker, D. S. Weiss, <u>Phys. Rev. Lett.</u>, 56, 823 (1986)

41. C. E. Moore, "Atomic energy levels", NSRDS-NBS 35 (1971)

42. M. Inguscio, K. R. Leopold, J. S. Murray; K. M. Evenson, <u>J. Opt. Soc. Am.</u>, B2, 1566 (1985)

43. D. A. Landman, A. Lurio, <u>J. Opt. Soc. Am.</u>, 60, 986 (1970)

44. W. L. Wiese, M. W. Smith, B. M. Miles, "Atomic transition probability", vol. II, NSRDS-NBS 22 (1969)

45. N. Beverini, P. Minguzzi, F. Strumia, <u>Phys. Rev.</u>, A4, 550 (1971)

Note added in proof

After the preparation of the manuscript, further experimental results have been obtained:

- the absolute wavelength and the isotope shifts of the Mg resonance line have been measured with sub-Doppler precision on a Mg atomic beam[1]: λ_{vac} = 285.2963(1) nm;

- new far-infrared laser lines have been discovered in $^{13}CH_2F_2$ and in CH_3OH (see ref. 2), whose frequency is very close to that of the Mg: $^3P_1-^3P_2$ and to the Ca: $^3P_1-^3P_0$ lines respectively. This laser could be phase locked to the atomic transitions if the efficiency of the sideband generator will be improved;

- the Ca R branching ratio has been measured[3]: R = 0.592×10^5 ($0.467 <$ R $< 0.81 \times 10^5$), and found to be in agreement with the estimated value thus confirming the possibility of a direct Ca laser deceleration.

REFERENCES

1. N. Beverini, E. Maccioni, D. Pereira, F. Strumia, G. Vissani, Wang Yu-zhi: AIP Proc. Third Int. Laser Scien. Conf., 1987, in press

2. N. Ioli, A. Moretti, D. Pereira, F. Strumia: Conf. Digest XII Int. Conf. Infrared MM Waves, IEEE cat N.87CH2490-1, p.61-62 (1987)

3. L. P. Lelluch, L. R. Hunter, <u>Phys. Rev.</u>, A36, 3490-3493, (1987)

DIGITAL LOGIC ELEMENTS FOR OPTICAL COMPUTING

F. A. P. Tooley

Department of Physics, Heriot-Watt University

Edinburgh EH14 4AS, UK

INTRODUCTION

Recently the field of optical computing has grown to encompass a new group of workers who are investigating the applications of digital optics. The foundation of this subject is a property that a broad range of material posses - optical bistability. The primary concern of this paper is the prospects for the use of arrays of optically bistable devices in optical computing. The aim of this work is to provide details of the necessary requirements of a digital logic element if it is to be suitable for inclusion in an optical computing system.

The latter part of the introduction will describe the basis of the interest in digital optical computing. In the second section the classification of optically bistable devices in outlined and the description in detail given of the properties of passive intrinsic devices utilising a refractive nonlinearity. The third section describes experimental results with such systems and provides a comparison of the performance of bistable devices currently under investigation. Finally, details of the construction of the first all-optical digital circuits in the form of a classical finite state machine will be presented.

Recently the adoption of optical techniques within computing systems - for interboard communication for example, has focussed considerable attention on whether an optical computer could be constructed. An up-to-date review of most of the main fields within optical computing has recently been published[1]. It is instructive to outline why optical computing is of interest at all. The communications problem from which electronic devices suffer is illustrated by an example presented by R. J. Keyes of IBM at the OSA Topical Meeting on Optical Computing in March 1987; transistors are capable of switching on or off in 10 picoseconds, a ring oscillator built from such devices might have a switching time of 30 ps, however when integrated into a chip the switching time increases to over 100 ps and when these chips are incorporated into a board 1 ns switching results. The minimum computer cycle time required for reliable operation may be

around 10 ns. The problem arises because of the need for synchronous operation and the domination of the 'RC' time constant associated with char-
ging the wire connection between gates, a time constant which does not
reduce as the size of circuit is decreased[2]. Light may therefore have a
role to play by allowing faster electronic gates to be used, the large
bandwidth of the visible or near infrared radiation used, playing a crucial part.

In addition to being constructed with gates which have improved switching speeds of around three orders of magnitude the six orders of magnitude increase in the processing power of present-day digital electronic
computers over those constructed 30 years ago is due to a change in architecture and in particular the degree of parallelism used. The communications problem within parallel electronic computers suggests the use of
optics since, for example, the capability of a single lens to resolve millions of points in its field of view can be regarded as the ability to
form millions of independent communication channels. This concept of massive parallelism of communication channels is a mainstay of the justification of optical computing. If a system is capable of processing millions
of bits and communicating the results simultaneously the potential processing power may be huge. However, the type of problem which such a computer
is able to deal with more effectively than conventional computers may be
restriced to those in which the data is already in the correct format such
as an image. Consequently most proposed optical computers are special purpose machines designed to solve problems such as image processing, pattern
recognition and matrix algebra[3]. Thus the architecture appropriate to an
optical computer will, in general, be similar to that used in a classical
finite state machine, one which allows simultaneous interrogation of all
the memory allowing parallel processing as opposed to the architecture
used in conventional machines which is based on von Neumann's invention
of address-orientated communication with memory.

A further problem which arises when electronic computers are run at
speeds which cause communication times to be dominant is that of clock-
skew which is the timing jitter caused by unequal signal path lengths
within a circuit. A lens' ability to image is a demonstration that optics
is in principle capable of a reduced degree of clock-skew since Fermat's
principle ensures that every path length will be identical.

The primary advantage of an optical computer is thus the high bandwidth of light giving large information capacity and the ability to form
non-interfering channels in massive numbers. However clearly other justification is required to adopt optics since although it is obvious that
beams of light have some advantages over currents in wires the generation
of arrays of beams all of which may have to be independently addressable
presents considerable technical problems. For example, two-dimensional
arrays of laser diodes are not yet commercially available and when they
are, will be unlikely to cost less than several orders of magnitude more
than arrays of transistors.

Foremost amongst this further justification required is the possibility for the adoption of swhitching systems and computer architectures

which are difficult to implement using wire-based communication. As three
examples, the perfect shuffle switching system, neural networks and symbolic substitution are illustrative. The perfect shuffle is a global interconnection network used in parallel computers to perform sorting and fast
Fourier transforms. If there are a number N of input and output lines, all
possible one-to-one interconnections can be achieved using $\log_2$ N perfect
shuffles and nearest-neighbour exchanges. As an example for 8 channels of
input numbered 1 to 8, from top to bottom, after a perfect shuffle the
order of channels would be 1,5,2,6,3,7,4,8, i.e. the lower four channels
have been interlaced with the upper channels. The interconnection then
proceeds by making four decision on 1 an 5; 2 and 6; 3 and 7 and 4 and 8,
whether the ultimate output channel to which this information should be
sent is toward the top or bottom, so they are either interchanged or not.
If this process is repeated three times any arbitrary arrangement of the
initial pattern is possible. This switching scheme can be implemented using
wires but all wire lengths must be padded out to make them equal to ensure
the signals arrive at the exchange/bypass logic simultaneously, for a large
number of channels the resulting complexity of wiring is prohibitively
expensive; particularly since it may be necessary to extend the shuffle
to two dimensions. Optics, however, can implement the 2D perfect shuffle
using either a beam splitter and two mirrors or four prisms and two lenses[4].

The optical implementation of neural networks is reviewed in a recent
article by Abu-Mostafa and Psaltis[5]. Neural networks operate using large
(≤ 100) and variable fan-in and fan-out of signals into each neuron, a situation which is difficult to mimic using electronic gates Optics may be
useful in the implementation of these networks because of the high space-
bandwidth available of the storage and interconnection media that can be
used, such as holographic materials.

The power of analogue optical computing is in its ability to process
all of the information in an image simultaneously, for example in the use
of matched filters in a Vanderlugt correlator. The same principle is used
in symbolic substitution[6]. Symbolic substitution however is proposed as a
means to perform highly parallel digital computations optically. In such a
system, information is distributed throughout an image in the form of
bright and dark pixels to form patterns, the change in position and transformation of these spacial patterns can be used to perform logic functions,
digital arithmetic and even the programming of an optical computer. It has
recently been shown that such an architecture can be used to produce a general purpose computer using regular interconnections between arrays of
logic gates[7].

The proposed scheme for optical computing produce a requirement for
the optical equivalent of the transistor. Symbolic substitution, for example, will be implemented using a configuration of masks, mirrors, beam
splitters/combiners and devices which determine whether the light level
is above a threshold. Further, since the architecture requires that this
new, threshold image is processed by the same scheme several times, there
is a requirement for an optical output and one which is at a power level
equal to that in the original processed image. The requirement is thus for

a logic gate with gain with optical signal input and output which can be fabricated in the form of a two-dimensional array of uniform response. The following section is a discussion of whether arrays of optically bistable devices can fill this role.

OPTICAL BISTABILITY

Optically bistable (OB) devices are optical elements which, over some range of light input powers, have two possible output states. Figures 1a and b show two typical input/output characteristics for such devices. The range of powers over which they are bistable corresponds to a region of hysteresis and is bounded by two discontinuities at which switching between the two states can occur. Such switching can be induced by holding close to one of these discontinuities and making an incremental change in the total light input. This power increment can derive from a completely independent (signal) input. It is possible to obtain a change in output larger than the signal input and hence achieve digital gain. Furthermore under the appropriate conditions these devices can have steep non-hysteretic characteristics (see Fig. 1c and d) such that analogue gain can be obtained - making them the optical equivalents to the transistor.

Since the first demonstration of devices of this sort in the late 1970's there have been a large number of experimental OB systems reported. In all cases switching occurs as a consequence of positive feedback: sometimes arising from an intrinsic nonlinear absorption process; sometimes provided by an optical cavity, and sometimes by an external electrical feedback mechanism.

Fig. 2 summarizes the way in which the many optically bistable devices can be sub-divided into four main classes: I-IV. The first division is between active and passive. Active devices are those where the total light output can be more than the total light input - i.e. those having an internal gain stage or light source. Passive devices have no light generating or amplifying components and consequently only produce signal gain by transferring power from one light beam to another. In these the power source is in the form of an optical input.

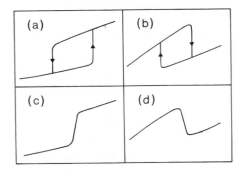

Fig. 1. (a) and (b): typical input/output characteristics for optically bistable (OB) devices
(c) and (d): input/output characteristics for nonlinear optical devices showing regions of analogue differential gain.

Both active and passive OB devices can be further divided into hybrid and intrinsic. A hybrid device is one in which the feedback required is provided electrically. This may involve some external circuitry or be achieved by the internal electrical characteristics of the device. The switching speeds of hybrid devices are always ultimately limited by the time constants of the electrical feedback. Intrinsic OB devices are constructed from optically nonlinear media and rely on either direct optical feedback or the nature of the nonlinearity itself. In this case switching speeds are determined by medium time constants or, in very fast devices, the optical feedback time.

A review[8] concentrating on the compact OB devices that fall within these four classification areas and excluding the larger devices based around nonlinear molecular gases; atomic vapour and gas discharge or optically pumped lasers concludes that Class IV devices (active hybrid) are only of technical interest in integrated form and if 2D arrays can be fabricated.

The Class III devices (active/intrinsic) of technical interest, are all based around laser diodes as the active components; sixteen distinct system which have been studied experimentally were found from a review of the literature, the system with the lowest operating energy is included in the comparison at the end of the section however until large arrays of diode lasers become available, it is difficult to see how these types of devices can influence the search for a system satisfying the requirement stated at the end of the first section.

Class II devices (passive/hybrid) are combinations of optical modulators and detectors, changes in the detected light level being transmitted are fed-back to the electrical control of the modulator. This class includes the many types of spatial light modulator presently being studied and used for optical computing, e.g. the Hughes liquid crystal light valve. Recently devices of this class have been proposed[9] and constructed[10] by the optical computing groups at AT&T Bell Laboratories and University College, London. These groups emphasise the advantages of operating these devices in

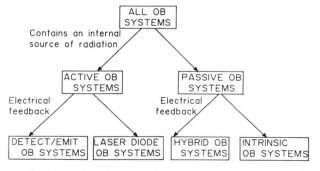

Fig. 2. All optically bistable devices can be subdivided into four main classes.

a mode in which they possess a thresholding characteristic with gain but not bistability.

The final division of Class I (passive/intrinsic) contains the systems studied by the author's institution and is thus discussed in the bulk of this paper. In the review, 38 systems in which optical bistability has been observed were identified, the radiation wavelengths used varied from 0.39 μm and 10.6 μm. The system currently (and since its first observation in 1978) receiving most attention is dispersive optical bistability in Fabry-Perot etalons. All bistable systems require positive feedback, which in this case is provided by the reflectivity of the etalon mirrors, and nonlinearity which for dispersive optical bistability is the intensity dependence of the optical thickness of the cavity.

The origin of this change in optical path length is almost always dominated by the refractive index change. In the system which operate at a sufficiently low energy to allow many gates to be operated simultaneously the origin of the nonlinearity is either a thermal effect (i. e. utilising the thermo-optic coefficient) or an electronic effect, caused by absorption of the incident radiation. The behaviour of one system utilising a thermal nonlinearity is outlined later in this section.

Several of the most interesting systems studied use the electronic nonlinearity in a semiconductor. The process which occurs is the absorption of radiation changes the distribution of carriers within and between the conduction and valence bands. The change in refractive index which results (n_2 is the change in refractive index with intensity) can be large and consequently the power required to observed switching is low, if the nonlinearity is resonant and the radiation wavelength chosen to optimise this enhancement. Resonance occurs at the band edge of semiconductors. Two examples of this bistable system are InSb illuminated by radiation of 5.5 μm wavelength[14] (either from a carbon monoxide laser or a lead salt diode laser) and GaAs/GaAlAs MQW material which is found to have virtually identical switching behaviour, at room temperature, to bulk GaAs) illuminated by radiation of 0.83 μm wavelength (either from a dye laser or a GaAlAs diode laser).

Not surprisingly, in addition to an increase in nonlinearity decreasing the required switch power, an increase in the strength of positive feedback acts likewise. In etalon devices a measure of the strength of feedback is the coefficient of finesse (F). If the front and back mirror reflectivities are R_F and R_B and the absorption coefficient of the nonlinear material (which is filling the cavity of physical thickness D) is α, then one obtains $R = (R_F R_B)^{\frac{1}{2}} \exp(-\alpha D)$ and $F = 4R/(1-R)^2$. Clearly, high absorption will decrease finesse, as will low mirror reflectivity. Without finite absorption however there will be no refractive index change for this form of nonlinear etalon[11]. The optimisation of cavity design thus depends on achieving the correct balance of absorbance and finesse and it is possible to obtain an expression for the lowest intensity at which bistability of this type can be observed[12,13].

Optical bistability occurs in a system with sufficient positive feedback and nonlinearity. In the case of dispersive optical bistability the role of positive feedback is contained in the dependence of the average internal intensity circulating in the etalon on its optical phase thickness (δ),

$$\frac{I_c}{I_o} = \frac{1+R_B \exp(-\alpha D) \ (1-R_F) \ (1-\exp(-\alpha D))}{\alpha D (1-R)^2} \ \frac{1}{1+F\sin^2\delta} \tag{1}$$

and

$$2\delta = 2\pi N + \delta_o + 4\pi n_2 I_c D \cos\theta/\lambda$$

where δ_o is the initial detuning of the etalon from the Nth order resonance, θ is the internal angle of propagation, λ is the radiation wavelength in vacuo and I_o is the incident intensity.

Consider the situation in which $\delta_o = \pi$, i.e. the etalon is exactly off resonance. As the incident intensity is increased, at every new value of I_o, the internal intensity I_c increases, causing the refractive index of the cavity to also increase. The phase thickness of the cavity is thus changed in a sense that increases the ratio of internal to incident intensity. If the increase in incident intensity is obtained by a series of steps of equal magnitude, the increase in internal intensity at every step get progressively greater since as each new value of internal intensity is established the etalon is nearer to on-resonance; the ratio I_c/I_o is increased since $\sin\delta$ is decreasing. Eventually (if the finesse is sufficiently high) the situation is reached when an increment in incident intensity causes a change in internal intensity which is big enough to change the magnitude of $\sin\delta$ such that the consequent change in internal intensity, for the new fixed value or I_o, is larger than the initial change in I_c. A runaway effect occurs and there is a consequent discontinuity in the internal intensity (and transmitted and reflected intensities which are related to it). The etalon resonance feature has been shifted by the change in optical thickness caused by the absorbed radiation. Switching occurs because of positive feedback: increasing internal density increases transmittance.

A particularly useful technique to obtain the characteristic which is easily implemented on a computer is the dummy-variable method and outlined in detail by Wherrett[14]. Characteristics like those presented in Fig. 1 are typical, the difference between Fig. 1a and c or Fig. B and d is obtained by a different choice of initial detuning (δ_o). In practice, this change is trivial to perform, a selection of the ways in which it is implemented include; use of a tunable laser, variation of angle of incidence, translation of a wedged sample, and temperatures change induced for example by resistive coupling.

The first system in which passive intrinsic optical bistability in a semiconductor was observed was in the form of an interference filter[15].

A simple interference filter has a general form similar to a Fabry-Perot etalon, being constructed by depositing a series of thin layers of transparent material of various refractive indices on a transparent substrate. The first layers deposited form a stack of alternating high and low refractive index all of optical thickness equal to one quarter of the operating wavelength. The next layer is a low integer (1-20) number of half wavelengths thick and finally a further stack is deposited to form the filter. The stacks have the property of high reflectivity at one wavelength thus forming a cavity. However, unlike a Fabry-Perot etalon, due to absorption in the spacer (which may be necessary to induce nonlinearity), matched (equal) stack reflectivities do not give the optimum cavity design to minimise switch power. A balanced design[13,16] which takes into account the decrease in back mirror reflectivity due to the double pass through the absorbing cavity is preferable and also results in greater contrast between bistable states. The balanced design $R_F = R_B \exp(-2\alpha D)$ is easily achieved by varying one or all of the available parameters; number of periods, thickness and refractive index of each layer within either stack. Another difference between an interference filter and a Fabry-Perot etalon is that the free spectral range of the former is given by a complicated combination of spacer thickness and stack design whereas in the latter it is simply given by the reciprocal of the optical thickness.

The recent interest in nonlinear interference filters by several groups[15,18] is due in part to three advantages it possesses over other systems.

1 - It relies on a thermal nonlinearity and thus the absorption can be introduced by any partially absorbing material forming part of the filter either in the cavity[15,18], as one of the mirrors[19] or external to the optical cavity[11]. Thus although a given filter will only work within a narrow range of wavelengths (with filters that operate in the visible the FWHM of the passband of a single cavity device is typically ~ 4 nm wide, multiple cavity designs reduce this), different filters of similar construction can be operated at whichever part of the spectrum there is a convenient laser source. The electronic-nonlinearity based systems are not this flexible.

2 - The requirement for an array of gates has as a consequence the need for high sample uniformity. The lack of information on the spatial uniformity of response for any bistable system other than filters is almost certainly indicative of the current unsatisfactory performance in this respect of most systems. Measurements on filters however show a systematic variation of threshold or typically 5%/cm which is negligible.

3 - The systems utilising an electronic nonlinearity also heat-up during operation. Typically, about a third of the incident power is absorbed and ultimately contributes to heating. Systems operating at mid-infrared wavelengths (e. g. InSb) have sufficiently high nonlinearities that the effects due to this heating are negligible. However as a consequence of the strong dependence of the electronic n_2 on operating wavelength, the nonlinearity in materials with a bandgap of around 1 eV, for example GaAs, is so

low that pulsed operation of a signal channel to minimise thermal effects is all that has so far been achieved.

The principle disadvantage of thermal-nonlinearity-based systems, including filters, is the present high switching energy required. However, in the case of filters, it seems possible that by operating with small device volume low enough switching powers will result (< 1 mW) to enable simultaneous operation of the order of 10^4 pixels (gates) and that switching times as low as $\sim 1\ \mu s$ will be possible[11,16]. These figures, if obtainable, would allow the first generation of useful computing systems to be constructed.

This extrapolation to small device volumes is based on an analysis which predicts the observed spot size dependence of switch power and time at large spot sizes (> 10 μm); power proportional to spot diameter and time proportional to spot area. (This is in contrast to the spot size dependence observed for electronic-nonlinearity-based devices in which, at large spot sizes, the power is proportional to beam area and the time is independent of spot size being determined solely by the carrier recombination rate). However the spot size scaling of power does not follow this dependence, in either nonlinearity case, if the spot size is reduced below a critical value. As an example of the loss of the linear dependence measurements performed using a nonlinear interference filter and radiation of wavelength 0.6333 μm showed that as the spot size used was reduced below 15 μm to 8 μm the power increased from 15 mW to 20 mW.

One reason for the departure from the dependence observed at large spot size is the effect of diffusion. In the case of electronic-based-nonlinearity devices, the photoexcited carrier population's transverse diffusion through the material will mean that those carriers outside the illuminated region will not contribute to the observed phase shift. Similarly, heat diffusing outside the illuminated region does not contribute in the thermal nonlinearity case. Hence at spot diameters smaller than the diffusion length this effect will become apparent. In InSb at 80 K the high mobility (10^6 cm^2/Vs) and long recombination time (200 ns) gives a diffusion length of 60 μm[20].

Another effect which influences the spot size scaling in thermal devices is due to the opposing phase shifts caused by the electronic and thermal contributions. As the spot size is decreased, the switching intensity is increased and since the electronic effect is intensity dependent, its contribution will be increased. Wherrett et al.[13] suggest that in some circumstances and at small spot sizes, the electronic contribution can be greater than the thermal.

The final effect considered here is that due to the small depth of focus of beams focussed to close to the diffraction limit. The confocal parameter of a Gaussian beam focussed to a diameter of 2λ is only $2\pi\lambda$. The effective optical thickness of cavity (L) is given by the photon lifetime[21], $L = D/(\alpha D - \ln\sqrt{R_F R_B})$. Typically, etalons of thickness 2λ are used, $R_F R_B \cong 1$ and $\alpha D = 0.1$ giving $L \cong 20\lambda$. Clearly this effective thickness should have been _less_ than the depth of focus if the power density at the focus is to be applicable. In addition, the finesse will decrease if the

depth of focus is of the same order of magnitude as L since diffraction
of the beam within the etalon will mean that, although the radiation angle
of incidence may be normal to the cavity mirrors, some of the light propagating back and forth within the cavity will be at an angle to the cavity
and consequently experience a different optical thickness. Atherton et
al.'s approximation for the effective finesse of an etalon[22] has been used
to obtain Fig. 3, which shows the reduction in the effective coefficient
of finesse (F_E) as the spot diameter is decreased (or the solid angle of
the cone of rays passing through a filter is increased). Typically in experiments a coefficient of finesse of greater than 30 is used, the reduction
in finesse and nonuniform power density will thus adversely effect the spot
size scaling at a spot diameter below $\sim 3\lambda$.

One solution to some of the problems experienced with etalons at small
spot sizes is the pixellation of samples into 2-D arrays of microresonators
each of which acts as a waveguide. An additional benefit of this approach
is that crosstalk between adjacent channels will be reduced to an acceptable
level. The approach the group at Heriot-Watt University is taking to achieving its aim is illustrated by Fig. 4. which shows the form a pixellated
filter might take. The lower thermal conctivity layer may be 100-200 μm
thick plastic (or glass) which is optically cemented to a high thermal
conductivity substrate (e.g. sapphire). After reticulation of the glass by
laser machining the filter is deposited to form the array of resonators.
The coupling losses between the modes in a waveguide of this type and free-
space have not yet been considered. An experiment by Jewell et al.[23] showed
the feasibility of this approach with GaAs etalons.

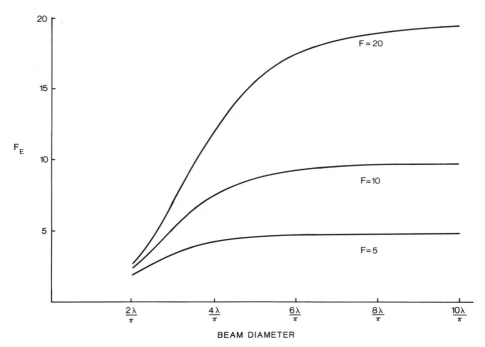

Fig. 3. The dependence of the effective coefficient of finesse (F_E) on
the spot diameter used, plotted for three etalons with F_E at
large spot diameters equal to 20, 10 and 5.

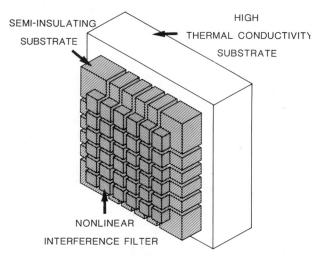

SEMI-INSULATING
SUBSTRATE

HIGH
THERMAL CONDUCTIVITY
SUBSTRATE

NONLINEAR
INTERFERENCE FILTER

Fig. 4. Schematic representation of the form a reticulated (pixellated)
array of thermal-nonlinearity-based waveguide microresonators
may take

The switching of a passive optically bistable device is a second
order phase transition and as such is critically slowed down (CSD). One
effect of CSD on the operation of logic gates in an optical circuit is
to give a switching time longer than that which might be expected given
the logic gate performance when irradiated by a high energy pulse. For
example compare the switching time obtained with NLIF's when used as a
looped circuit (next section)[16] ($\sim$ 10 ms) with that obtained when an
identical gate is irradiated by a picosecond duration pulse[17] ($\sim$ 1 ns).
The switching time is dependent on the degree of 'overswitching' used,
i.e. the amount by which the minimum is exceeded. For example, Janosey et
al.[24] have shown that if an NLIF with a switching power of 10 mW is biased
by a holding beam of 6 mW and switched with a signal power of 4 mW (15 mW)
the switching time is 0.25 s (0.5 ms).

It is instructive to compare the observed switching power/time of
the varoius optically bistable devices currently receiving most attention.
In Fig. 5, seven systems are compared by plotting their position on a graph
of switching power against time. In the plot CSD is ignored, the switching
time used is that which is obtained when a large degree of overswitching
is used. The SEED is a passive hybrid device, the diode laser an active
intrinsic device, the other all passive intrinsic. In the case of the
latter, the dimensions included refer to the spot size used. The SEED
performance is determined by device area and lower energy operation should
be obtainable when smaller devices are grown. The variation in SEED performance is obtained by variation of the electrical feedback parameter. The
higher power entry for the diode laser is the electrical input necessary.

413

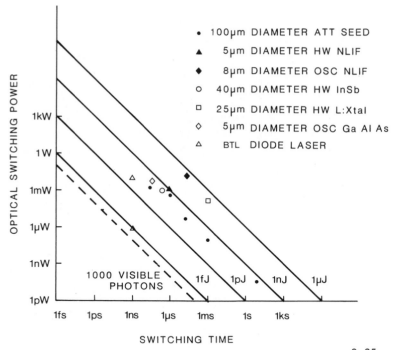

Fig. 5. Comparison of optically bistable devices: AT+T SEED[9,25] - HW NLIF -
Heriot-Watt University nonlinear interference filter[13,16,11],
OSC NLIF - Optical Sciences Center, University of Arizona[18,26],
HX InSb[20] - HW L:Xtal - liquid crystal[19], OSC GaAlAs[27] -, BTL
diode laser - British Telecom Laboratories[28]

The figure shows that the performance of nonlinear filters in an unpixelled
form and with an unoptimised cavity is comparable to other present technologies.

OPTICAL CIRCUITS

To use passive intrinsic optically bistable devices as logic gates
it is necessary to hold the nonlinear system at some point on its characteristic where a small increase in the switching beam will cause a larger
increase in the output (the reflected or transmitted part of the holding
beam). In practice reflective operation is preferable due to the greater
change in output compared to that which occurs in transmission and the
advantage of the possibility for signal introduction onto an absorbing
layer[28,11] or a coupled cavity[29]. However, the gain available even with
this mode of operation is still small and largely determined by the uniformity of the characteristics in the array. The lack of uniformity may
be due to variations in the sample or in the array of beams with which it
is illuminated. The Heriot-Watt group presently favour the use of hologra-
phically written lenslet arrays for beam generation which give 95% effi-

ciency and uniformity, in arrays of up to 10 by 10 beams[16].

The use of this array generator in conjunction with the nonlinear filters has allowed the construction of the first all-optical digital circuits. All the circuits constructed are looped so that every gate in each circuit is capable of restoring the logic level to a standard, thus proving cascadability. One of the first circuits constructed is shown in Fig. 6. In this configuration the three holding beams to the optical gates are controlled by three acousto-optic modulators (AOM). The transmissivity of the AOM is set at one of two levels. The lower level allows an amount of power to fall on the gate which is less than the switch off power, i.e. reset level. The higher level gives a power level which minimises the difference between the switching power and holding power, i.e. enables the gate. These six levels and the amount of time they persist are independently controlled on a microcomputer.

The fanning together of signal beam (transmitted holding beam of preceding gate) and holding beam was performed by the 2.5 cm focal length lens used to focus the beams onto the filter by ensuring that the two collimated beams were incident on the lens when parallel to each other. The transmitted holding beams are recollimated by identical lenses for relaying to the succeeding gate which can consequently be placed at any arbitrary convenient position. The holding beams were incident on the filter at an angle of 30 ± 2°, the variation was deliberate to ensure that all three

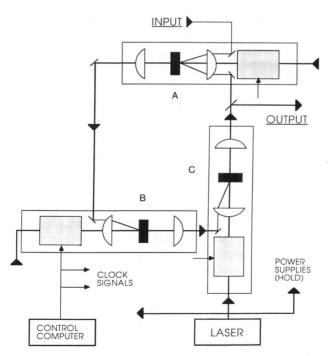

Fig. 6. The optical configuration used for the demonstration of a looped circuit.

415

filters had the same switching power (∼100 mW). This was achieved to within ∼ 25% by varying this angle and the position of the filter in the focussing beam (i.e. using the dependence of switching power on beam diameter).

Emphasis was placed on ensuring efficient transfer of the output power to the next gate to ensure that the full output power change was being utilised efficiently. The maximum total loss between gates is around 25%. However, more important than this is the effective utilisation of this power. The angle of incidence of the holding power was determined by the position of the passband of the filter and the required characteristic, i.e. to be around 5 nm (FWHM of filter) on the long wavelength side of the feature. The signal beam could be incident at any angle and this was chosen to give minimum reflection of the signal; i.e. at the passband transmission maximum, thus allowing the signal to be used with the greatest effect[28,29].

The circuit also featured a fourth beam which derived directly from the laser and was controllable using the microcomputer. The power level of this beam could be set at any one of three levels; 1) close to zero and a level equal to the transmitted output of the logic gates when 2) off and 3) on. The beam may be considered as the input to the circuit. It was incident on one of the gates, A, along with the other two beams.

The operation of the circuit successfully was achieved by clocking the holding beam power levels in the sequence shown in Fig. 7a. The repetitive sequence is A alone with holding power level high, AB, B, BC, C, CA, and so on. A process called 'lock and clock'. In addition when A and C are 'gate-high' (enabled) simultaneously, the input is also present either as an off or on signal. The state of the three gates is also monitored and this is shown in Fig. 7b. The state sequence shows that the logic circuit only responds to the high input bit and transfers the information completely around the loop.

The switching time of the devices in this demonstration was 50 ms, i.e. 150 ms cycle time. The time is a consequence of the large spot size used (100 μm) and critical slowing down. CSD is important here because the change in output when the gates switching is only ∼15% of the switch power. The difference in the holding and switching power was around 5% (to overcome the effect of the 3% amplitude noise on the laser). The remaining 10% gives only a small overswitch. The effect of CSD is apparent in Fig. 8 as the delay between when switching is initiated and when it occurs.

This circuit was later modified to that shown in Fig. 8. In this configuration the output of gate A is the reflected holding beam. Thus one of the gates is performing negation or alternatively acting as a NOR gate. With this configuration the logic gates changed state after each cycle, there was no requirement to 'lock and clock' the gates.

The circuit could, however, be operated with more reliability with a lock and clock approach. The sequence of holding power levels used is shown in Fig. 9a. The transmitted output states are shown in Fig. 9b. Note again the effect of CSD is to give a delay btween when switching is initiated and when it occurs. Because such a circuit automatically inverts the circu-

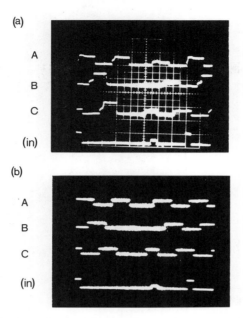

Fig. 7. (a) The 'lock and clock' sequence of reset and enable levels to
the three gates in Fig. 5.
(b) The output sequence obtained using the optical configuration
in Fig. 6 and the timing sequence in Fig. 7a.

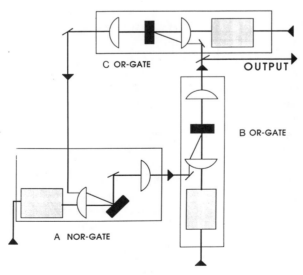

Fig. 8. The NOR/OR/OR gate circuit.

(a)

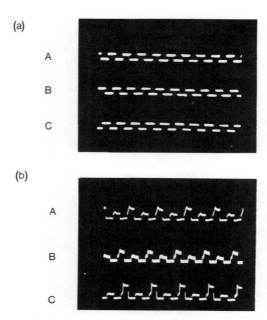

A

B

C

(b)

A

B

C

Fig. 9. (a) The gate enable/reset sequence to 'lock and clock' the circuit shown in Fig. 8.
(b) The response of the NOR/OR/OR circuit to the gate enable sequence above.

lating datum on each cycle, no external input is needed to test for proper operation.

These configurations do not readily lend themselves to expansion to arrays of logic gates. It was also difficult to keep the circuit in operation for extended periods because of the small output power change available in the transmitted output ∿ 15%. The same factor is also partly responsible for the slow switching speeds obtained (CSD). For these reasons the circuit shown in Fig. 10 was designed. Only the reflected output power is used to interact the logic gates, the output power change was ∿ 40% of the switching power. This led to the configuration being more stable than one involving transmission. The interconnection between logic gate arrays is one-to-one mapping. The system is thus degenerate, the channels being completely independent. Interaction could readily be provided by incorporating other optical components such as a dove prism, into the system.

The interconnection is performed using a two-element condenser lens. This was chosen to minimise aberration which was found to be negligible. A system that operates with arrays can only use a conjugate ratio of unity, i.e. one-to-one imaging, since anything else would involve (de)magnification and thus destroy the exact register of signal beams with holding beams required. The focal length of the composite lens (two plano-convex lenses vertex to vertex) was 5 cm and the distance between arrays thus fixed at 20 cm. The principle virtue of this configuration is that its operation is

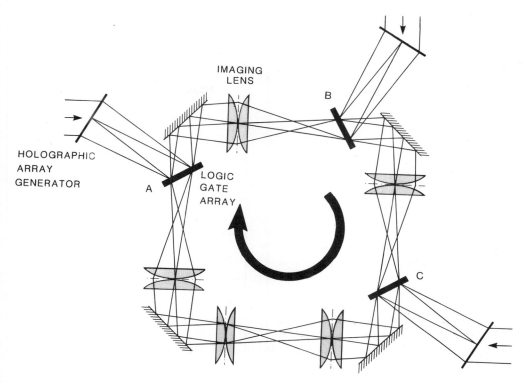

Fig. 10. A parallel optical circuit using imaging of arrays of beams.

almost completely independent of the precise nature of the gate array
(e.g. spot size of elements number of elements, focal length of lenses
used to produce elements) since it merely images one plane to another. If
very small spot sizes or (physically) large arrays are used the off-axis
aberration of the lens and its aperture will start to limit performance.
However the chosen spot size of $\sim$ 100μm in the shape of an equilateral
triangle with edge 2 nm is unaffected by aberration.

This spacing of elements was a consequence of the generation method:
three circular holographic lenses of 2 nm aperture which did not overlap.
A consequence of this array generation method was that cross talk due to
thermal diffusion between logic elements was not apparent. The focal length
of the three holographic lenses was 10 cm thus the lenses were not required to produce near diffraction-limited performance. Parts of this circuit
have been operated. A single channel, i.e. 3 logic gates all reflection-
coupled, has been shown to work. In addition this single channel could
also operate with 50% attenuation inserted between gates that is with an
effective fan-out of 2, the minimum required.

More significant than this was the demonstration of one stage of the
loop, the interaction of three logic gates which three similar gates. This
experiment was performed by first operating each set of three gates indi-

vidually and ensuring that all three had near-identical switching powers. Then the power level to set B (say) was set at a level which was insufficient to make it switch. As the power to the preceding set (A) was increased the power reflected to set B increased until at a critical level sufficient power was incident to detune set B to a point where it was switched by the (previously too low) holding power supplied. As the holding power level to set A was increased further, set A switched and the reflected power from all three gates in set A dropped. The detuning of set B produced by this lower level was insufficient to allow the fixed holding power level to maintain set B in their on-resonance state. So simultaneously with set A switching to on-resonance, set B switched to off-resonance. When, now, the level to set A was reduced, set A switched to their high-reflectivity state and set B switched back to on-resonance in synchronisation.

The essential point of this demonstration is that it is the first demonstration of the interaction of arrays. More than that, the way in which the demonstration was performed ensured that once again restoring optical logic was being proven. The change in output of set A when they switched was used to make set B, which had similar switch powers, also switch.

This circuit was later modified to that shown in Fig. 11 which is a multiple channel flip-flop. This complete circuit was operated successfully with fan-out of 2 between each gate-pair and with an array of 2 by 2 beams incident on each filter. An external signal incident on any channel was able to change the state of each flip-flop channel in turn confirming independent operation.

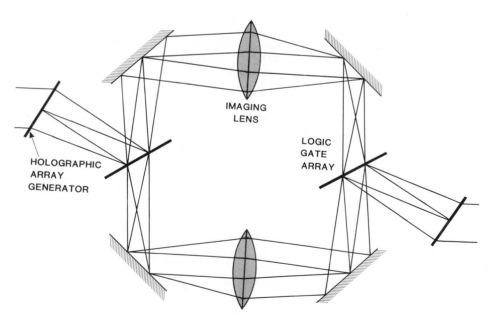

Fig. 11. A parallel optical circuit in the form of two sets of gate arrays coupled to form a flip-flop.

CONCLUSION

Recently the optical counterpart to the transistor has become available. This paper has attempted to indicate what role this device might play
in the construction of an optical computer. The performance of these digital gates is presently at a stage at which the first optical circuits can
be successfully operated. With the predicted improvement in the performance
of these devices, the switching energies may eventually be low enough to
allow the use of arrays of these devices in commercially viable systems.

ACKNOWLEDGEMENTS

Regular discussions with all about 30 members of the optical computing group at Heriot-Watt University is acknowledged. In addition to work
acknowledged in the references, invaluable input and assistance has been
forthcoming from Mr. D. Hutchings, Mr. N. Craft, Professor B. S. Wherrett,
Professor S. D. Smith, Dr. A. C. Walker, Dr. J. G. H. Mathew and Dr. M.
Taghizadeh.

REFERENCES

1. T. E. Bell, Optical Computing: A Field in Flux , IEEE Spectrum, 23:
    34 (1986)
2. A. Huang, Architectural Consideration Evolved in the Design of an
    Optical Digital Computer , Proc. IEEE, 72:780 (1984)
3. "Technical Digests", OSA Topical Meetings on Optical Computing (1985-
    1987)
4. A. W. Lohmann, What Classical Optics can do for the Digital Optical
    Computer , Appl. Opt., 25:1543 (1986)
5. Y. S. Abu-Mostafa and D. Psaltis, Optical Neural Computers ,
    Scientific American, 256:3,66 (1987)
6. K. H. Brenner, A. Huang and N. Streibl, Digital Optical Computing
    with Symbolic Substitution , Appl. Opt., 25:3054 (1986)
7. M. J. Murdocca, Digital Optical Computing with One-Rule Cellular
    Automata , Appl. Opt., 26:682 (1987)
8. A. K. Kar, C. R. Paton, F. A. P. Tooley and A. C. Walker, "General
    Review of Optically Bistable Switching Devices", First interim
    report for Project 1019 by Heriot-Watt University (1986)
9. D. A. B. Miller, D. S. Chemla, T. C. Damen, T. H. Wood, C. A. Burrus,
    A. C. Gossard and W. Wiegmann, Quantum Well Self-Electro-Optic
    Effect Device: Optoelectronic Bistability and Oscillation and
    Self-Linearised Modulation , IEEE J. Quant. Elect., QE-21:1462
    (1985)
10. P. Wheatley, P. J. Bradley, M. Whitehead, G. Parry, J. E. Midwinter,
    D. Mistry, M. A. Paste and J. S. Roberts, Novel Nonresonant Opto-
    electronic Logic Device , Elect. Lett., 23:2,7 (1987)
11. A. C. Walker, Reflection Bistable Etalons with Absorbed Transmission ,
    Opt. Commun., 59:145 (1986)
12. B.S. Wherrett, Fabry-Perot Bistable Cavity Optimisation - On
    Reflection , IEEE J. Quant. Elect., QE-20:646 (1984)

13. B. S. Wherrett, D. Hutchings and D. Russel, Optically Bistable Interference Filters: Optimisation Considerations , J. Opt. Soc. Am. B, 3:351 (1986)

14. B. S. Wherrett, Semiconductor Optical Bistability: Toward the Optical Computer , in: "Nonlinear Optics: Materials and Devices", C. Flytzanis and J. L. Oudar, eds., Springer-Verlag, Berlin (1986)

15. F. V. Karpushko and G. V. Sinitsyn, The Anomalous Nonlinearity and Optical Bistability in Thin-Film Interference Structures, Appl. Phys. B, 28:137 (1982)

16. S. D. Smith, A. C. Walker, F. A. P. Tooley and B. S. Wherrett, The Demonstration of Restoring Digital Optical Logic, Nature, 325: 6099 (1987)

17. J. Y. Bigot, A. Daunois, R. Leonelli, M. Sence, J. G. H. Mathew, S. D. Smith and A. C. Walker, Appl. Phys. Lett., 49:844 (1986)

18. M. T. Tsao, L. Wang; R. Jin, R. W. Sprague, G. Gigioli, H. M. Kulche, Y. D. Li, H. M. Chou, H. M. Gibbs and N. Peyghambarian, Simbolic Substitution using ZnS Interference Filters, Opt. Eng., 26:41 (1987)

19. D. C. Hutchings, A. D. Lloyd, I. Janossy and B. S. Wherrett, Theory of Optical Bistability in Metal Mirrored Fabry-Perot Cavities Containing Thermo-Optic Materials, Opt. Comm., 61:345 (1987)

20. D. J. Hagan, H. A. MacKenzie, J. J. E. Reid, A. C. Walker and F. A. P. Tooley, Spot Size Dependence of Switching Power for an Optically Bistable InSb Element", Appl. Phys. Lett., 47:203 (1985)

21. A. Yariv,"Introduction to Optical Electronics", Holt, Rinehart and Winston, New York (1976)

22. R. Atherton, N. K. Reay, J. Ring and T. R. Hicks, Tunable Fabry-Perot Filters, Opt. Eng., 20:806 (1981)

23. J. L. Jewell, A. Scherer, S. L. McCall, A. C. Gossard and J. H. English, GaAs-A1As Monolithic Microresonator Arrays, Paper PDP1 at the OSA Topical Meeting on Photonic Switching, Lake Tahoe, March 1987

24. I. Janossy, J. G. H. Mathew, E. Abraham, M. R. Taghizadeh and S. D. Smith, Dynamics of Thermally Induced Optical Bistability, IEEE J. Quant. Elect., QE-22:2224 (1986)

25. G. R. Olbright, N. Peyghambarian, H. M. Gibbs, H. A. MacLeod and F. Van Milligen, Microsecond Room-Temperature Optical Bistability and Crosstalk Studies in ZnS and ZnSe Interference Filters with Visible Light and Milliwatt Powers, Appl. Phys. Lett., 45:1031 (1984)

26. S. S. Tarng, T. Venkatesan and W. Wiegman, Use of a Diode Laser to Observe Room-Temperature, Low-Power Optically Bistability in a GaAs-AlGaAs Etalon", Appl. Phys. Lett., 44:360 (1984)

27. M. J. Adams, H. J. Westlake, M. J. O'Mahony and I. D. Henning, A Comparison of Active and Passive Optical Bistability in Semiconductors, IEEE J. Quant. Elect., QE-21:1948 (1984)

28. F. A. P. Tooley, Fan-Out Considerations of Digital Optical Circuits, Appl. Opt., May 1987

29. N. C. Craft and S. D. Smith, Highly Cascadable Optically Bistable Device for Large Fan-Out Optical Computing Applications", Paper TaC3 in Technical Digest of the OSA Topical Meeting of Optical Computing, Lake Tahoe, March 1987.

INVESTIGATION OF ATMOSPHERIC PROPERTIES WITH LASER-RADIATION

W. Steinbrecht, K.W. Rothe and H. Walther

Sektion Physik, Universität München and

Max-Planck Institut für Quantenoptik, Garching, F.R.G.

INTRODUCTION

Survey on the Laser Methods

In this contribution we will focus on the investigation of atmospheric properties using laser radiation ranging from the ultra-violet to the infrared spectral region. The laser is an ideal instrument for this purpose due to its outstanding properties such as large spectral brightness and small angular divergence. In recent years lasers with discrete and continuous wavelength tunability got available. They can now be used to measure specific gaseous components of the atmosphere. So the laser got also important for the remote sensing of pollutants and other trace constituents. In the following a survey of methods used to date to investigate the atmosphere will be given.

Scattering of laser light in the lower atmosphere is dominated by Mie scattering caused by aerosol, clouds, dust and other particles. Rayleigh scattering from the molecular constituents of the atmosphere is two orders of magnitude smaller. It becomes dominant above heights of 30 km, where virtually no aerosols exist. It is therefore possible to derive from the scattering intensity at these heights information about the gas density. From these, one can determine the pressure and temperature distributions; even seasonal variations of these parameters were studied in RADAR-like experiments[1,2].

For the analysis of gases in the lower atmosphere, fluorescence or Raman scattering can be measured. Absorption measurements yield the highest sensitivity and are very simple, but give concentrations only integrated along the light path[3]. To cover a larger area by the laser beam, mirrors can be used. A very simple setup uses topographic targets for reflection of the light back to its source for detection. By the use of mirror arrays absorption measurements can be performed like in tomography and can yield the spatial distribution of gas constituents[4].

Normally, absorption measurements can be performed with low-power laser (e.g. diode lasers). Even arc lamps with spectral filtering can be applied. Such a setup was very successful in detecting SO_2, N_2O, O_3, NO_2 and NO_3 with very high sensitivity[5].

A remarkable improvement of the detectivity in absorption measurements becomes possible by the use of the heterodyne technique, where a tunable laser is used as local oscillator and a photodetector as a mixer[6]. The signal at the intermediate frequency is amplified by a suitable amplifier of narrow bandwidth. This technique (most advantageous in the infrared spectral range) yields an increase in sensitivity of several orders of magnitude, allowing appreciable lengthening of the absorption paths. With heterodyne detection, fast detectors used at room temperature (like pyroelectric detectors) are equally sensitive as cooled infrared detectors.

For direct detection of the radiation emitted by a pollutant, the heterodyne method is very useful, too. Of course, it works only if the temperature of the gas under study is higher than the ambient temperature (as it is the case e.g. in the exhaust of a chimney). The heterodyne method is also being used in satellite or balloon-borne experiments for detecting atmospheric gases (e.g. O_3) by measuring the absorption of the skylight serving as light source[7].

It is a particular advantage of the laser that intense light pulses can be generated, this allows time resolved measurement of the light back-scattered from the atmosphere (as in RADAR); this allows to evaluate the spatial distribution of the scatterers. Since light replaces the radio waves, these methods are known under the name LIDAR (Light Detection And Ranging). Table I summarizes the properties of the atmosphere that can be measured by means of the different scattering processes.

The relative intensity of the different scattering processes is given in Table II. From these values it is obvious that Raman scattering can only be used for the detection of gases with higher concentrations. For most pollutants, the ratio Np/Na is of the order of 10^{-6}. It follows that the Raman signal of the pollutant is 9 orders of magnitude lower than the Rayleigh signal, to which all gases contribute.

Raman scattering was particularly used for the measurement of H_2O[8]. Typically the mixing ratio of H_2O and N_2 molecules is measured by means of the relative intensities of the first Stokes (vibrational) Raman LIDAR returns. From measurements of rotational Raman bands temperature profiles have been determined[9]. The most recent work has concentrated on high spatial resolution[10] and longterm, high accuracy profiling of H_2O[11].

The use of fluctuations of the refractive index for anemometry and the study of turbulences has been investigated by several authors. An analysis of these methods is given in Ref.(12). For anemometry for example two parallel beams of a He-Ne laser are transmitted through the atmosphere, at a distance of 0.5 to 1 m. The motion of the air causes a correlation between the intensity fluctuations of the two beams; the component of the wind velocity in the plane of the beams and perpendicolar to them can be

Table 1. Properties of the atmosphere that can be studied by Lasers

Process	Information
Rayleigh scattering	total gas density distribution, temperature and pressure distributions
Mie scattering	aerosols, clouds, dust, smog
Raman scattering	molecules, temperature
Resonance scattering	atoms, molecules, temperature
Resonance absorption	atoms, molecules, temperature
Fluctuations of refractive index	turbulence, wind velocity

Table 2. Comparison of the relative intensity of backscattered signals

Scattering process	Relative signal	Remarks
Rayleigh scattering	1	
Mie scattering	80	at a visibility distance of a 5 km
Raman scattering	10^{-3} Np/Na	Np, Na = densities of pollutant and of total atmosphere respectively
Fluorescence scattering	80	for a scattering cross-section of 10^{-16} cm^2 (electronic transitions of molecules) and on the assumption that 1% of the absorbed energy is reemitted as radiation, concentration 0.3 ppm

evaluated from the correlation time and the distance of the beams.

Anemometry can also be performed by measuring the frequency shift caused by the Doppler effect. The frequency shift of the back-scattered signal (Mie scattering) versus laser frequency allows the determination of the velocity component parallel to the laser beam. Information about the wind velocity can also be derived measuring the photon correlation of the backscattered light[12].

Lidar measurements using Mie scattering have often been performed e.g. to measure the smoke plume from chimneys or the heights of inversion and cloud layers. Furthermore from the variation of the backscattered signals in fog or haze the actual visibility can be determined. As mentioned before pure Rayleigh scattering signals which were received from higher altitudes (30-150 km) allow to deduce gas density profiles. These measurements which were performed at many places on earth with high power lasers are standard (for reviews see Refs. (1) and (2)).

In order to determine gaseous constituents in the atmosphere fluorescence scattering seems to be a much better method than Raman scattering. However, in the lower atmosphere problems arise from the collisional quenching of the fluorescence: in general only 1% of the molecules radiate their excitation energy, wereas the majority transfer their energy to other molecules by collisions. The value of the scattering cross section used in Table 2 belongs to electronic transitions of molecules, it is 4 to 5 orders of magnitude smaller for vibrational transitions in the infrared so that the relative backscattering is, in general, considerably smaller. An additional disadvantage not considered is that for electronic transitions of molecules, the fluorescence is distributed over a wide range of the spectrum since the ground state is split up into vibrational levels. In order to detect sufficient fluorescent light the receiver system has to be broadbanded. This, of course, lowers the ratio between fluorescent and background signals.

The detection of atoms or molecules in the upper atmosphere by fluorescence radiation is, however, much easier: a LIDAR experiment of this kind was performed by Sandford et al. well in the pioneer days of dye lasers; Na atoms were detected in these measurements[13]. To minimize the background, the first experiments could only be performed at night. As more powerful lasers are now available, it is also possible to measure during the day. The temporal variation of the atomic density can now be observed around the clock.

Dye lasers with pulse energies of up to 100 mJ were used. The laser is tuned to the Na D_2-line; the fluorescence radiation is collected by a large aperture mirror. A (generally double-structured) layer about 15 km thick was found at an altitude of 90 km, the Na atom density being of the order of 10^3 cm^{-3}. The laser experiments showed clearly that the concentration of the sodium atoms changes with the abundance of meteorites, proving that the sodium atoms originate from evaporating meteorites. For sodium to persist in atomic form a certain ratio of atomic oxygen and ozone is necessary; this is only guaranteed in an altitude of about 90 km.

The work on Na in the mesosphere has been extended to other elements, such as K, Li, Ca, Ca$^+$ and to monitoring the sodium layer as an indicator of mesospheric temperature, atmospheric waves, and sudden changes in the ionosphere (see e.g. Ref. (14) and (15)). Experiments have also begun on using the LIDAR-induced sodium fluorescence beam spot in the upper atmosphere as a light source for adaptive optics corrections for astronomical imaging[16].

Fluorescence measurements in the stratosphere have also been performed with balloon-borne LIDAR-instruments[17,18] and methods have been proposed[19] to use OH fluorescence LIDAR to measure mesopheric temperature and pressure.

THE DIFFERENTIAL ABSORPTION METHOD

As mentioned above, an absorption measurement usually does not yield
the density distribution of a gas. This disadvantage can be overcome by
using Mie scattering (which is very strong in the lower atmosphere) as a
"mirror". The position of the "mirror" can be determined by the time elaps-
ing between the emission of a laser pulse and the detection of the backscattered signal. The dependence of the Mie scattering on the wavelength is
small. It is therefore possible to eliminate the local variation of the
scattering by measurements at two different wavelenghts. Only one of these
has to be absorbed by the gas to be detected. Both wavelengths must be
sufficiently close together to ensure that their Mie scattering is in fact
equal.

This method, called "differential absorption", was first applied by
Rothe et al.[20] for measuring air pollutants. Theoretical studies[21] had
shown before that it is the most sensitive one of all LIDAR methods which
can be used to determine trace constituents.

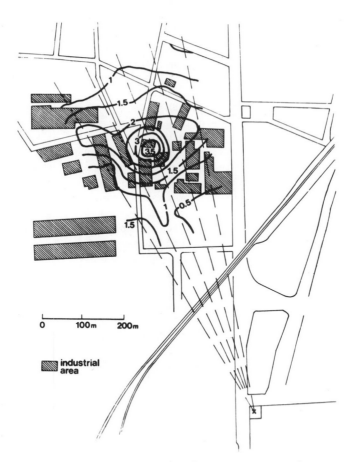

Fig. 1. Distribution of NO_2 concentration above a chemical plant. Concentrations are given in ppm. The results were obtained by averaging about 40000 laser shots[20].

For the first laser measurements performed with this method[20] the apparatus was equipped with a tunable dye laser. The concentration of NO_2 near a chemical factory was measured. With only 1 mJ of pulse energy, concentrations as low as 0.2 ppm could be detected at distances of 4 km. The distribution of the NO_2 concentration above the plant was determined by varying the direction of the laser beam. Connecting points of equal concentration measured in five different directions, the map of isolines shown in Fig. 1 was obtained. It becomes evident from this picture which building is the source of the pollutant. By means of a dye laser the differential absorption method can also be applied to SO_2 and O_3. SO_2 measurements were carried out e.g. by Svanberg et al.[22]. The O_3 measurements will be discussed later in this paper.

If the differential absorption method is to be extended to a large number of pollutants or trace constituents of the atmosphere, measurements have to be carried out in the infrared range of the spectrum. The most universal setup for this purpose would contain a continuously tunable laser. As such lasers, at present, are still too complicated to be used in field measurements, molecular lasers with many emission lines (as e.g. DF, HF, CO, CO_2 and N_2O lasers) must be employed and the measurements must rely on accidental coincidences of laser lines with absorption lines of the trace constituents[2]. Several pollutants can even be detected simultaneously by a multi-line measurement or sequential measurements using different emission lines.

In the following, some results obtained from a setup equipped with a multi-gas laser are reported. Details of the apparatus are published elsewhere[3]. The first experiments to be discussed here were obtained at the cooling tower of the power station in Meppen. The objective of these measurement was to determine the concentration of water vapour around the cooling tower in order to obtain data comparable with computational results of the distribution. The measurements were made from two different points located outside of the power plant; the concentration was determined in different planes of varying elevation. This yielded a three-dimensional distribution of the H_2O concentration. The laser was operated with CO_2. At a pulse repetition rate of 70 Hz, a measuring time of 10 minutes was necessary to obtain the distribution in one plane, as shown in Figs. 2a and 2b. Figure 2a shows the contour lines of constant concentration, while Fig. 2b displays the concentration on the vertical axis.

As next example, measurements of the ethylene concentration around an oil refinery near Ingolstadt are discussed. The purpose of these in-situ measurements was to extend the measurements to an organic gas and to test the improvement of the detectivity into the ppb-range.

The ethylene detected leaks out of the distillation plant. These measurements were also made with a CO_2 laser placed at a distance of about 500 m from the refinery. An example of the measurements is displayed on Fig. 3a and 3b.

428

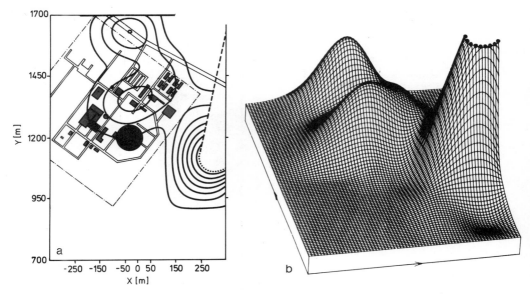

Fig. 2. a) Plan of the Meppen power plant with the cooling tower (round
shaded area). Contour lines of the water vapour concentration begin
with 2 Torr at the outermost line, the steps between the lines
being also 2 Torr. The dotted line shows the contour of the visible
plume. The plane of the measurement is inclined to the surface; it
touches the surface at the position of the apparatus (x=0, y=0) and
is at an altitude of about 100 m above the cooling tower[20]; b) two-
dimensional distribution of water vapour concentration (vertical
axis) above the power plant[20].

MEASUREMENT OF THE STRATOSPHERIC OZONE LAYER WITH A DIFFERENTIAL ABSORPTION LIDAR

As mentioned above ozone (O_3) is one of the most interesting atmospheric
trace gases at the present time. Its maximum concentration, up to 500 $\mu g/m^3$,
is reached at an altitude of approximately 20 to 25 km. The ozone layer –
absorbing in the UV – acts as a shield against the harmful UV radiation of
the sun; the dissipation of the absorbed energy on the other hand strongly
influences the temperature balance of the atmosphere.

The ozone distribution around 35 km is of special interest because
theoretical models predict that it is particularly sensitive to anthropogenic
perturbations: chlorofluorocarbons are expected to cause maximum ozone reduction at that height. These substances are used, for example, as propellant
gases in spray cans or as cooling liquids in refrigerators. At ground level
they are inert, but after diffusing into the upper stratosphere they are
dissociated by the solar UV irradiance. The free chlorine atoms thus generated act on the ozone, destroying it catalytically. A depletion of the upper
stratospheric ozone content is therefore expected. The influences of other
species such as nitrogen oxides and carbon dioxide have also been taken into
account in the calculations.

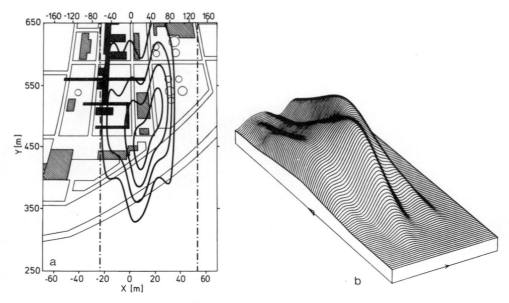

Fig. 3. a) Map of the refinery and contour lines of the ethylene concentration. The lines start with 20 ppb, the inward increase
being 20 ppb between the lines[20]; b) ethylene concentration
(vertical axis) above the refinery[20].

The increasing carbon dioxide content is especially of great importance.
Its main consequence is a change in the atmospheric temperature. The reaction equilibrium of temperature-dependent chemical cycles, involved in
models of ozone chemistry, therefore changes and has to be considered in
the calculations. Recent computer models predict an ozone increase due to
these species in the troposphere and lower stratosphere, by about the same
amount as the depletion in the upper stratosphere.

The two effects together, therefore, leave the total ozone content
nearly unchanged. Nevertheless, further attention to the problem is needed
because the described shift of the ozone layer to lower altitudes changes
the temperature distribution in the atmosphere. As a consequence, influences
on the weather, which at present cannot be predicted in detail, are expected.

The complexity of these computer models is illustrated by the fact that
more than a hundred chemical reactions, some exhibiting nonlinear behaviour
and some being strongly temperature dependent, have to be taken into account.
In addition, many of the reaction constants are not yet known accurately
enough. With the observation of the ozone depletion over Anctartica the
ozone problem gained new and dramatic interest (for a recent review see
e.g. ref. 23).

Only precise long-term measurements of the height-resolved ozone concentration can lead to a final proof of the phenomena involved. Therefore
new methods for these investigations are necessary.

First measurements of the ozone layer with lasers were performed by Gibson et al.[24,25] and later by Megie et al.[26]. The apparatus has been improved since then; today useful results on the ozone layer can be obtained from laser experiments[27]. As Mie scattering can be neglected in the upper atmosphere, Rayleigh scattering was used as "mirror" in these experiments.

In these laser measurements time intervals in the range of several hours were required to get an accuracy of 30%. This also holds for the measurement of Uchino et al.[28]; they used a XeCl laser at 308 nm without having a second laser line necessary as a reference for the differential-absorption technique. Instead they used the data of a balloon sonde, launched the next day, as a reference.

In Munich we have constructed a new setup for ozone measurements in the stratosphere taking into account the requirements for a high power laser as well as for a proper reference line. We use a XeCl laser with a pulse energy of 130 mJ and a repetition rate of up to 40 Hz. Its radiation is focussed into a high-pressure methane or hydrogen cell to generate the reference line by stimulated Raman scattering (see fig. 4). Typical conversion rates are 30% at a pressure of 35 atm for methane or 8 atm for hydrogen. This is more than sufficient since the Raman shifted reference line at 338 nm for methane and 353 nm for hydrogen is absorbed significantly less by ozone than the unshifted XeCl laser line at 308 nm.

The backscattered light is focussed by a spherical mirror (60 cm diameter, focal length 240 cm). The two wavelengths are separated by a dichroic beam splitter, interference and Fabry-Perot filters (the latter were used for daytime measurements in the Arctic). The two photomultipliers must be protected against the very strong intensities backscattered from the first few kilometers; a chopper wheel therefore rotates in front of the multipliers, covering them during and shortly after emission of the laser pulse. The photons backscattered from the stratosphere are counted, and the corresponding count-rates stored and afterwards transferred to a computer for further data evaluation.

The major advantage of this setup is that simultaneous measurements at both wavelengths are possible, thus eliminating all problems associated with rapidly changing atmospheric conditions (turbulences).

One disadvantage of all ground-based stratospheric LIDAR measurements is the rather strong decrease of the light intensity in the first few kilometers primarily due to Mie scattering, and in the UV spectral region also due to Rayleigh scattering. The measurements with our ozone LIDAR were therefore made from the summit of the Zugspitze in the Alp Mountains (altitude 3 km). This difference in altitude gives rise to an increase of the intensity of the stratospheric backscattering by a factor of about three for clean air, and of twenty for hazy air. The whole apparatus was installed in a small container (3 m x 2 m) and transported by helicopter to the summit of Zugspitze.

The LIDAR station has been in operation from October 1982 to September 1986[29,30]. The inherent height resolution is 200 m. In the evaluation pro-

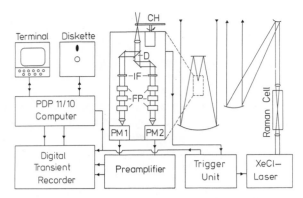

Fig. 4. Scheme of the ozone LIDAR. The transmitter is shown on the right and the receiver on the left part of the figure.

cedure the data are smoothed so that the final resolution is about one kilometre. It takes about 15 minutes averaging time to obtain profiles up to 30 km with less than one per cent statistical error. At an altitude of 40 km, however, several hours integration time are needed to get an accuracy of the same order. Information on still higher altitudes requires a whole night of laser operation at a pulse repetition rate of 50 Hz.

Fig. 5 shows all ozone profiles measured in September 1983. Each single profile represents one night observation. The fluctuations are only due to natural changes; in comparison with these the measurement error is negligibly small at all altitudes.

The variability of the ozone concentration above 30 km is relatively small because photochemical reactions dominate the ozone concentration in the upper stratosphere. On the contrary, the ozone content in the lower stratosphere is highly variable. This is due to transport mechanisms, taking place mainly at these altitudes.

The LIDAR profiles agree quite well with measurements performed by chemical ozone sondes, launched from Hohenpeissenberg, in about 30 km distance from Zugspitze[31]. There Brewer/Mast electrochemical sondes are launched on a routine basis two or three times a week since about fifteen years. (For details of the experimental setup see Ref. 32). The sondes are prepared very carefully, a fact which is reflected in the low correction factors. They reach maximum heights of up to 35 km. The accuracy is about five percent and decreases with altitude in the stratosphere.

Early 1987 the improved LIDAR system, capable of daytime measurements, was placed on board of the research vessel and ice-breaker Polarstern. Measurements were performed during a cruise from Bremerhaven to Tromsoe via Spitzbergen (May 14 to July 1, 1987). Details of these measurements will be published elsewhere . Measurements could be performed whenever there was a clear sky.

A survey of the major results is given in Fig. 6. Typical ozone profiles for various latitudes are shown. A systematic decrease of the ozone con-

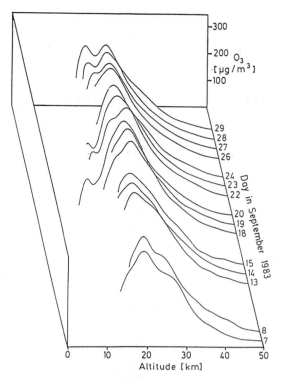

Fig. 5. Ozone density profiles measured in September 1983. Each profile represents the mean ozone concentration measured in one night[30].

centration — especially for altitudes from 10 to about 25 km — can clearly be seen for increasing geographic latitude. This confirms the global variation of the total ozone concentration as found by satellite measurements.

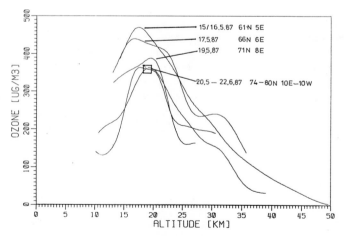

Fig. 6. Results of the ozone measurements during a cruise of the research vessel Polarstern May to July 1987. The concentrations are given in $\mu g/m^3$.

In addition, the height dependence strongly supports the assumption that transport phenomena in the stratosphere rather than photochemistry dominate the variation of the ozone concentration.

This very systematic decrease with increasing latitude is superimposed by short-term fluctuations of the ozone concentration at all altitudes up to about 35 km, reflecting the sensitive dependence of the ozone content on various meteorological parameters.

CONCLUSION

In this review we have discussed the application of laser radiation for the determination of trace constituents in the atmosphere. The few examples show that lasers are a very good tool for remote sensing allowing to achieve a very high detection sensitivity. In special cases also measurements over large distances are possible. The ozone problem we described in the paper requires long term measurements; especially for this purpose the laser is superior to all other detection methods used so far.

REFERENCES

1. D. Hinkley, ed. in "Laser Monitoring of the Atmosphere", Topics in Applied Physics, Vol. 14, Springer Verlag, Berlin, Heidelberg, New York (1976)
2. D. K. Killinger and A. Mooradian, eds., in Workshop on Optical and Laser Remote Sensing, Springer Series in Optical Sciences, Springer Verlag, Berlin, Heidelberg, New York (1982)
3. K. W. Rothe, H. Walther in: "Tunable Lasers and Applications", A. Mooradian, T. Jaeger, P. Stokseth, eds., Springer Series in Optical Sciences Vol. 3, Springer Verlag Berlin, Heidelberg, New York (1976)
4. R. L. Byer, L. A. Stepp, Optics Lett. 4, 75 (1979)
5. U. Platt, D. Perner, H. W. Patz, Journ. Geophys. Rev. , 84, 6329 (1979); D. Perner, U. Platt, Geophysic. Rev. Lett., 6, 917 (1979); U. Platt, D. Perner, A. M. Winer, G. W. Harris, J. N. Pitts, Jr. Geophys. Rev. Lett. 88, 112 (1983)
6. R. T. Menzies, M. S. Shumate, Science, 184, 570 (1974)
7. R. T. Menzies in "Laser Spectroscopy III", J. L. Hall and J. L. Carlsten, eds., Springer Series in Optical Sciences Vol. 7, Springer Verlag Berlin, Heidelberg, New York (1977)
8. D. A. Leonard, Nature, 216, 142 (1967)
9. J. A. Cooney, Opt. Eng., 22, 292 (1983)
   Y. F. Arshinov, S. M. Bobrovnikov, V. E. Zuev, V. M. Mitev, Appl. Optics, 22, 2984 (1983)
10. J. Cooney, K. Petri, A. Salik, Appl. Optics, 24, 104 (1985)
11. S. H. Melfi, D. Whiteman, Bull. Am. Meteor. Soc., 66, 1288 (1985)
12. V. E. Derr, M. J. Post, R. L. Schwieso, R. F. Calfee, G. T. McNice, "A Theoretical Analysis of the Information Content of LIDAR Atmosphere Returns", NOAA Technical Report ERL 296 - WPL 29
13. M. R. Bowman, A. J. Gibson, M. C. W. Sandford, Nature 221, 456 (1969);

M. C. W. Sandford, A. J. Gibson, J. Atmosph. Terr. Phys., 32, 1243
(1970), A. J. Gibson, M. C. W. Sandford, Nature, 239, 509 (1972)

14. G. Megie, F. Bos, J. E. Blamont, M. L. Chanin, Planet Space Sci., 26,
    27 (1978); C. Gramer, J. P. Jegon, G. Megie, Geophys. Res. Lett.,
    12, 655 (1985)
15. K. H. Fricke, U. von Zahn, J. Atmos. Terr. Phys., 47, 499 (1985)
16. R. Foy, A. Labeyrie, Astron. and Astrophys., 152, L29 (1985)
17. W. S. Heaps, Appl. Optics, 19,243 (1980)
18. W. S. Heaps, T. J. McGee, J. Geophys. Res., 90, 7913 (1985)
19. T. J. McGee, T. J. McIlrath, Appl. Optics, 18, 1710 (1969)
20. K. W. Rothe, U. Brinkmann, H. Walther, Appl. Phys., 3, 114 (1974)
    and K. W. Rothe, U. Brinckmann, H. Walther, Appl. Phys., 4, 181
    (1974)
21. R. L. Byer, M. Gaburny, Appl. Optics, 12, 1496 (1973)
22. S. Svanberg in "Surveillance of Environmental Pollution and Resources
    by Electromagnetic Waves - Principles and Applications", T. Lund,
    ed., Nato Advanced Study Institute Series, D. Reidel Publishing
    Company, Dordrecht, Holland (1978)
23. R. J. Cicerone, Science , 237, 35 (1987)
24. M. C. Sandford, A. J. Gibson, J. Atmosph. Terr. Phys., 32, 1423 (1970)
25. A. J. Gibson, L. Thomas, Nature, 256, 561 (1975)
26. G. Megie, J. Y. Allain, M. L. Chanin, J. E. Blamont, Nature, 270,329
    (1977)
27. M. S. Shumate, R. T. Menzies, W. B. Grant, D. S. Dougal, Applied Optics,
    20, 545 (1981)
28. O. Uchino, M. Maeda, M. Hirono, IEEE, QE-15, 1094 (1979)
29. J. Werner, K. W. Rothe, H. Walther, Appl. Phys. B32, 113 (1983)
30. J. Werner, K. W. Rothe, H. Walter, in: "Atmospheric Ozone", Proc. of
    the Quadrennial Ozone Symposium, ed. C. S. Zerefos and A. Ghazi,
    Chalkidiki, Greece, Reidel Dordrecht (1985)
31. W. Attmannspacher, R. Hertmannsgruber, J. Werner, K. W. Rothe, H.
    Walther, in: "Atmospheric Ozone", Proc. of the Quadriennal Ozone
    Symposium, ed. C.S. Zerefos and A. Ghazi, Chalkidiki, Greece,
    Reidel Dordrecht (1985)
32. W. Attmannspacher, H. U. Dütsch, Ber. Dt. Wettered, Nr. 120 (1970)
    and Nr. 157 (1978), Offenbach am Main, Verlag des Deutschen
    Wetterdienstes

LASER TECHNIQUES FOR MEASURING ATMOSPHERIC TRACE CONSTITUENTS

G. Benedetti-Michelangeli, M.Fazi and C. Santoro

SELENIA S.p.A.

Via dei Castelli Romani, 2, Pomezia, Italy

1. INTRODUCTION

In the past twenty years many LIDAR systems have been developed all over the world. These are systems which gain a wealth of data about static and dynamic characteristics of the atmosphere by sending a laser beam into the atmosphere and then measuring the signal back-scattered by its constituents. This instrumentation has been developed by scientists for scientific purpose and has all the characteristics of such instruments: it usually only functions when the scientist himself or his technicians make it work, and moreover it is made to work only in typical laboratory conditions. If an industry should produce it without major modification, its cost would probably be prohibitive and its performance would hardly satisfy the end user.

The results which have been obtained and the experience gained after so many years of scientific activity make it feasible for industry to begin designing and producing such systems. We are not concerned here with what LIDAR can do, but with the problems presented by its industrialization.

2. LIDAR APPLICATIONS

Let us now have a look at the reasons why an industrial LIDAR is necessary and at the functions it should perform. In order to reduce the serious effects produced by air pollution on the quality of human life, three major corrective actions can be taken:

1) Modify the manufacturing processes of industry to reduce the impact on the environment.
2) Adjust the quantity of pollutants produced, in accordance with the local atmospheric and meteorological situation, to keep concentrations below the hazard threshold.
3) Perform "environmental police" action to enforce the law.

The measuring techniques at present provided for by the law (in Italy, and presumably in the rest of the industrialized countries) are based on point detectors which sample the local atmosphere, distributed all over the territory. The choice of the location of these detectors must take into account local meteorology and its evolution during the year. One possible approach to environmental monitoring in a limited part of the territory can be based on the use of a mathematical model of the local atmosphere, which, given a set of meteorological and atmospheric data, can, in real time, forecast its evolution and give early warning on possible dangerous situations.

In the framework of the activities which are necessary to organize the structure and operate the local network of sensors, the LIDAR has six relevant functions:

1) Gathering local meteorological and atmospheric data, to build up a valid predictive mathematical model.
2) Finding the most effective locations to install the ground sensors.
3) Continuous monitoring of the evolution of the atmosphere and of its relevant constituents to perform the necessary forecast to control the production of pollutants, with the help of the mathematical model.
4) Characterizing new pollution sources to help update the model.
5) Performing "environmental police" action to enforce the law.
6) Helping in emergency actions in case of disasters, e.g. fire, ecological accidents, etc., by taking advantage of its remote-sensing capabilities.

As far as its technical aspects are concerned, the LIDAR must be able to measure the needed quantities with the right range, within the right temporal and spatial resolution, and within the recurring and non-recurring costs allowed for by the particular situation. A LIDAR system can assume different configurations to comply with the requirements of different applications. It can be installed in fixed, transportable or mobile ground stations, on aeroplanes, on balloons, on board satellites or spacecraft, on robots, to remotely gain information about the presence of toxic or explosive atmosphere, inside factories, to control industrial processes, on underwater or surface systems, to monitor biological processes dynamics in lakes or in the sea, along methane pipe-lines to detect the presence of leaks, etc.

We are here concerned with the most relevant aspects of the development of a transportable ground LIDAR to be used in monitoring and controlling an industrial area in a feedback loop. The network is intented to be run by local health authorities.

3. HOW A LIDAR WORKS

A laser beam, sent into the atmosphere, interacts with its constituents, modifying and being modified by them in various way. The most relevant interaction effects, in terms of atmospheric remote-sensing applications, include:

A) Scattering from particles whose dimensions are comparable with the wavelength of the impinging radiation (typically aerosols - Mie scattering).

B) Scattering from particles whose dimensions are small compared with the wavelength of impinging radiation (typically molecules - Rayleigh scattering).

C) Absorption from gaseous species which possess absorption spectral lines at the wavelength of impinging radiation.

D) Emission from gaseous species of radiation spectrally different from the impinging one (Raman effect).

E) Scattering from moving aerosols (radial component of the wind) of radiation spectrally shifted with respect to the impinging one (Doppler effect).

F) Scattering from molecules, of radiation spectrally wider than the impinging one, due to thermal agitation (Doppler effect).

LIDAR systems get information about the atmosphere by measuring the characteristics of backscattered laser radiation which has been modified by the interaction with its constituents. The main information obtained are:

1) The aerosol distribution profile.
2) The distribution profile of concentration of gaseous species.
3) The longitudinal component of wind velocity (or of turbulence).
4) The molecular temperature.

This list does not include those quantities which can be inferred indirectly (e.g. atmospheric pressure, obtained by absorption measurements on a spectral doublet).

A single LIDAR cannot be designed to perform all the functions above unless its structure is very complex. In effect, the LIDAR is assembled in the configuration suitable to perform only the needed functions. To this end it needs to be made up of modules.

4. FUNCTIONAL STRUCTURE OF A LIDAR SYSTEM

A typical LIDAR is made up of the following blocks:

1) laser source
2) beam expander
3) beam steering systems
4) receiver optics
5) spatial filter
6) spectral filter
7) detector
8) gate generator
9) electronic control system
10) data processor
11) display unit

The laser source emits short pulses. The divergence of the transmit-

ted beam is controlled by a beam-expander. A beam-steering system allows
the pointing of the expanded beam in the chosen direction. The backscattered signal is gathered by receiver optics, is spectrally and spatially
filtered to reduce the background noise, and is finally sent to the detector. The electrical signal coming from the detector, subdivided in temporal intervals controlled by an electronic gate (corresponding to atmospheric
longitudinal intervals), is processed to extract the useful information. A
control system synchronizes the laser pulses and the temporal gates, and
controls the beam-steering movements. A display unit makes the processed
data available to the end user.

Transmitter and receiver must respectively illuminate and receive from
the same atmospheric volume. The two optical axes can be non-coincident
(bistatic system) or coincident (monostatic system).

To minimize background noise contribution, the receiver field of view
and the transmitter divergence must be as small as possible, so far as they
can be kept in alignment.

5. INDUSTRIAL REQUIREMENTS OF A LIDAR SYSTEM

The difference between industrial LIDARS and those developed for scientific purposes become evident when four general aspects are taken into consideration in the design phase:

1) Environment/machine interface
2) Man/machine interface
3) Differences in functional requirements
4) Design goals of subassemblies and of the system

5.1. Environment/machine interface

The intended application of the LIDAR requires that its working does
not generate interference problems with other human activities which exist
in the area where it must operate. The interference concerns possible laser
radiation effects on men, electromagnetic disturbances which it could generate or by which it could be disturbed, and the problems of location as
regards the point of observation, the easiness of access, and the possible
spoiling of the landscape.

5.2. Man/machine interface

The people working with the LIDAR are those who are in charge of in-
stalling and running it and those who use information provided. The former
must be able to choose the best configuration among those made available,
to suit particular needs. The latter must be able to define the needs and
to interpret the results obtained from the measurements.

To meet the requirements of the former, the design must take into account all those procedures which will allow them to control, run, and maintain the instrument, keeping in mind their level of technical expertise.

440

The design goals should include reliability, easiness of use, and maintainability.

So far as the end user is concerned, it is necessary that data are adequately processed to make them directly useful.

5.3. Differences in functional requirements

A capability which might not be so important for scientific applications but which is fundamental in our case, is that of using LIDAR by daylight. This requirement makes the transceiver design particularly difficult.

Another important aspect is its versatility, that is, its ability to adapt itself to different needs and situations by its modularity. The easy interchangeability of modules allows diversifying the functions of the LIDAR.

5.4. Design goals of subassemblies and of the system

In designing an instrument for scientific research, generally the main goal is the attainment of the highest possible performance in each subassembly through state-of-the-art technical solutions. These solutions are often of low reliability because of the short experience which has been gained. The designer of an industrial instrument must choose those solutions which, even if they cannot give the best performance, give the necessary confidence and a good reliability and producibility within the cost frame dictated by the market. The economic aspect of an industrial product has a great importance due both to the production cost and to the logistic which is to be provided to the customer.

Now, we will discuss some of the specific design features, within the requirements mentioned above, of an industrial LIDAR system.

5.5. Eye safety

LIDAR makes use of laser radiation projected into the atmosphere, usually in densely populated area. Safety concern, mainly regarding the eyes of people who may be within reach of the laser beam, is the first requirement of an industrial LIDAR and needs an adequate design effort.

The radiation is dangerous in proportion to the absorbed energy per unit volume and therefore to the raising of the tissue temperature. From this point of view the spectrum can be divided into three parts:

1) The first one includes those wavelengths which are strongly absorbed by the cornea or by other parts of the eye between cornea and retina; the thickness of the affected layer is small and the surface is large. The total volume can widely vary between hazard and safe situations. This group includes ultraviolet (absorbed by the lens) and far infrared.

2) The second group includes those wavelengths to which the eye is

transparent. These wavelengths are focussed on the retina, producing spots of very small volume. This group, which includes visible and near infrared light, is the most dangerous one. The retina obviously absorbs this radiation very well, in accordance with the well known photopic and scotopic curves.

3) The third group includes those wavelengths which are absorbed in the whole of the eye, and the absorbing volume is large. See, for istance, D.B. Judd (*).

There are two spectral zones outside the visible to near infrared region which are available as eye-safe spectral bands; they are situated between the high transparency and high absorption spectral regions of the eye, where the absorption volume is large. Available regulations and literature confirm that the spectral region between 0.315 and 0.4 microns, and wavelengths longer than 1.4 microns, allow safe exposure levels much higher than those in the visible.

A safe LIDAR system must project into the atmosphere the lowest possible laser power possibly at a wavelength within one of the safe spectral regions. The energy gathered by the eye depends on the ratio of its pupil to the laser beam diameter: it is therefore advisable to use large diameter laser beams. To keep the emitted power low, the receiver sensitivity must be high. The receiver sensitivity depends on the optics input pupil area, on the quantum efficiency of the detector, and on the transmissivity of the optical system at the laser wavelength. The receiver optics pupil diameter must be large, and due consideration must be given to the efficiency of the optical system.

5.6. Daylight capability

In the daytime, unlike night, solar radiation increases the background noise level. Higher laser power is therefore needed, making the attainment of eye-safety, life, reliability and cost goals more difficult. To minimize this problem, two approaches are possible:

A) The reduction of the receiver field of view.
B) The reduction of the receiver spectral band.

The receiver field of view is determined by dimensions of a spatial filter positioned on the optics focal plane. In order to avoid losses in the signal, the angle and the orientation of the field of view must be well matched with the divergence and the pointing orientation of the transmitter: the field of view of the receiver must contain only the volume of the atmosphere illuminated by the laser beam. Small receiver fields of view and small laser divergences are not too difficult to achieve, but it is not easy to keep them aligned under all environmental conditions. In the design phase this is a demanding requirement, which leads one to select: monostatic solutions rather than bistatic ones; high laser stability, that is, sources with low beam-wander; and highly stable mechanical structures,

---

* D. B. Judd, Natl. Bur. Standard (U.S.) Circ. N. 478, p. 2 (1950)

to keep the two channels mutually aligned and the optical system simultaneously thermally stabilized with an appropriate modulation transfer function (M.T.F.).

In a monostatic structure, a realistically feasible receiver field of view and the corresponding laser beam divergence are not smaller than about 0.1 mrad.

The receiver spectral bandwidth is determined by the filter used. Interference filters, which are the most usual solution of this application, must satisfy a compromise among spectral bandwith divergence of impinging radiation (and consequently, diameter), temperature stability, and cost. The typical result is a minimum bandwidth of the order of 10 Angstroms. In a daylight system, more selective filtering is likely to be necessary, and Fabry-Perot interferometers are a promising solution.

6. MODULARITY

An industrial LIDAR system must be flexible enough to cope with highly diversified operational situations. This flexibility can be achieved through modularity of its subassemblies; this means that every interface between subassemblies must be defined in all respects, and the characteristics of each module must carry out well defined and circumscribed functions, all of which must be considered as a unit when the module is replaced. A well designed module, both form the point of view of the functions performed and the interface characteristics, opens the door to future developments, because it facilitates the introduction of new technologies and new functions.

7. CONCLUSIONS

Research and industrial LIDARS, which are similar with respect to their capabilities to gather physical quantities characterizing the atmosphere, differ both in their objectives and in their design approaches, because they must satisfy significantly different requirements.

The designer of an industrial system begins his job by making an analysis of what the end user really needs, drawing on concepts and ideas which have lead the most outstanding results of the experience accumulated in the scientific research environment, and selecting those items which allow him to assemble the most suitable system for the market, among the materials, the technologies, and the instruments which the industry makes available. This undertaking can be pursued only with collaboration among scientists, users, and industry, through a continuous flow of information. The designer has the duty of harmonizing the user's desires with the state-of-the-art achievements of research to make a satisfactory industrial product available to the market.

Degeneracy factor, 168
Degenerate electron gas, 297
Delfin, 303
Depletion instability, 23
Dermatology, 273
Detection efficiency, 344
DFB laser, 99, 134, 139
DHE, 234
Diagnostic equipment, 138
DIAL, 137, 138
Dielectric breakdown, 152
Differential absorption method, 427
Diffusion length, 173
Digital logic elements, 403
Digital optical computing, 403
Diode laser, 95
Direct laser writing, 137
Direct writing, 204
Discharge excitation, 4
Dispersion relation, 296
Dispersive mirror, 103
Displacement operator, 45
Displacement polarizability, 156
Dissociation attachment, 10
Distributed feedback lasers, 99, 134
Di-Hematoporphyrin Ether, 234
DNA, 264
  biosynthesis, 223
  synthesis, 221
DOPA-melanin, 274
Doppler
  detection, 138
  effect, 388
  free resolution, 339
  shifted absorption, 189
Dove prism, 418
DRAM
  chips, 143
  manufacturing, 144
DRAW, 95
Drilling, 145
Drilling treatments, 132
Driving current, 44, 46
Droplet model, 360
Drude relaxation time, 156
DT fuel, 297, 302
Duffing's equation, 322
Dummy-variable method, 409
DY isotopes, 351
Dynamic random access memory
  chips, 143

Eczema, 260
EEC international program RACE, 139
EFRAW, 95
Eigenvalues, 110
Einstein coefficients, 8
Electric field eigenmodes, 88
Electric power, 143
Electric power cables, 143
Electric quadrupole moment, 339
Electron beam
  excitation, 5
  pumping, 95
Electron
  cyclotron frequency, 317
  excitation, 156
  hole plasma, 156, 166, 167
  phonon scattering, 161
  plasma waves, 311
  positron colliders, 309
  pressure, 120
  thermalization, 156
  trapping, 116, 123, 124
Electronic nonlinearity, 410
Electro-optic sensors, 137
Electro-optic systems, 138
Endoscopic applications, 276
Endoscopic PDT, 278
Endoscopic photocoagulation, 132
Endoscopic surgery, 132
Endoscopic techniques, 237
Endoscopic therapy, 277
Energy balance, 296
Energy production, 129
Entropy waves, 19
Envelope solitons, 91
Environment control, 138
Environmental monitoring, 438
Environmental police, 438

Epitaxial layers, 134
Eu isotopes, 346
Eureka, 139
Eureka Eurolaser project, 27, 139
Excimer fluid dynamics, 18
Excimer laser, 2, 125, 136, 188
Excitation mechanisms, 155
Excitation method
Explosive instability, 334
Eye safety, 441
F_2, 23

HVD, 234
Hybrid fusion fission scheme, 294
Hybrid OB devices, 407
Hybrid reactor, 305
Hydrodynamic efficiency, 298
Hydrodynamical effects, 155
Hydrossiethylvinil
  deuteroporphyrin, 234
Hyperbilirubinemia, 217, 265
Hyperfine splitting, 339
Hyperfine structure, 339
H-like ions X-ray lasers, 33, 37

ICF, 293
IFEL, 332
Image processing, 404
Impact ionization, 169
Implosion, 297
Incoherent laser array, 108
Incoherent radar systems, 148
Index-guided stripe lasers, 97
Indirect band semiconductors, 161
Induced emission, 342
Industrial LIDAR, 437
Industrial applications, 129
Industrial chemical processes, 197
Industrial laser, 1
Industrial metrology, 130
Industrial photochemistry, 197
Inertial confinement fusion, 293
  reactor, 305, 306
Inertial fusion, 337
Informatics, 138
Information
  printing, 130
  reading, 130
Infrared detectors, 424
Infrared differential absorption
  LIDAR, 137
Injection laser, 95
Injection pumping, 95
Integrated circuits
  manufacturing, 143
Integrated CO_2 laser, 79
Integrated optic technologies, 138
Intelligent sensors, 138
Interaction time, 145
Intercarrier scattering, 162
International market, 131
Intersecting storage rings, 310
Intracavity doubling, 388
Intravessel endoscopy, 245

Intrinsic OB devices, 407
Inverse Cerenkov regime, 330
Inverse free electron laser, 332
Ion pressure, 120
ISOLDE, 347
  facility, 339, 350
Isomeric states, 358
Isotones, 358
Isotope
  production, 363
  separation, 115, 201, 363
  shift, 339
ISR, 310
K constant, 180
Klystrons, 334
KrF, 2, 5, 25
KrF lasers, 335

LAAV, 290
Ladar, 151
Laguerre-Gaussian modes, 179
Landau mechanism, 320
Laparoscopic laser surgery, 270
Large electron positron
  collider, 309
Large high aspect ratio target, 302
Laryngeal microsurgery, 272
Laser
  ablation plasma, 298
  angioplasty, 243
  annealing, 155
  applications, 129
  assisted vasal anastomosis, 286
    vasovasostomy, 288, 290
  beams summing, 138
  biostimulation, 217
  cooling, 367, 383, 386
  cutting, 155
  deceleration, 391
  detection and ranging, 151
  diagnostics, 187, 255
  disk, 95
  displays, 95
  drilling, 137, 155
  driven plasma, 312
  driven plasma waves, 318
  fusion, 293
  gain, 8
  heating, 155
  imaging, 138
  induced
    acceleration, 328

MHD
  approximation, 116, 119
  equations, 119
  model, 120
MICF target, 302
Microelectronics, 133, 143
Microoptic components, 134
Microsurgery, 74, 132, 271
Mie scattering, 189, 235, 423, 425
Military applications, 154
Mishen, 303
MOCVD, 205,
Mode
  discrimination, 103, 104, 105
    effect, 103
Modulation transfer function, 443
Molecular lasers, 428
Molecular photobiology, 217
Molecular photomedicine, 217
Molecular structure, 3
Molecular temperature, 439
Momentum-exchange collisions, 119
Monomode laser diodes, 139
Mostek, 144
Motion equation, 119
MPI, 115
  experiments, 121
  plasma, 120
  produced plasma, 127
MTF, 443
Multi quantum well, 134
Multiphoton absorption, 157
Multiphoton dissociation, 202
Multiphoton excitation, 188
Multiphoton ionization, 115
Multiphoton
  ionization experiments, 125
Multiphoton processes, 201
Multiple epitaxy techniques, 133
Mycosis, 260

National defense, 138
Na-isotopes, 340
Near-field
  intensity profiles, 110
  profiles of phase-locked
    array, 110
Neodymium laser, 138
Neonatal jaundice, 265
NEP, 149
Neural networks, 405
Neutrality condition, 121

Ne-like ions X-ray lasers, 33, 37
NLIF, 413
NMR tomography, 202
Noise-equivalent power, 149
Nondestructive
  inspecting technique, 152
Nonlinear etalon, 408
Nonlinear interference filters, 410
Nonlinear propagation, 83
Nonlinear refractive index
  coefficients, 88
NOVA, 303
Nuclear binding energy, 360
Nuclear fusion, 129
Nuclear fusion gain, 295
Nuclear moments, 358
Nuclear technology, 202
Numerical simulation, 155

OB devices, 406
Octal, 303
Odd group coupling model, 358
Ohm equation, 120
Omega-X, 303
Optical amplifiers, 139
Optical bistability, 403, 406
Optical circuits, 414
Optical communication, 99
Optical components, 177
Optical computing, 403
  system, 403
Optical data storage, 154
Optical disk, 98, 106, 154
Optical feed-back, 133
Optical fiber, 83
  catheter, 134
  communications, 95
  sensing, 95
  technology, 243
Optical Kerr effect, 88
Optical knife, 269
Optical loops, 139
Optical modulators, 407
Optical pumping, 95
Optical Ramsey interrogation, 378
Optical reading, 95
Optical recording, 108
Optical resonance excitation, 342
Optical signal processing, 134
Optical storage, 97
  technology, 98
Optical writing, 95

451

Pressure waves, 18
Propagation constant, 109
Proton-antiproton collider, 309
Protoporphyrin, 234
PSK mode, 98
Psoralens, 264
Psoriasis, 260, 264
PuF_6, 202
Pulse
  broadening, 92
  chirping, 326
  compression, 92, 137
Pulsed angioplasty, 250
Pulsed dye laser, 125
Pulsed laser, 2
Pumping mechanisms, 4
Pumping technique, 95
Pusher, 299
PUVA
  therapy, 265
  treatment risks, 265
PVC, 198
Pyroelectric detectors, 424
Pyrolysis, 203
PWS, 273

Quadratic Zeeman effect, 371
Quarks, 310
Quasi Fermi levels, 159
Quenching electron, 5
Quenching two body, 6
Q-switching technique, 135

Ra isotopes, 351
Radar, 151
  systems, 143
  technology, 148
Radiactive detection
  of optical pumping, 339
Radio detection and ranging, 151
RADOP, 356
  method, 339
Raman
  back scattering, 335
  cascade, 323
  scattering, 88, 161, 189, 425
  shifted excimer laser, 137
Ramsey lineshape, 374
Rangefinder, 148
Rare gas halide, 2, 3
Rayleigh
  range, 181

Rayleigh (continued)
  scattering, 189, 423, 425
Reactor laser, 301
Recombination, 168
Refractive nonlinearity, 403
Relativistic detuning, 323
Relativistic electrons, 313
Relativistic equations of motions, 312
Relaxation rates, 341
Remote sensing, 129, 130, 423
  of pollutants, 137
REMPI, 192
Resonance
  absorption, 297, 425
  frequency, 343
  ionization mass spectrometry, 348
  ionization spectroscopy, 340, 345
  scattering, 425
Resonant beat waves, 321
Resonant beating, 312
Resonant multiphoton ionization, 192
RF excitation, 79
Rhodamine 6G, 388
Rhodamine B dye laser, 236
Richardson-Dushman relation, 164
RIMS, 348
Ring Dye laser, 388
RIS, 343, 345
RNA, 264
  biosynthesis, 223
Robot systems, 137
Robotized systems, 177
Rocket model, 300
Roots-blowers, 79
Runge-Kutta integration, 118

SASE, 334
Scaling laws, 305
Scattering
  matrix, 102, 103
  parameters, 102, 160
Scribing treatments, 132
Second harmonic distortion, 381
Second order Doppler effect, 381
SEED, 413
Self amplification of spontaneous
  emission, 334
Self-generated electric field, 120
Semiconductor laser, 95, 108
  phase-locked array, 108
SENRI I, 306, 307
Sheath instability, 24

VC, 198
Vibration detection, 138
Video disks, 95, 154
Vinyl chloride, 198
VLSI
  devices, 133
  microstructures, 204
Vulcan laser facility, 38, 303
VUV FEL's, 63

W₀ radius, 180
Wave coupling, 329
Wavelength multiplexer, 134
Wave-breaking condition, 315
Welding, 132, 145
White noise, 376
W.K.B. approximation, 324
Wind velocity, 439
Work function, 164

X-band radar, 149, 150
XeCl, 2, 5, 25
  laser, 431
XLPE, 151
X-ray
  amplifier, 35
  angiography, 245
  FEL's, 63
  gun, 16
  lasers, 33
  mirrors, 35
  preionization, 10, 12, 16

YAG lasers, 106

Z₀ distance, 180,
Zeeman
  optical pumping, 386
  splitting, 357
ZnSe lens, 178